The Reluctant Patron
Science and the State in Britain 1850–1920

PETER ALTER

The Reluctant Patron
Science and the State in Britain
1850–1920

Translated from the German by
ANGELA DAVIES

BERG
Oxford / Hamburg / New York
Distributed exclusively in the US by
St. Martin's Press, New York

First published in 1987 by
Berg Publishers Limited
Market House, Deddington, Oxford OX5 4SW, UK
Schenefelder Landstr. 14K, 2000 Hamburg 55, W.-Germany
175 Fifth Avenue/Room 400, New York, NY 10010, USA

Copyright © German Historical Institute 1987

This translation is based on *Wissenschaft, Staat, Mäzene. Anfänge moderner Wissenschaftspolitik in Grossbritannien 1850-1920* (Stuttgart, 1982) but has been revised and updated by the author for this edition.

All rights reserved.
No part of this publication may be reproduced in any form or by any means without the prior permission of Berg Publishers Limited.

British Library Cataloguing in Publication Data

Alter, Peter
   The reluctant patron: science and the state in Britain 1850–1920.
   1. Science—Social aspects—Great Britain 2. Science—Great Britain—History—19th century 3. Science—Great Britain—History—20th century
   I. Title
   306'.45'0941   Q175.52.G7
   ISBN 0-907582-67-2

Library of Congress Cataloging-in Publication Data

Alter, Peter.
   The reluctant patron.
     Rev. translation of: Wissenschaft, Staat, Mäzene.
     Bibliography: p.
     Includes index.
     1. Science and state—Great Britain—History.
     2. Research—Great Britain—History. 3. Engineering—Research—Great Britain—History. I. Title.
   Q127.G4A6413 1986   338.94106   86–31770
   ISBN 0–907582–67–2

Printed in Great Britain by Redwood Burn Limited, Trowbridge, Wiltshire

Contents

Introduction Science and the State in Britain before 1918   1

1. The Spectrum of Science Patronage in
Nineteenth-Century Britain   13
    Specialists and Amateurs: Learned and Scientific
        Societies   14
    The Natural Sciences become Academically
        Established   22
    The Era of Patronage   35
    Indifference and Improvisation:
        The Role of the State   60

2. Science and the Awareness of Crisis   75
    The British Science Lobby in the Nineteenth and Early
        Twentieth Centuries   76
    'Competition between Nations': The Arguments
        and Interests of Science   98

3. Science Policy Decisions in Britain: Institutional
Innovations after 1900   138
    The National Physical Laboratory   138
    Imperial College of Science and Technology   149
    The Medical Research Committee   172
    Spokesmen for Science   177

4. Policies for Science during the First World War   191
    'The War of Chemists and Engineers'   191
    The Organisation of Science by the State: The
        Founding of the Department of Scientific and
        Industrial Research   201

5. Science as a Profession: From Amateur to Outsider?   214
    Science in Victorian England   214
    The Social Situation of British Scientists in the
        Nineteenth Century   235

Conclusion Science in a Liberal Industrial State   246

Appendices
    1. Foundation of learned and scientific societies in all fields in
        the United Kingdom during the nineteenth century   256
    2. Scientific societies and institutions in London, 1883   257
    3. Membership figures of the Royal Society of London
        1660–1935   259

Bibliography   261

Index   283

Tables

1.1. Percentage of students from within a radius of 30 miles of their university — 29
1.2. Universities in the United Kingdom in 1910 — 31
1.3. State expenditure on science, 1850–1900 — 68
1.4. State expenditure on science, 1869–1914 — 69
2.1. Average annual growth in industrial output in the three largest industrial nations, 1860–1913 — 106
2.2. Iron and steel production in the leading industrial nations, 1870–1910 — 108
2.3. Number of patents granted for synthetic dyes in Britain, 1856–1913 — 110
2.4. World production and consumption of synthetic dyes, 1913 — 111

Introduction
Science and the State in Britain before 1918

'During the past few years much has been done by the State to provide facilities for research', wrote *Nature* in 1920, surveying the support which science in Britain had received since the turn of the century. This tribute to what had been achieved was quickly followed by criticism: 'But it is not too much to say that even now neither the public nor our statesmen understand the debt they owe to the peculiar and rare geniuses to whom the greatest discoveries are due, or that any attempt has been made to discharge it'.[1]

The relationship between science and the state in Britain is the subject of this book. It deals with the promotion and organisation of science in Britain in the late nineteenth and early twentieth centuries, a period which, in historical perspective, assumes particular significance for the relationship between science and the state in this country. The study attempts to survey the change that government policy towards science underwent during this period. *Laissez-faire* policies hitherto pursued gave way to interventionist policies, which aimed to coordinate scientific research and to influence its direction, overtly or covertly, by making money available and tying it to specific purposes. In comparison to Germany, for example, this change in Britain took place relatively late, in the two decades between the Boer War (1899–1902) and the end of the First World War. The conceptual and material foundations of a modern public science policy in Britain were laid at this time. The book therefore focuses on two questions, which to a certain extent are interrelated. Firstly, what factors precipitated state intervention in the science sector, which until then had largely been left to its own devices? And secondly, what was the basic shape of the British science policy laid down at this time? Before we attempt to answer these questions, it is necessary to make some general points concerning terminology.

In this book the term 'science' refers only to the natural sciences.

1. *Nature*, 105 (1920), p. 189.

Introduction

It encompasses both the pure sciences and the whole spectrum of the 'applied sciences'. The distinction between pure and applied sciences is of relatively recent origin, although at the turn of the century its justification was already increasingly being questioned. The large degree of overlapping that exists between the two categories makes the distinction seem artificial. Basically, it can be justified only in terms of objectives: the term 'applied science' relates to scientific research undertaken primarily in order to solve problems which have practical application. This study also takes into account medical science and biology as they relate to specific points. The humanities and social sciences, on the other hand, are largely excluded. To do otherwise would have been to take on a subject of unmanageable proportions. Where it seems appropriate, however, they are included in the discussion.

Science policy in the widest sense here covers all long-term measures undertaken by governments, informal groups of scientists or politicians, and science organisations, to develop, and provide financial aid for, the country's scientific resources in order to maximise their effectiveness for the general good as well as for the national interest.[2] The term 'scientific resources' includes universities, non-university research institutions, and scientific staff. This restricted definition is based on the now generally accepted argument that science, scientific productivity and scientific progress require more than the availability of financial resources; equally important are a favourable institutional framework and an attractive career structure for scientists. From this definition it follows that science cannot be regarded in isolation from its national context; science cannot be considered supra-national. On the contrary, the element of 'national interest' in the definition points to the fact that a policy *for* science clearly aims to develop science within a particular country. By promoting science, governments frequently hope to achieve economic and industrial superiority over other countries. At the beginning of the twentieth century this strong national element was especially pronounced in 'policies for science', as the second chapter of this book will show in detail for Britain. Nowadays science policy is still generally based on the assumption that in a modern industrial society, science and technology must be encouraged if economic efficiency and political influence are to be

---

2. On the definition of science policy, see Frank R. Pfetsch, *Zur Entwicklung der Wissenschaftspolitik in Deutschland 1750–1914* (Berlin, 1974), p. 30; Hilary Rose and Steven Rose, *Science and Society* (Harmondsworth, 1971), p. 131; and J.B. Poole and Kay Andrews (eds.), *The Government of Science in Britain* (London, 1972), p. 1.

Introduction

achieved within a competitive international system.[3] From the scientist's point of view, 'the organisation of scientific research is an important task for every modern state'.[4] Scientists and spokesmen for science argue (in order to justify, often uncritically, their financial demands), that the development of science and technology therefore requires special public support. According to this line of argument, scientific research is one, perhaps even the decisive, precondition for, indeed is the driving force behind, economic growth and social change. The introduction of new technologies and products, it is claimed, will lead to increased industrial production and productivity.

For long the proposition that there is a close connection between science and economic growth was universally accepted without challenge. Only recently has it been questioned — and not only by historians of science.[5] One important factor in stimulating this process of questioning has been that only a few, albeit spectacular, cases have provided convincing proof of any direct connection between state support for science, economic efficiency and economic growth, such as has been postulated since the beginnings of industrialisation. Also, the self-evident assumption that economic growth is the primary aim of economic policy is increasingly being regarded with scepticism. One commentator, for example, writes:

> We have discovered that the Faustian urge for unlimited growth and an unlimited increase in power is a process that is as dangerous as it is profitable. Growth commonly exacts a high price. The uncontrolled and uncoordinated forces – all of which function separately and with an excessive measure of power – unleashed by modern science, modern technology and modern politics, in fact only contribute to damaging the healthy balance of the organism as a whole.[6]

3. See Norman J. Vig, *Science and Technology in British Politics* (Oxford, 1968), p. 2; Q.M. Hogg, *Science and Politics* (London, 1963), pp. 11–12; Roy MacLeod and Kay MacLeod, 'The Social Relations of Science and Technology 1914–1939' in Carlo M. Cipolla (ed.), *The Fontana Economic History of Europe: The Twentieth Century* (Glasgow, 1976), p. 301; Sanford A. Lakoff, 'Scientists, Technologists and Political Power' in Ina Spiegel-Rösing and Derek J. de Solla Price (eds.), *Science, Technology and Society. A Cross-Disciplinary Perspective* (London and Beverley Hills, 1977), pp. 355–91.
4. Translated from Werner Heisenberg, *Tradition in der Wissenschaft. Reden und Aufsätze* (Munich, 1977), p. 102.
5. See Wolfgang Pohrt (ed.), *Wissenschaftspolitik – von wem, für wen, wie?* (Munich, no date [1977]), p. 199; Joseph Ben-David, *The Scientist's Role in Society: A Comparative Study* (Englewood Cliffs, NJ, 1971), pp. 15–16; C.F. Carter and B.R. Williams, *Industry and Technical Progress: Factors Governing the Speed of Application of Science* (London, 1957); and Pfetsch, *Wissenschaftspolitik in Deutschland*, pp. 129–92.
6. Translated from Robert Sinai, 'Was uns krank macht. Die Zivilisation ist am Ende', *Der Monat*, vol. 31, no. 3 (1979), p. 8.

Introduction

Since the late 1970s the threat to a 'healthy balance', as well as alarm among some sections of the public about certain developments in science and industrial technology and their consequences, have produced growing hostility towards science and technology in the USA and Western Europe. Ultimately, a rejection of science paves the way for a denial of any connection between scientific and social progress[7] — whatever the real meaning of these two terms may be. The development of science and technology has obviously reached a point where their 'blessings' are perceived less clearly than the dangers associated with them.

In short, the criteria by which Western industrial societies judge science and its results have changed. The almost unboundedly optimistic and often rather naive faith in science which firmly re-established itself after the Second World War and was reflected in rapidly growing government expenditure on research, has been under threat since the beginning of the 1970s. Discussion now focuses on the limitations of science rather than on its potentialities. Since at least the late eighteenth century the promise held out by modern science has been closely associated with the central idea of progress.[8] Now, however, this promise seems to have been transformed into a threat to human existence; never before have the consequences of science been so fundamental and far-reaching. The new hostility towards science can also be seen at institutional level. In Britain, for example, where an old tradition can easily be evoked, this hostility is expressed in the closure or cutting of research and teaching institutions, both within the university system and outside it, some of which were established at great expense as recently as the 1960s.

The term 'science policy' also requires further explanation. Coined only a few decades ago, it was unknown before the First World War. Frank R. Pfetsch, a German historian of science, has

---

7. D. Nelkin, 'Technology and Public Policy' in Ina Spiegel-Rösing and Derek J. de Solla Price (eds.), *Science, Technology and Society: A Cross-Disciplinary Perspective* (London and Beverly Hills, 1977), pp. 393–441; and Jean-Jacques Salomon, 'Science Policy Studies and the Development of Science Policy', in ibid., pp. 64–6. Also Hermann Lübbe, 'Relevanz contra Curiositas. Über die anwachsende Wissenschaftsfeindlichkeit' in idem, *Wissenschaftspolitik. Planung, Politisierung, Relevanz* (Zurich, 1977), pp. 7–29; and Walter L. Bühl, *Einführung in die Wissenschaftssoziologie* (Munich, 1974), pp. 9–19.

8. Reinhart Koselleck, 'Fortschritt' in Otto Brunner, Werner Conze and Reinhart Koselleck (eds.), *Geschichtliche Grundbegriffe. Historisches Lexikon zur politisch-sozialen Sprache in Deutschland* (Stuttgart, 1975), vol. 2, pp. 351–423; Alfred Heuss, 'Das Problem des "Fortschritts" in den historischen Wissenschaften', *Zeitschrift für Religions- und Geistesgeschichte*, vol. 31 (1979), pp. 132–46; Edgar Zilsel, 'Die Entstehung des Begriffs des wissenschaftlichen Fortschritts' in idem *Die sozialen Ursprünge der neuzeitlichen Wissenschaft* (Frankfurt, 1976), pp. 127–50.

Introduction

pointed out[9] that the earliest documented instance of its use in Germany seems to be that by Karl Griewank in 1927.[10] Presumably the term 'science policy' was coined at the same time in Britain, but we do not yet have an exact reference for its first use. The synonymous term 'scientific policy' (which is occasionally still found today), appears in a leader in *Nature* in 1924.[11] A few years earlier, the terms 'policy of research' and 'research policy' occur in the correspondence between scientific organisations and high government officials.[12] Circumlocutions such as 'a definite policy of encouraging scientific research' can be found in scientific journals in the early years of the First World War.[13] These link up with expressions which had been used since the end of the nineteenth century to describe the same thing: 'encouragement of research', 'endowment of science', 'the furtherance of scientific work', 'the organisation of scientific research'.

Whenever this book refers to the 'state', this means specifically the central government in London. Any activities undertaken by local authorities and corporations to promote science in the nineteenth century are excluded from the study. This applies equally to county and city councils and to local scientific societies. A lack of detailed local and regional studies as yet makes it difficult to generalise about their activities.[14] But the term 'central govern-

9. Pfetsch, *Wissenschaftspolitik in Deutschland*, p. 29.
10. 'Cultural policy involves more than strengthening the functional aspects of science and technology; it also aims to unite a people and the state by means of a common culture and promoting cultural values through the state. In this context a "science policy" ["Wissenschaftspolitik"] with a dual aspect proves to be necessary: science policy is not only the activity of the state in the service of science, but at the same time it is the activity of science in the service of the state as the embodiment of the nation' (translated from Karl Griewank, *Staat und Wissenschaft im Deutschen Reich. Zur Geschichte und Organisation der Wissenschaftspflege in Deutschland*, Freiburg, 1927, p. 9). Bernhard vom Brocke, however, points out that Adolf Harnack had already used the term in 1908; see 'Der deutsch-amerikanische Professorenaustausch. Preußische Wissenschaftspolitik, internationale Wissenschaftsbeziehungen und die Anfänge einer deutschen auswärtigen Kulturpolitik vor dem Ersten Weltkrieg', *Zeitschrift für Kulturaustausch*, vol. 31 (1981), pp. 128–82.
11. It reads: 'By wise and prudent direction, by advancing . . . "with daring caution", the Department of Scientific and Industrial Research may give us an accepted national scientific policy, definite, continuous, and consistent' (*Nature*, 114, 1924, p. 707).
12. The earliest examples of its use date from the first half of 1915. See, for instance, Medical Research Council Archives, P.F. 2/1 (2 March 1915) and Department of Education and Science, Ed. 24/1576 (21 June 1915), Public Record Office (PRO).
13. See, for example, *Nature*, 94 (1914–15), p. 548 ('State Aid for Science', 14 January 1915).
14. Robert H. Kargon, *Science in Victorian Manchester. Enterprise and Expertise* (Manchester, 1977); Robert E. Schofield, *The Lunar Society of Birmingham: A Social History of Provincial Science and Industry in Eighteenth-Century England* (Oxford, 1963); and Ian Inkster and Jack Morrell (eds.), *Metropolis and Province: Science in British*

ment' must also be used with great care, because central government was never a monolith with a single and uniform attitude towards science. The ministries and offices of central government showed differing degrees of open-mindedness towards issues connected with science, its organisational problems and financial demands. It is obvious that certain ministries within central government (and individuals within ministries) had a greater, or especially towards the end of the century, a more rapidly growing, interest than others in encouraging science. Ministries with traditionally close ties to science included the Treasury,[15] the Colonial Office, the Board of Trade and, after its establishment in 1902, especially the Board of Education. A Ministry for Science was not created in Britain until 1959. No previous recommendations that such a ministry be established, including one made as early as 1875 by a Royal Commission chaired by the Duke of Devonshire, were acted upon.

It is a commonplace among historians of science, but rarely sustained in detail, that the period between 1900 and 1920 marked 'the origins of the organisation of scientific research as we now understand it'.[16] Particular significance is generally attributed to the First World War and the social and economic upheavals accompanying it. In this view, 1914 represents a caesura not only in political history but also in the relationship between the state and science in Britain. The war was both a turning point and 'the germ of national science planning'.[17] During it, Britain, and in similar fashion the USA, experienced 'the first significant wave of government intervention in scientific affairs'.[18]

The present study also covers the period between 1900 and 1920, but 1914 is not considered to be the major turning point. Rather, it is the thesis of this book that between the Boer War and the First World War a change took place in the political, economic and social status of science in Britain. This change was accompanied by the establishment of several scientific institutions, which created the institutional framework for a blossoming of British science in the

---

Culture, 1780–1850 (London, 1983) – all are pioneering studies in this field. On methodological problems see Robert E. Schofield, 'Histories of Scientific Societies: Needs and Opportunities for Research', *History of Science: An Annual Review of Literature, Research and Teaching*, vol. 2 (1963), pp. 70–83.
 15. See Roy M. MacLeod, 'Science and the Treasury: Principles, Personalities and Policies, 1870–85' in G.L'E. Turner (ed.), *The Patronage of Science in the Nineteenth Century* (Leiden, 1976), pp. 115–72.
 16. Poole and Andrews (eds.), *Government of Science*, p. 28.
 17. Ibid., p. 2.
 18. Vig, *Science and Technology*, p. 1.

Introduction

1920s and 1930s.[19] At the same time, in the two decades prior to 1918, an administrative structure was created for science which remained essentially unchanged until after the Second World War. To what extent a causal relationship did in fact exist between the foundation of new institutions and the take-off in individual disciplines after the First World War is a problem which remains to be solved by the historiography of science. However, there is much to suggest that at least a correlation can be traced between Britain's leading position in certain sciences (such as medicine, nuclear physics, chemistry, physiology, molecular biology and statistics) since the 1920s, and increasing public support for science.

Exactly when the government began to give more support to science in Britain, especially in financial terms, has not yet been unequivocally settled. Similarly, the reasons for the government's increased readiness to support science, and the organisational and administrative forms this support took, have not yet been fully investigated by historians of science. This is partly due to the fact that so far only limited access has been available to government documents related to decision-making in the area of science policy. *The Government of Science in Britain* edited by J.B. Poole and Kay Andrews provides a useful selection of key documents relating to British science policy since the late nineteenth century.[20] But in spite of their enormous significance for the development of science since the seventeenth century, scientific organisations, societies and institutions have largely only been dealt with in celebratory anniversary monographs; analyses which take account of the modern historiography of science are still needed. Pioneering work has been done in this field by Morris Berman, Robert E. Schofield and Robert H. Kargon, whose works are also methodologically innovative.[21]

On the whole, it is largely since the 1960s that the history of science has developed as a special discipline within history, es-

---

19. It is a matter of controversy whether the advancement to 'scientific centres' of the USA and Britain in the 1920s was paralleled by a decline in Germany's position. In any case, the humanities and sciences certainly declined rapidly in Germany after 1933, when most Jewish academics emigrated and National Socialist ideology penetrated various scientific disciplines. See Ben-David, *The Scientist's Role*; Rainald von Gizycki, 'Centre and Periphery in the International Scientific Community: Germany, France and Great Britain in the Nineteenth Century', *Minerva*, vol. 11 (1973), pp. 474–94; Herbert Mehrtens and Steffen Richter (eds.), *Naturwissenschaft, Technik und NS-Ideologie. Beiträge zur Wissenschaftsgeschichte des Dritten Reichs* (Frankfurt, 1980).
20. Poole and Andrews (eds.), *Government of Science*.
21. Morris Berman, *Social Change and Scientific Organisation: The Royal Institution, 1799–1844* (London, 1978). For references to the works by Schofield and Kargon, see note 14.

pecially in the USA and Western Europe.[22] Historians of science in Britain since the 1960s have been able to look back to an unbroken tradition. As early as the 1930s John D. Bernal was writing about the history of science in a social context.[23] Later, W.H.G. Armytage, D.S.L. Cardwell, Eric Ashby and Roy M. MacLeod dealt with important aspects of the history of science in Britain, and developed a methodological framework. But there is as yet no comprehensive analysis of modern science policy in Britain, locating it between the three poles of state, industry and science. D.S.L. Cardwell, Norman J. Vig and Ian Varcoe provide useful overviews, Cardwell concentrating on the nineteenth century and Vig and Varcoe on the post-1945 period.[24]

This book is a contribution both towards an as yet unwritten history of the organisation and politics of science in Britain in the nineteenth and twentieth centuries, and towards the study of the nature and changing role of the modern state. Around the turn of the century the functions of the state within society were being re-defined in Britain as elsewhere. The growing obligations of the state in the most varied areas resulted in a greater readiness to grant it and its organs a growing sphere of responsibility and ever more powers. Lloyd George was not the only person in Britain calling for 'efficiency in all the departments of the State'.[25] In the past few years historians have paid more attention to problems associated with this issue, which must be considered in connection with the

---

22. See Roy M. MacLeod, 'Changing Perspectives in the Social History of Science' in Ina Spiegel-Rösing and Derek J. de Solla Price (eds.), *Science, Technology and Society: A Cross-Disciplinary Perspective* (London and Beverly Hills, 1977), pp. 149–95; Frank R. Pfetsch, 'Wissenschaft als autonomes und integriertes System. Tendenzen in der neueren Literatur zur Wissenschaftspolitik und -soziologie', *Neue Politische Literatur*, vol. 17 (1972), pp. 15–38; Thomas S. Kuhn, 'The History of Science' in *International Encyclopedia of the Social Sciences* (New York, 1968), vol. 14, pp. 74–83, reprinted in idem, *The Essential Tension: Selected Studies in Scientific Tradition and Change* (Chicago and London, 1977), pp. 105–26; Ina Spiegel-Rösing, 'The Study of Science, Technology and Society (SSTS): Recent Trends and Future Challenges' in Thomas S. Kuhn and Derek J. de Solla Price (eds.), *Science, Technology and Society: A Cross-Disciplinary Perspective* (London and Beverly Hills, 1977), pp. 7–42.
23. John D. Bernal, *The Social Function of Science* (London, 1939, reprinted Cambridge, Mass., and London, 1967); idem. *Science in History*, 2nd edn (London, 1957). On Bernal's life (1901–71) and political work see Paul G. Werskey, *The Visible College: A Collective Biography of British Scientists and Socialists of the 1930s* (London, 1978); idem, 'British Scientists and "Outsider" Politics, 1931–1945', *Science Studies*, vol. 1 (1971), pp. 67–83; and Maurice Goldsmith, *Sage: A Life of J. D. Bernal* (London, 1980).
24. D. S. L. Cardwell, *The Organisation of Science in Britain*, 2nd edn (London, 1972, reprinted 1980); Vig, *Science and Technology*; Ian Varcoe, *Organizing for Science in Britain: A Case-Study* (London, 1974).
25. Speech held on 4 April 1903 in Newcastle (David Lloyd George Papers, A/11/1/26).

Introduction

decline of liberalism, particularly relating to the advent and development of the interventionist state. More peripherally, this study also looks at why Britain, which was the leading and most dynamic of the industrial economies until after the middle of the nineteenth century, lost its relative advantage in the technical and industrial sector towards the end of the century, when its industrial capacity fell further and further behind that of its rivals in Europe and North America. Eric J. Hobsbawm regards this development as the crucial question of nineteenth-century British economic history.[26]

The issue which lies at the heart of this study, therefore, concerns the political and institutional conditions under which a public science policy involving systematic encouragement of science based on long-term planning developed in Britain after the end of the nineteenth century, building on and extending previous attempts in this direction. What was the state's attitude towards the problems of science and the position of science in society? What factors contributed to the state's increasing readiness, between 1900 and 1918, to intervene in the science sector? Along what lines did state intervention proceed; what motives and ideas governed it? Did the state consider itself capable of promoting science? To what extent did scientists demand, or reject, state intervention and the increasing entanglement of the state with science at institutional level, and what arguments did they use in putting their case? How did decision-making processes function in the developing area of public science policy? Is there any evidence to suggest that innovations in the organisation of science undertaken in Britain during this period were modelled on foreign patterns?[27] If so, how were these models integrated into Britain's own political and social system? Although a great deal of detailed research into these questions remains to be done, the main objective of this study is to synthesise the present state of research by using an inductive approach.

The methodology of this study is different from that used by Thomas Kuhn.[28] In recent years Kuhn's work has been in the forefront of the theoretical debate in the historiography of science, and it has also influenced work in other areas.[29] Kuhn's main

26. Eric J. Hobsbawm, *Industry and Empire: An Economic History of Britain since 1750*, 4th edn (London, 1973), p. 149. For a discussion of the concept of modernisation, see Hans-Ulrich Wehler, *Modernisierungstheorie und Geschichte* (Göttingen, 1975).
27. The term 'innovation', which is generally used in many different contexts, is here reserved for institutional innovations.
28. Thomas S. Kuhn, *The Structure of Scientific Revolutions*, 2nd edn (Chicago, 1970).
29. MacLeod, 'Changing Perspectives', p. 157; T.W. Hutchison, *On Revolutions and Progress in Economic Knowledge* (Cambridge, 1978). On Kuhn's influence on the

Introduction

interest is devoted less to the organisation of science and the reciprocal relations between state and science, than to an analysis of qualitative advances in scientific knowledge. His work centres on the sort of scientific revolution which is associated, for example, with the names of Copernicus, Newton, Lavoisier, Darwin and Einstein. Kuhn analyses discontinuities of a revolutionary nature: abrupt transitions within a discipline, accompanied by critical upheavals, from a received theoretical model ('paradigm') to a new one, as the result of revolutionary advances in research.[30] According to Kuhn, scientific revolutions are 'episodes . . . in which a scientific community abandons one time-honored way of regarding the world and of pursuing science in favor of some other, usually incompatible, approach to its discipline'.[31] This method is generally described in the historiography of science as an 'internal approach'.[32] The 'external approach' which this study adopts emphasises the institutional, organisational, political and social context of scientific activities, as well as external influences which affect the development of science.

This study is for the most part organised chronologically. The greatest emphasis is on representative changes within the institutional infrastructure of science in Britain, and on turning points in the relationship between science and the state. The first chapter takes stock of the scientific organisations which worked for science and for its financial support in nineteenth-century Britain; it also looks at the relative shares of private and public patronage in the founding and financing of scientific enterprises and institutions. The second chapter analyses the public discussion in Britain about the country's technical and scientific backwardness, the reasons for it and the conclusions arrived at. In the second half of the nineteenth century there was already some awareness that Britain was falling behind other countries in the science sector. Occasionally

---

theoretical discussion within historiography over the last ten years, see George G. Iggers, *Neue Geschichtswissenschaft. Vom Historismus zur Historischen Sozialwissenschaft. Ein internationaler Vergleich* (Munich, 1978); and David A. Hollinger, 'T. S. Kuhn's Theory of Science and its Implications for History', *American Historical Review*, vol. 78 (1973), pp. 370–93.

30. The discovery of the structure of the DNA molecule provides a well-documented recent example of a paradigm shift. See James D. Watson, *The Double Helix: A Personal Account of the Discovery of the Structure of DNA* (London, 1968; reissued Harmondsworth, 1976); Horace F. Judson, *The Eighth Day of Creation: Makers of the Revolution in Biology* (London, 1979); Robert C. Olby, *The Path to the Double Helix* (London, 1974).

31. Thomas S. Kuhn, 'The Essential Tension: Tradition and Innovation in Scientific Research' in idem, *The Essential Tension*, p. 226.

32. George Basalla (ed.), *The Rise of Modern Science: External or Internal Factors?* (Lexington, Mass., 1968).

Introduction

public controversies flared up about the failure of British policy in this field. But reform was not discussed with greater urgency until after the turn of the century, when Britain had experienced the Boer War and growing economic competition from other countries, especially Germany and the USA. The discussion continued until well into the First World War.

The third chapter of the book analyses the planning and foundation of specific scientific institutions between 1890 and 1914, against a background of social and economic crisis, symptoms of which were becoming more and more obvious in Britain after the turn of the century. Chapter 3 also attempts to establish more precisely the part played by state authorities in the making of important science policy decisions. In addition, it looks at what categories of people were advocating more state support for science in Britain after the turn of the century, and who the active decision-makers were in the field of science policy. Chapter 4 concentrates on the years after 1914 when, under the exceptional conditions created by the war in Europe, arguments which had been put forward by the 'science lobby' in the reform debate prior to 1914 suddenly appeared in a new light. This chapter traces the creation of new, central organisations for science, a centralised financing policy for research outside the university system, and growing understanding on the part of the public, industry and central government for the demands of scientists. The concluding fifth chapter deals with the social status of scientists in Britain. It attempts to establish to what extent scientific professions opened up prospects for a career, and how the profession of 'scientist' has changed since the second half of the nineteenth century, seen against the background of specialisation and 'professionalisation' taking place in the sciences.

Finally, here are some brief remarks concerning the sources on which this study has drawn. In addition to newspapers and journals which provide insights into the public debate on science policy issues in Britain, the files of the Board of Education are of particular significance. They provide information about the role played by the government in establishing new organisations for science, and in taking measures to encourage science. But the total holdings of relevant official documents for the period are not extensive. It appears that science policy was discussed only rarely in cabinet. This cannot, however, be established with certainty, because minutes of cabinet meetings were not kept until December 1916, when Lloyd George's government introduced a cabinet secretariat. An extremely valuable supplement to the official documents is

Introduction

provided by the numerous reports of the various royal commissions, departmental committees and select committees set up either by specific government offices or by Parliament itself to investigate clearly defined problems relating to the organisation of science. In the reports, not yet fully evaluated, these problems are generally treated painstakingly. Often detailed and lengthy, in many cases they formed the basis for state intervention in the science sector. The motives and objectives governing these measures often emerge much more clearly from the reports than from the surviving official documents or from the records of debates and question times in the House of Commons. On the whole, this institution devoted little time and attention to science policy.[33] Another extremely useful source is the leading science journal of the time, *Nature*.[34] Its well-informed, knowledgeable and committed reportage of issues relating to the organisation and support of science in Britain proved invaluable for this study. Material in the archives of scientific societies such as the Royal Society of London, and in those of scientific institutions such as Imperial College of Science and Technology, the National Physical Laboratory and the Medical Research Council, was consulted to complete the analysis of decisions relating to science policy in Britain. The private papers of the leading protagonists in the science policy debate of the time proved to be of relatively little use, although in some cases, such as the papers of Beatrice and Sidney Webb, and of Richard Haldane, they yielded important information.

---

33. See below, pp. 185–90.
34. The first edition of *Nature* was published in November 1869. See the special edition, 224 (1969), which was put out to mark the journal's 100th anniversary on 1 November 1969.

1
The Spectrum of Science Patronage in Nineteenth-Century Britain

In an early study of science and society since the eighteenth century, Hilary and Steven Rose argue that before 1914 the British government adopted a *laissez-faire* attitude towards the support and organisation of science.[1] This view, encountered frequently in the literature on the subject, provides the background to the following discussion. It is important to remember in this context that science can be promoted in a variety of ways, and by different bodies. The 'state' is only one among many such authorities; nowadays it has become by far the most important patron of science, largely because of the resources available to it. This was by no means generally the rule in the nineteenth century, with the probable exception of the German states after the Congress of Vienna and the German Reich after 1871. Only there, for reasons which are beyond the scope of this study, did the state take a large part in supporting science in the nineteenth century, thereby anticipating a development that in other countries took place only gradually after the turn of the century.

The promotion of science does not follow a uniform pattern which is more or less repeated in all countries. To take an image from physics, the patronage of science can be compared to a spectrum in which specific national conditions and traditions allow certain bands to stand out clearly, while others are only dimly visible, or indeed, invisible.[2] In Germany, for example, private endowments for science were not nearly as important as they were in Britain in the late nineteenth century. The entry under *Stiftungen* (endowments) in the popular German encyclopedia, *Brockhaus' Con-*

---

1. Rose and Rose, *Science and Society*, p. 36. On the problems associated with the use of the term '*laissez-faire*', see Arthur J. Taylor, *Laissez-faire and State Intervention in Nineteenth-Century Britain* (London and Basingstoke, 1972).
2. W. H. Brock uses the image of the spectrum in 'The Spectrum of Science Patronage' in Turner (ed.), *Patronage of Science*, pp. 173–206.

versations-Lexikon, gives an insight into the German situation at the time. It refers the reader to *Milde Stiftungen* (charities) and *Moralische Personen* (people of high moral standing).[3] Endowments were seen as relating only to religious or charitable purposes. *Mäzen* (patron) and *Mäzenatentum* (patronage) do not appear in the 1886 edition at all. This suggests that, hypothetically, every country has a typical spectrum of science patronage. If this is accepted, the spectrum of science patronage in nineteenth-century Germany was different from that characteristic of France, the USA or Britain. This chapter attempts to describe its general outline in Britain.

Specialists and Amateurs: Learned and Scientific Societies

Science was promoted essentially through four main channels in Britain during the nineteenth century: the universities, private patrons, public bodies in the widest sense and independent learned and scientific societies.[4] Scientific societies represent the oldest form of patronage, which in the nineteenth century was still incomparably more important in Britain than, for example, in Germany. Many local and national learned and scientific societies in Britain owe their origins to private initiatives. The best known of these societies are the Royal Society of London (1660),[5] the Literary and Philosophical Society in Manchester (1781) and in Liverpool (1793), the Royal Society of Edinburgh (1783) and the Lunar Society in Birmingham (1768);[6] specialist societies include the

---

3. This edition was published in 1886 in Leipzig. On this topic, see Peter Stadelmayer, 'Das schwierige Geschäft, Gutes zu tun. Von Theorie und Praxis der Stiftungen' in Rolf Hauer et al. (eds.), *Deutsches Stiftungswesen 1966–1976. Wissenschaft und Praxis* (Tübingen, 1977), p. 312.

4. References are provided by Cardwell, *Organisation of Science*, and Turner (ed.), *Patronage of Science*.

5. In the following, the dates in brackets indicate the year in which the society was founded. The Royal Society of London for Improving Natural Knowledge was founded in 1660, and received a royal charter from Charles II in 1662. It celebrated its 250th anniversary in 1912.

6. According to the statistics available, there were about thirty learned and scientific societies in Britain at the end of the eighteenth century. The majority saw themselves as a 'Royal Society in Miniature', like the Northampton Philosophical Society, for example. See Peter Mathias, 'Who Unbound Prometheus? Science and Technical Change, 1600–1800' in idem (ed.), *Science and Society, 1600–1900* (Cambridge, 1972), p. 62; and Douglas McKie, 'Scientific Societies to the End of the Eighteenth Century' in Allan Ferguson (ed.), *Natural Philosophy Through the 18th Century*, 2nd edn (London, 1972), pp. 133–43. *The Record of the Royal Society of London for the Promotion of Natural Knowledge*, 4th edn (London, 1940), has a large appendix of documents. Margery Purver, *The Royal Society: Concept and Creation* (London, 1967); R. H. Syfret, 'The Origins of the Royal Society', *Notes and Records of the Royal Society of London*, vol. 5 (1948), pp. 75–137; Douglas McKie, 'The Origins and Foundation of the Royal Society of London', *Notes and Records of the*

London Geological Society (1807), the Royal Geographical Society (1830), the Chemical Society (1841) and the Physical Society (1874), founded somewhat later than the others. Even today the specialist societies are not generally orientated towards research. They tend to be devoted to one particular science and promote the scientific, social and political interests of their members. Like similar societies in other European countries, the majority were established in the nineteenth century, but as a rule, British societies were founded earlier, and had more members, than their German counterparts.[7] The Linnean Society of London, a botanical society founded in 1788 and still in existence today, appears to have been the first such specialist society in England.[8] The speed with which new societies sprang up from the early nineteenth century onwards is a clear indicator of the expansion and specialisation which was taking place in the sciences at this time. It is also symptomatic of a growing public interest in science and in discussing learned problems. This interest was obviously not being catered for by the universities, for example. If we leave the growing number of scientific journals aside for the moment,[9] these societies provided the only forum in which learned questions could be discussed and the latest research results disseminated, often in the form of the 'public lecture' so popular in the nineteenth century.

According to the *Year-Book of the Scientific and Learned Societies*, published for the first time in 1884, there were just over 500 learned and scientific societies and institutions for all disciplines in the whole of the United Kingdom in 1883.[10] In the same year, London,

---

Royal Society of London, vol. 15 (1960), pp. 1–37; Kargon, *Science in Victorian Manchester*; D. S. L. Cardwell, 'The Patronage of Science in Nineteenth-Century Manchester' in Turner (ed.), *Patronage of Science*, pp. 95–113; Asa Briggs, *Victorian Cities* (Harmondsworth, 1977), ch. 3: 'Manchester, Symbol of a New Age', pp. 88–138. Steven Shapin, 'Property, Patronage, and the Politics of Science: The Founding of the Royal Society of Edinburgh', *British Journal for the History of Science*, vol. 7 (1974), pp. 1–41; Schofield, *Lunar Society*. According to J. D. Bernal, 'the effective centre of scientific thought in England at the end of the eighteenth century was not the Royal Society but the Lunar Society' (*Social Function of Science*, p. 25). On the general context, see Jürgen Voss, 'Die Akademien als Organisationsträger der Wissenschaften im 18. Jahrhundert', *Historische Zeitschrift*, vol. 231 (1980), pp. 43–74; also Fritz Hartmann and Rudolf Vierhaus (eds.), *Der Akademiegedanke im 17. und 18. Jahrhundert* (Bremen and Wolfenbüttel, 1977). This collection of essays does not deal with the Royal Society.

7. Cardwell, *Organisation of Science*, pp. 99–100; Pfetsch, *Wissenschaftspolitik in Deutschland*, pp. 343–45. See also Appendix 1 in this volume, p. 256.

8. McKie, 'Scientific Societies', p. 139; A.T. Gage, *A History of the Linnean Society of London* (London, 1938); David Elliston Allen, *The Naturalist in Britain. A Social History* (Harmondsworth, 1978), pp. 46–7.

9. A. J. Meadows, *Communication in Science* (London, 1974), pp. 66–86.

10. *Year-Book of the Scientific and Learned Societies of Great Britain and Ireland* (London, 1884), p. III. On the state of research, see Roy M. MacLeod, J. R. Friday

which dominated the scientific scene and the organisation of science, boasted fifty-four societies for the natural sciences (of widely varying significance), in addition to seventeen literary and historical societies, ten archaeological and thirty-six medical societies. The scale stretched from small specialist societies such as the Mathematical Society with 170 members or the Physiological Society with 50, to the large supra-regional societies such as the British Association for the Advancement of Science with 4,000 members, the Chemical Society with 1,300, and the Pharmaceutical Society of Great Britain with about 6,000.[11]

In almost all cases the core of the membership was made up of scientists, who were either private scholars or employed in universities or industry. In individual cases, suitability for membership was ascertained by an entrance examination, like that run by the Institute of Chemistry, for example, or the Pharmaceutical Society.[12] But such strict admission procedures were exceptional in the nineteenth century and already pointed towards the transition from specialist society to professional association which certain societies were undergoing. This development was to become a characteristic feature in the twentieth century.

By far the majority of learned societies in the nineteenth century provided a setting as much for social as for scientific activities. The fact that some of them accepted women members, or at least allowed women to attend public meetings, encouraged this trend.[13] For the bourgeoisie, an interest in science was often an entertainment or a diversion. In many cases a society's public prestige depended on the brilliance of its social activities and the social status of its president or patron. As institutions catering for bourgeois sociability, most learned societies, especially those outside London, had many lay members, even as late as the end of the century. Given a lack of financial support from public funds, membership subscriptions were an important source of income — sometimes the only one. But even taking the social aspect into account, it is noteworthy that membership numbers in all learned and scientific societies rose sharply, in line with the development of the sciences,

---

and Carol Gregor, *The Corresponding Societies of the British Association for the Advancement of Science 1883–1929: A Survey of Historical Records, Archives and Publications* (London, 1975).
   11. See Appendix 2 in this volume, p. 257–9.
   12. Lutz F. Haber, *The Chemical Industry 1900–1930: International Growth and Technological Change* (Oxford, 1971), p. 36; *Year-Book*, pp. 52–3.
   13. David E. Allen, 'The Women Members of the Botanical Society of London, 1836–1856', *British Journal for the History of Science*, vol. 13 (1980), pp. 240–54.

after the middle of the century. Roy MacLeod has demonstrated this convincingly for London. According to his figures, the thirteen most important scientific societies in the capital had a total of 5,000 members in 1850, around 10,000 in 1870 and 20,000 in 1910.[14] But by this time 'interested lay people' were increasingly being excluded from learned societies, though this development has not been completed to the present day. Learned societies were becoming professional associations in the strictest sense.

In 1899 the total number of learned and scientific societies in the United Kingdom fell to 453, because dissolutions and amalgamations exceeded new foundations. It is difficult to say how accurately this figure reflects the actual situation. The 1900 *Year-Book of the Scientific and Learned Societies*, from which this information comes, points out that many local medical, legal and literary-philosophical associations were not included in the statistics. A learned name alone was not a sure indication that a society was primarily orientated towards goals of a scientific rather than a social nature. Naturalists' Field Clubs, for instance, which flourished all over the United Kingdom between 1850 and 1914 and are thought to have had 50,000 members by the end of the century,[15] were overwhelmingly orientated towards social activities. For this reason, the year-books do not list them except in exceptional cases. Nevertheless, information in the year-books is important as an indicator of, and frame of reference for, the evolution of learned societies. In spite of statistical errors, it makes clear to what a large extent scientific societies were a phenomenon of the nineteenth century, appearing against the background of an unprecedented development, fragmentation and specialisation of science. Only 18 of the 453 societies in the United Kingdom at the end of the nineteenth century had been founded before 1800. According to available (but incomplete) statistics, the vast majority, numbering 149 societies, was founded between 1865 and 1888.[16]

14. Roy M. MacLeod, 'Resources of Science in Victorian England: The Endowment of Science Movement, 1868–1900' in Mathias (ed.), *Science and Society*, p. 114. The membership figures of the Chemical Society of London, for example, increased as follows:

1860	323 members
1880	1,034 members
1900	2,292 members
1920	3,721 members

(Meadows, *Communication in Science*, p. 13).

15. Allen, *Naturalist in Britain*, p. 170. Allen's informative book is the first comprehensive study of the organisational forms and the scientific and social activities of British naturalists and nature lovers from the seventeenth century to the present. He discusses the reasons for the decline in Naturalists' Field Clubs after the First World War (pp. 244–51).

16. See Appendix 1 in this volume, p. 256.

The most important of the British scientific societies was the Royal Society of London, which was also the oldest, having been established in the seventeenth century. It dealt with all the natural sciences, and therefore ranked above individual specialist societies as a sort of 'national academy of sciences', though it lacked a historical-philological section.[17] But in spite of its tradition and elevated social position compared with other scientific organisations, the Royal Society's scientific reputation was in dispute by the early nineteenth century.[18] This was largely because many of its members were not, in the strict sense of the word, professional scientists. Like other societies, the Royal Society had many interested lay members in the first half of the nineteenth century. Those who patronised science because of their social position, financial means or the idea that it was the thing for a 'cultured gentleman' to do were also strongly represented. In fact, since its inception, the Royal Society had followed a deliberate policy of accepting this sort of member for financial reasons. In the Royal Society's early years, at most a third of its Fellows could be considered scientists,[19] and by the beginning of the nineteenth century non-scientists were still in a clear majority. In 1830 it counted among its members ten Anglican bishops, of whom only one had a scientific publication to his credit. None of its sixty-three aristocratic Fellows had produced one. Eventually, in 1830 the Duke of Sussex was elected president of the Royal Society, the only consideration in his nomination being his social position.[20]

17. For the humanities, The British Academy for the Promotion of Historical, Philosophical, and Philological Studies was founded in 1902. On its foundation, see *Record of the Royal Society*, pp. 69–72; Frederic G. Kenyon, *The British Academy: The First Fifty Years* (London, 1952); Peter Alter, 'The Royal Society and the International Association of Academies 1897–1919', *Notes and Records of the Royal Society of London*, vol. 34 (1979–80), pp. 241–64. For the Royal Society, see the references in note 6. In addition, see Henry Lyons, *The Royal Society 1660-1940: A History of its Administration under its Charters* (Cambridge, 1944); E. N. da Costa Andrade, *A Brief History of the Royal Society* (London, 1960); Robert K. Merton, *Science, Technology and Society in Seventeenth Century England*, (Bruges, 1938, reprinted New York, 1970); P. M. Rattansi, 'The Intellectual Origins of the Royal Society', *Notes and Records of the Royal Society of London*, vol. 23 (1968), pp. 129–43; Christopher Hill, 'The Intellectual Origins of the Royal Society — London or Oxford?', *Notes and Records of the Royal Society of London*, vol. 23 (1968), pp. 144–56; Rupert A. Hall and Marie Boas, 'The Intellectual Origins of the Royal Society — London and Oxford', *Notes and Records of the Royal Society of London*, vol. 23 (1968), pp. 157–68; Christopher Hill, *Intellectual Origins of the English Revolution* (Oxford, 1965).
18. Dorothy Stimson, *Scientists and Amateurs: A History of the Royal Society* (New York, 1948), pp. 179–96; L. Pearce Williams, 'The Royal Society and the Founding of the British Association for the Advancement of Science', *Notes and Records of the Royal Society of London*, vol. 16 (1961), pp. 221–33.
19. *Record of the Royal Society*, pp. 55, 58.
20. Everett Mendelsohn, 'The Emergence of Science as a Profession in Nine-

This state of affairs, which could have resulted in the Royal Society degenerating into merely another London club like so many others, was a cause for concern during the first decades of the nineteenth century. Uneasiness was expressed not only by the scientifically aware public but also within the Royal Society itself. A rather cautious reform of the statutes in 1830, however, did not produce a fundamentally new membership structure. Not until 1846–7, when the statutes were altered to limit the number of new members who could be elected each year and to make scientific achievement a qualification for membership, did the situation gradually change. As the Royal Society became an organisation of scientists during the second half of the nineteenth century, the total number of Fellows fell.[21] 'By 1860 the reformers had succeeded in completely transforming the Society; for the first time since its foundation the scientific Fellows were more numerous than their colleagues.'[22] In line with this development, the Royal Society once again devoted itself more to scientific and organisational tasks. Within a generation, claims the official history of the Society, it had changed from 'a body of well-educated and cultivated men' into a 'scientific institution of the highest rank'.[23]

All of Britain's learned and scientific societies were dependent on regular membership subscriptions to finance their activities. This explains why the support they could offer to scientific enterprises in the nineteenth century was as a rule extremely limited. Since the middle of the century the larger specialist societies had received government subsidies while maintaining organisational independence. But the amounts they received were generally very small, usually enough only to contribute to day-to-day running costs.[24] The Royal Geographical Society, which had almost 3,400 members in 1883, making it one of the largest specialist societies in the nineteenth century, had received £500 annually from the government since 1854, as had the Royal Zoological Society of Ireland. The Royal Society of Edinburgh had been given £300 annually since 1834 and the Royal Scottish Meteorological Society a modest

---

teenth-Century Europe' in Karl Hill (ed.), *The Management of Scientists* (Boston, Mass., 1964), p. 28; Lyons, *The Royal Society*, pp. 232–4.

21. While the number of scientists in general continued to increase, the number of Fellows declined to 456 in 1900 after a peak of 769 in 1841. See Appendix 3 in this volume, p. 259.

22. Lyons, *The Royal Society*, p. 270.

23. *Record of the Royal Society*, p. 58. See also the chapter 'A Scientific Society: 1860–1900' in Lyons, *The Royal Society*, pp. 272–97.

24. See the lists of subsidies for 1856 to 1906 in *Scientific Societies (Government Grants): Return to an Order of the Honourable The House of Commons, dated 25 July 1906* (London, 1906), pp. 2–3.

£100 since 1893.[25] Very few societies possessed assets of their own, acquired, for example, through endowments. Thus the Chemical Society, which expanded from 1,300 members in 1883 to 3,100 in 1911, had only £220 to allocate for research at the beginning of the twentieth century. According to the president of the Society at the time, this was 'a very modest amount considering the number of claims and the activity of our workers'[26] — an opinion also shared by others.

The Royal Society of London, which had increasingly taken on the role of an unofficial scientific advisory body to the government since the second half of the nineteenth century,[27] had greater financial resources at its disposal because of the proceeds from its own assets and endowments. But as recent studies have shown,[28] it too could give only limited financial support to scientific projects. If we leave aside the Society's subsidising of scientific publications, one of its traditional activities, then before the middle of the century it had practically no money available to support actual research. Not until 1828 did some of its Fellows establish a Donation Fund, the proceeds of which were eventually to finance scientific projects. By 1900 the Society had seven similar funds tied to specific purposes. Most of them had been bequeathed by members during the second half of the nineteenth century. In 1900 the annual proceeds of these funds was just under £1,400.[29] The Society did not receive more-substantial endowments until after the First World War.

In recognition of its prominent position amongst scientific organisations, the Royal Society received the largest share of public

25. Ibid., pp. 2–3; *Nature*, 94 (1914–15), p. 552 ('State Aid for Science'); Roy M. MacLeod and E. Kay Andrews, *Selected Science Statistics Relating to Research Endowment and Higher Education, 1850–1914*, Mimeograph (University of Sussex, 1967), p. 10. On the membership figures of scientific societies in Britain, see also Appendix 2 in this volume, p. 257–9.

26. President's speech at the annual meeting of the Chemical Society on 22 March 1907. An abridged version is printed as 'The Position and Prospects of Chemical Research in Great Britain', *Nature*, 76 (1907), pp. 231–5, quotation on p. 234.

27. The Royal Society, which came to see itself as 'the Government organ for science' (*Year-Book of the Royal Society of London 1902*, London, 1902, p. 183), was commissioned from time to time by various ministries and public bodies to carry out or supervise scientific enquiries or projects. Examples include research into tropical diseases in 1896 and 1904, and into volcanism in the Caribbean in 1902 (*Record of the Royal Society*, p. 81; 'The Royal Society: A Retrospect of 250 Years', *The Times*, 16 July 1912). On the Society's role in other parts of the empire, see Roy M. MacLeod, 'Scientific Advice for British India: Imperial Perceptions and Administrative Goals, 1898–1923', *Modern Asian Studies*, vol. 9 (1975), pp. 343–84.

28. See Roy M. MacLeod, 'The Royal Society and the Government Grant: Notes on the Administration of Scientific Research, 1849–1914', *Historical Journal*, vol. 14, (1971), pp. 323–58.

29. *Record of the Royal Society*, pp. 65–7, 78 and 120–5. In 1842 the proceeds were £120; by 1912 they had risen to £2,300.

money. Since 1849 it had administered a fund, allocated by Parliament, which initially amounted to £1,000 per annum. But after the publication of the Devonshire Commission's Report in 1876, when the number of applications for research grants increased rapidly, the Grant-in-Aid for Scientific Investigations reached £5,000. In 1882 it was reduced to £4,000 and stayed at this level until 1920. Even by the yardsticks of those times, this amount was small, but it enabled the Society both to subsidise publishing costs and to provide financial support for the more expensive research projects proposed by individual scientists. Until the end of the nineteenth century this grant represented the largest regular contribution made by the state towards funding science; between 1850 and 1914 almost 1,000 scientists and more than 2,300 projects were subsidised out of it. The total amount spent during this period was about £179,000.[30]

The Royal Society supported a range of projects including all the natural sciences, with special emphasis on astronomy (relatively costly in the nineteenth century because of the expensive instruments required), physics, chemistry, mathematics and geology, and towards the end of the century, biology, which was developing extremely rapidly at that time. The Royal Society rarely supported applied sciences both because the costs involved were high and because it tended to concentrate on the pure sciences. The largest proportion of grants the Royal Society administered, seldom exceeding £200 per project between 1850 and 1914, went to its own members, though their share fell in the last third of the century as a wider circle of experts was consulted. Nevertheless, 52.5 per cent of those who received grants between 1905 and 1909 were members of the Royal Society. Although grants were unevenly distributed, members of the Royal Society and especially scientists in London being favoured, this represented the beginnings of indirect state support for science. Of course, the actual sums involved were very small, even during the period from 1876 to 1882 when the grant-in-aid was temporarily increased. The Royal Society, which administered this regular government subsidy as well as the proceeds from its own assets and funds, became the most important patron of British science in the nineteenth century. Its significance lay in the fact that it supported the projects of individual scientists; private patrons and, later, state authorities preferred to fund research institutions such as universities, laboratories and libraries.[31]

30. MacLeod, 'The Royal Society', p. 324; MacLeod, 'Resources of Science', p. 146; *Record of the Royal Society*, pp. 79–80 and 185–91.
31. See below, pp. 35–68 in this volume.

Not until after the First World War did the practice of giving direct grants for scientific research become more widely established.

The Natural Sciences Become Academically Established

Only from the second half of the nineteenth century have universities had a larger share in supporting science, but since then they have played an increasingly important part. In the nineteenth century there were relatively few universities in the United Kingdom, and they differed from each other in origin, organisation and social prestige to a far greater extent than was the case in Germany or France. British universities enjoyed a large degree of autonomy from the state, including financial independence, and this placed many obstacles in the way of any standardisation of organisation and courses until very recently. In the nineteenth century, British universities were incorporated under royal charter or Act of Parliament and governed their own internal affairs. The state granted them responsibility for academic education, but it reserved the right to set up royal commissions and to pass laws based on their recommendations. But no reorganisation of the whole university system was ever attempted.[32]

In terms of social prestige, Oxford and Cambridge were undoubtedly the top universities in the nineteenth century, but the Scottish universities had an equal, if not superior, scientific reputation, at least until the middle of the century. In the early nineteenth century, Oxford and Cambridge did not fully participate in developments which made the Prussian universities in particular (as a result of Wilhelm von Humboldt's reforms) into centres of scientific research. For reasons which are beyond the scope of this study, until the end of the century Oxford and Cambridge were not primarily places of research where students received a specialised academic training from teachers who also engaged in research. Rather, they were establishments providing a higher education for the children of the social and political elite. According to Matthew Arnold, who published a comparative study of European universities in 1868, Oxford and Cambridge were not universities in the continental European sense; they were 'places where the youth of the upper-class prolong to a very great age, and under some very admirable influences, their school education . . . . They are in fact

---

32. On this see Joseph Ben-David, *Centers of Learning: Britain, France, Germany, United States* (New York, 1977), pp. 52–3.

still schools'.[33] The French historian, Hippolyte Taine, called the two English universities 'in many respects a club for young men of the nobility and gentry, or at least of wealth'.[34] According to Michael Sanderson, Oxford and Cambridge were 'traditionally ... the natural resort for the complete education of the Anglican gentleman and the normal prerequisite for preferment in Church and public life'.[35] Regulations which permitted only members of the Church of England to enter the two English universities were not changed until 1854 in Oxford and two years later in Cambridge. Until this time, sons of the prosperous, Nonconformist industrial bourgeoisie from the north of England and the Midlands were excluded and had to turn to London and the Scottish universities. Not until 1871, however, could non-Anglicans hold college fellowships at Oxford or Cambridge.[36]

Although close contacts existed between individual scientists, scientific societies and industry before and during the industrial revolution, as A. E. Musson and Eric Robinson have demonstrated for Manchester, and Robert E. Schofield for Birmingham,[37] industrialisation in Britain took place almost entirely independently of the universities. According to Eric Ashby, author of a study of the relationship between universities and the scientific and industrial revolution in Britain since the seventeenth century, Britain's 'industrial strength lay in its amateurs and self-made men: the craftsman-inventor, the mill-owner, the iron-master'.[38] This as-

33. Quoted from Eric Ashby, *Technology and the Academics: An Essay on Universities and the Scientific Revolution* (London, 1958, reprinted London and New York, 1966), p. 13.
34. Hippolyte Taine, *Notes on England* (London, 1957), p. 118, trans. from the French, *Notes sur l'Angleterre* (Paris, 1872).
35. Michael Sanderson, *The Universities and British Industry 1850–1970* (London, 1972), p. 31. See also D. C. Coleman, 'Gentlemen and Players', *Economic History Review*, vol. 26 (1973), pp. 99–100. Sanderson (p. 5) quotes J. S. Mill's definition of what a university should *not* be: 'It is not a place of professional education. Universities are not intended to teach the knowledge required to fit men for some special mode of gaining a livelihood. Their object is not to make skilful lawyers and physicians or engineers, but capable and cultivated human beings' (Inaugural Address delivered to the University of St Andrews, 1867). On the ideal of the gentleman, see also Martin J. Wiener, *English Culture and the Decline of the Industrial Spirit 1850–1980* (Cambridge, 1981).
36. V. H. H. Green, *Religion at Oxford and Cambridge* (London, 1964), pp. 297–305; A. J. Engel, *From Clergyman to Don: The Rise of the Academic Profession in Nineteenth-Century Oxford* (Oxford, 1983), pp. 56–81.
37. A. E. Musson and Eric Robinson, *Science and Technology in the Industrial Revolution* (Manchester, 1969); A. E. Musson (ed.) *Wissenschaft, Technik und Wirtschaftswachstum im 18. Jahrhundert* (Frankfurt, 1972); Rose and Rose, *Science and Society*, pp. 17–19; Schofield, *Lunar Society*; Robert E. Schofield, 'Die Orientierung der Wissenschaft auf die Industrie in der Lunar Society von Birmingham' in Musson (ed.), *Wissenschaft*, pp. 153–64.
38. Ashby, *Technology and the Academics*, p. 50. See also Eric Ashby, 'On Universities and the Scientific Revolution' in A. H. Halsey et al. (eds.), *Education,*

sessment must be modified in the light of more recent research which has shown that the Industrial Revolution also had a scientific basis. The men of the Industrial Revolution — the inventors, innovators and industrialists — were not, as a rule, academically trained scientists. They were technicians, well-grounded in empiricism. But they frequently had a highly developed interest in science which found expression in, among other things, their membership of national and local scientific societies — something that has long been overlooked. Recently, however, economic historians have drawn attention to it, unearthing numerous examples.[39]

It remains an open question as to whether the entrepreneurs of the Industrial Revolution were aware of the long-term problem: industrial innovations are not one-off events which provide the basis for production over an unlimited period of time. In many branches of modern industry, innovations and inventions are tied closely to scientific research and form an integral part of a continuous process. The full implications of this problem were not perceived during the first half of the nineteenth century; probably they were obscured by British industry's dominant position at this time. It was only the appearance of foreign competitors towards the end of the century that triggered off the well-known debate about the weaknesses of British industry, the conservatism of British entrepreneurs and the inadequacies of the British educational system.[40] Among the universities, it was only Cambridge that encouraged the sciences around the middle of the nineteenth century. Large physical and chemical research laboratories on the European model were not set up until the early 1870s at the two English universities. The Clarendon Laboratory in Oxford dates from 1872, and the Cavendish Laboratory in Cambridge from 1874. Famous physicists such as James Clerk Maxwell (1831–79), Lord Rayleigh (1842–1919), Ernest Rutherford (1871–1937) and Joseph J. Thomson (1857–1940) worked at the Cavendish Laboratory, which attained international fame after the First World War for the work done there in nuclear physics and biochemistry.

Until well into the nineteenth century, Oxford and Cambridge remained the only two universities in England, in spite of their

---

Economy and Society (New York, 1961), pp. 466–76; and R. M. Hartwell, *The Industrial Revolution and Economic Growth* (London, 1971), pp. 146–9, 233, 243.
  39. In addition to the works referred to, see Mathias, 'Who Unbound Prometheus?', pp. 62–3.
  40. Coleman, 'Gentlemen and Players'; David Ward, 'The Public Schools and Industry in Britain after 1870', *Journal of Contemporary History*, vol. 2, no. 3 (1967), pp. 37–52; Wiener, *English Culture*, pp. 16–24. Also see below, pp. 98–137 in this volume.

adherence to pre-industrial educational ideals, and in spite of their shortcomings as far as science was concerned. Various reforms since the middle of the century had attempted to remedy these, influenced to some extent by developments taking place in German universities.[41] This situation did not change until the 1830s, when two new universities were founded: Durham University in 1832 and the University of London in 1836. The University of London was established both because it was felt that the metropolis of the empire needed a university, and because of dissatisfaction with the Anglican character of Oxford and Cambridge. Non-Anglicans were not admitted as students until 1854 at Oxford and 1856 at Cambridge, and could not teach there until 1871. The University of London was preceded by University College, which was founded as a limited company in 1826 on the private initiative of Jeremy Bentham and John Stuart Mill. It was set up as a secular institution, modelled on the universities of Berlin and Bonn, in order to give non-Anglicans the chance of an academic education. The Jewish community in England in particular benefited from the establishment of the College, and subsequently became one of its benefactors. King's College, founded in 1829 by a group of Anglicans, including the Duke of Wellington, Robert Peel and the Archbishop of Canterbury, was set up in direct competition to the 'godless institution in Gower Street' — 'godless' because theology was not taught there. Despite the denominational origins of King's College, its organisation, curricula and conditions of admission were very similar to those of University College.[42] But professors at King's College had to be members of the Anglican Church — one of the main reasons, incidentally, why Justus von Liebig did not get a chair there in 1845.

After 1836 the two rival colleges, both strong in the natural sciences, together made up the University of London. The new university was not a teaching institution; it was basically an examining body for the two London colleges, and had the right to confer degrees. Since 1839 it had received the modest sum of £5,000 from the government for carrying out this function. But its responsibilities in this respect were far greater than those of Oxford and

41. Gizycki, 'Centre and Periphery', pp. 486–8; MacLeod, 'Resources of Science', pp. 115–21; W. H. G. Armytage, *The German Influence on English Education* (London, 1969); Harald Husemann, 'Zu den deutsch-englischen Universitätsbeziehungen während der letzten hundert Jahre' in Hartmut Boockmann et al. (eds.), *Geschichte und Gegenwart. Festschrift für Karl Dietrich Erdmann* (Neumünster, 1980), pp. 459–90.
42. H. Hale Bellot, *University College, London, 1826–1926* (London, 1929); F. J. C. Hearnshaw, *The Centenary History of King's College London 1828-1928* (London, 1929).

Cambridge, because it was an examining body not only for University College and King's College but for all the academic institutions in the United Kingdom which it recognised as such. This was a completely new and unparalleled role for a university. In 1849 its jurisdiction was extended to include the whole of the empire, which meant that the University of London played an important part in the founding and development of tertiary institutions not only in England but throughout the colonies and dominions. Its examination requirements set the standards for education in universities being created all over the empire. Not until 1903 was the University of London reorganised along lines which provided the foundation for its present form. It became a modern university with teaching commitments, but its rather unusual organisational structure was retained. On the eve of the First World War the University of London comprised eighty-nine colleges, schools, institutes and hospitals.[43]

The period during which the University of London was reorganised was one of great significance in the history of British universities because such a large number were founded at this time. The period between 1900 and 1910 compares only with that from 1950 to 1970, when for the second time this century the number of universities in Britain increased sharply, in line with international developments. There were several reasons for the first wave of new foundations, which resulted in the creation of the modern British university system. The most obvious was an increasing public awareness of the backwardness of the British educational system in comparison with Germany and the USA.[44] In 1901, for example, it was noted that the United Kingdom had fewer universities than any other western country, and that it had considerably fewer students than comparable countries.[45] The proportion of university students in Britain was not maintained as the population increased

---

43. Thomas Lloyd Humberstone, *University Reform in London* (London, 1926); Sanderson, *Universities*, pp. 106–7; Michael Sanderson, 'The University of London and Industrial Progress 1880–1914', *Journal of Contemporary History*, vol. 7, nos. 3 and 4 (1972), pp. 243–62. See also Abraham Flexner, *Universities: American, English, German* (New York, 1930, reprinted London, 1968).
44. On this see below, pp. 98–104 in this volume.
45. Figures given for the number of British students do not tally. According to A. H. Halsey, there was a total of 20,000 students in Britain in the academic year 1900/1 (A. H. Halsey, ed., *Trends in British Society since 1900: A Guide to the Changing Social Structure of Britain*, London, 1972, p. 206). Haber calculates the number of students in the United Kingdom as 21,000 in 1900, and 25,000 in 1914 (*The Chemical Industry 1900–1930*, p. 52). According to F. R. Pfetsch, there were just under 34,000 students at universities and just under 10,800 at *Technische Hochschulen* in Germany in 1900 (*Wissenschaftspolitik in Deutschland*, p. 186).

during the nineteenth century.[46] Added to this was the knowledge that the industrial efficiency of Britain's European competitors and of the USA was increasing. Finally, the Boer War had revealed the need for reform in various areas, and had sparked off a public campaign for 'national efficiency'.

During the nineteenth century only three new universities had been founded in Britain: Durham University in 1832, with strengths in the humanities,[47] London in 1836 and Victoria University in 1880. Thus before 1900 there was a total of five universities in England. But a large number of regional colleges was founded in the second half of the nineteenth century, under the academic patronage of the University of London or, to a lesser extent, Oxford or Cambridge. These new colleges concentrated mainly on science and technology, which were taught at university level. Students were examined by the University of London and were awarded its degrees. Owens College in Manchester (1851), Yorkshire College of Science in Leeds (1847), Firth College in Sheffield (1879) and Mason College in Birmingham (1875) were all founded as local educational and training establishments of this sort. Unlike Oxford and Cambridge, whose social prestige was by no means threatened by these new institutions,[48] the regional colleges were orientated strongly towards the demand of local trade and industry for people with academic training. Colleges were therefore often founded as private initiatives, or by municipal administrations or county councils. Central government in London was not involved.

Academic standards and syllabuses varied widely between the provincial colleges. The development from small educational establishments, which were likely to be private institutions, to full

46. See Sanderson, *Universities*, p. 3. According to one estimate, in 1900 there were 12.8 students per 10,000 of the population in the USA, 7.9 per 10,000 in Germany and 5 per 10,000 in the United Kingdom (Cardwell, *Organisation of Science*, p. 204).
47. The foundation of Durham University was a special case in the nineteenth century. It was founded by the cathedral chapter, 'which, in its alarm over the prospect of church reform and possible confiscation of ecclesiastical revenues, hastened to invest some of its accumulated surplus in a college' (David Owen, *English Philanthropy, 1600–1960*, Cambridge, Mass., 1964). As a university which modelled itself on Oxford and Cambridge, it was not a success. In 1863 it had a total of only forty-four students (Ward, 'Public Schools', p. 51).
48. 'As a result a two tier structure emerged in the early twentieth century. Oxford and Cambridge were national universities connected with the national elites of politics, administration, business and the liberal professions. The rest were provincial, all of them, including London, taking most of their students from their own region and training them in undergraduate professional schools for the newer technological and professional occupations created by industrialism such as chemistry, electrical engineering, state grammar school teaching and the scientific civil service' (A. H. Halsey, 'The Changing Functions of Universities' in A. H. Halsey et al., eds., *Education*, p. 461).

teaching universities with strengths in the natural sciences, had progressed furthest in the north of England and the Midlands by the end of the nineteenth century. In the 1880s Owens College in Manchester, University College in Liverpool (founded in 1881) and Yorkshire College of Science in Leeds amalgamated to form Victoria University. Like the University of London, Victoria University was essentially a coordinating umbrella organisation and an examining body with the right to confer degrees. It existed until 1903, when it was disbanded largely on the instigation of Liverpool and Manchester. Its constituent colleges became independent universities; Manchester and Liverpool became full universities in 1903, while Leeds, which had always been the weakest of the three, especially in financial terms, did not receive full university status until 1904. Birmingham, thanks to the initiative of Joseph Chamberlain, a former mayor, had had its own university since 1900. It had developed from Mason College, founded in 1875 during Chamberlain's period of office. Birmingham University was the first civic university in the country. Its creation undoubtedly encouraged the colleges in Manchester and Liverpool to aim for the dissolution of Victoria University. Sheffield received university status in 1905, and Bristol in 1909.

The organisation of the new provincial universities basically resembled that of German universities. They did not develop along the lines of Oxford and Cambridge and, to a certain extent, London — that is, by the addition of new colleges — but by the expansion of existing faculties, departments, chairs and institutes, and the creation of new ones. At the new universities, much greater emphasis was placed on scientific and technological subjects, which had been inadequately represented at British universities. The influence of the European *Technische Hochschulen* on this development is unmistakable. Speaking in Birmingham in 1905, Chamberlain expressly rejected any competition between the new universities and Oxford and Cambridge in the humanities. The *raison d'être* of Birmingham University, he said, was 'to teach, as it had never been taught before, science, and also to proceed in the work of scientific research'.[49] Chamberlain could refer back to the intentions of Josiah Mason, founder of the university college in Birmingham. Mason had rejected the overwhelming preponderance of the humanities in the two old English universities, and had therefore intended excluding them from the institution he was founding.[50] In Leeds, too, influential voices were raised in support of excluding the humani-

49. *The Times*, 15 May 1905.
50. Sanderson, *Universities*, p. 6.

Table 1.1. Percentage of students from within a radius of 30 miles of the university at selected British universities 1908–56

University	1908/9	1948/9	1955/6
Birmingham	—	56	38
Bristol	87	39	26
Leeds	78	60	40
Liverpool	75	62	55
Manchester	73	59	48
University College London	66	53	43

Source: A. H. Halsey, 'The Changing Functions of Universities' in A. H. Halsey, Jean E. Floud and C. Arnold Anderson (eds.), *Education, Economy and Society* (New York, 1961), p. 462.

ties from the curriculum and concentrating on science and technology. Even today these origins are visible in British universities founded since the end of the nineteenth century. In Liverpool, for example, the ratio of arts to science students was forty-two to fifty-eight in the late 1970s.[51]

The foundation of six new civic universities within ten years meant that England had made up its obvious deficit in the tertiary sector, at least in quantitative terms. The centuries-old monopoly of Oxford and Cambridge had been broken. The particular significance of the new universities was not only that they gave the natural sciences a privileged position, but also that they cooperated with local industry, establishing a relationship between science and industry which was new in England. Initially, they drew most of their students from the immediate vicinity (Table 1.1). Manchester in particular flourished in terms of science. This rapidly growing industrial city developed a lively scientific and cultural scene,[52] which since the eighteenth century could have been matched outside London only in Edinburgh and perhaps Birmingham and Dublin. Against this background, Manchester earned itself a high

---

51. Ngaio Crequer, 'Rising from the Ruins: Liverpool's Success Story', *Times Higher Education Supplement*, 1 February 1980, p. 8.
52. The Hallé Orchestra, for example, which still exists today, was founded as early as 1857 by Carl Hallé in Manchester. See Briggs, *Victorian Cities*, pp. 88–138 (ch. 3, 'Manchester, Symbol of a New Age'), and Cardwell, 'Patronage of Science', pp. 95–113. 'The endowment of science in Manchester over the years 1800 to 1914 is perhaps the best instance of self-endowed, or self-supporting science on record. The profits made from the great Manchester cotton trade, and from the industries to which it gave rise, to a large extent endowed Manchester science' (Cardwell, p. 108).

reputation in the natural sciences, impressively underlined by the appointment and work of the physicist Ernest Rutherford and his circle (Niels Bohr, James Chadwick, H. G. J. Moseley).[53] Within a few decades Manchester, whose social and administrative problems had made it the 'shock city' of the 1840s,[54] had become 'a world centre for scientific research'.[55]

Its rapidly acquired status in the sciences made Manchester one of the few exceptions on the educational scene in the late nineteenth century. According to Raphael Meldola, president of the Chemical Society, the contribution made by British universities to basic research, even until early this century, was 'far below the standard, both of quality and quantity, which might be expected and which we should like to see attained'. Referring to his own subject, he concluded that 'many of our universities are distinct failures as centres of chemical research, and that the total output of work from university laboratories is by no means worthy of the great traditions of this country as a pioneering nation in scientific discovery'.[56] We can only speculate as to why Meldola's negative assessment, even of his own subject, did not apply to Manchester University. Scientific interest, traditionally strong among members of the well-to-do middle class, many of whom were also willing to provide financial support, undoubtedly created favourable conditions for the establishment of a university in Manchester. In addition, as Cardwell points out, 'the high technology of local industry helped to direct attention towards science and the scientific problems that related to industry'.[57] It remains an open question whether these factors, fortuitously combining, provide a satisfactory explanation for Manchester's remarkable development in the context of the new universities.

The wave of new foundations during the last third of the nineteenth century took the number of universities in England to ten by 1910 (Table 1.2). Apart from these, there were four Scottish universities, founded in the fifteenth and sixteenth centuries, whose

---

53. J. B. Birks (ed.), *Rutherford at Manchester* (London, 1962); Kargon, *Science in Victorian Manchester*; Mario Bunge and William R. Shea (eds.), *Rutherford and Physics at the Turn of the Century* (New York, 1979).
54. Briggs, *Victorian Cities*, p. 56.
55. Kargon, *Science in Victorian Manchester*, p. 234. For example, Manchester had a chair of organic chemistry since 1874. For years it was the only one in the United Kingdom. The first professor to hold this chair had been Carl Schorlemmer (1834–92).
56. Speech given to the annual meeting of the Chemical Society on 22 March 1907. A shortened version was printed in *Nature*, 76 (1907), pp. 231–5, quotation on p. 232.
57. Cardwell, 'Patronage of Science', p. 108.

Table 1.2. Universities in the United Kingdom in 1910

University	Year of foundation
England	
Oxford	End of twelfth century
Cambridge	Beginning of thirteenth century
Durham	1832
London	1836
University College	1826
King's College	1829
London School of Economics and Political Science	1895
Imperial College of Science and Technology	1907
Birmingham	1900
Manchester	1903
Liverpool	1903
Leeds	1904
Sheffield	1905
Bristol	1909
Scotland	
St Andrews	1410
Glasgow	1451
Aberdeen	1495
Edinburgh	1583
Wales	
University of Wales	1893
Aberystwyth	1872
Bangor	1883
Cardiff	1884
Ireland	
Trinity College Dublin	1592
National University of Ireland	1908
Dublin	1908
Cork	1845
Galway	1845
Belfast	1908
Queen's College	1845

organisational and teaching structures resembled those of the German universities. In the context of the United Kingdom they were in many respects anomalous.[58] Their intellectual roots were Calvinist–Presbyterian, and their admissions policy was not restrictive but aimed to educate as broad a section of the population as possible. In 1850 England and Wales had four universities for a population of 18 million; Scotland had four universities catering for a population of 2.8 million. According to Sanderson's calculations, Scotland had one university to one million people in 1885, while the figure for England was one university to six million people.[59] The education received by students at Scottish universities was much more vocationally orientated than that provided in England. Scottish universities had medical faculties, and the natural sciences were well represented in the eighteenth and nineteenth centuries. The connection between the universities and the pioneers of industrialisation in Scotland are well known. The state had a large say in the running of Scottish universities — they received state subsidies from the early nineteenth century because private assets and endowments did not provide enough to finance them.[60] With the exception of University College Dundee (1880), which was integrated into the University of St Andrews in 1897, no new universities were founded in Scotland in the nineteenth century. The lack of new stimuli and reforms may have contributed to the stagnation which affected Scottish universities in the second half of the nineteenth century.

Wales had had its own university since 1893. From the day of its foundation it was regarded as a 'powerful symbol of popular achievement and of national status'.[61] But like the University of London and Victoria University, it was only an examining body and administrative centre for colleges in Aberystwyth, Bangor and Cardiff. After 1883 they received a government grant from London, which increased rapidly in subsequent years.[62]

58. See Sanderson, *Universities*, pp. 146–83 (ch. 3, 'The Scottish Universities 1850–1914'); J. B. Morrell, 'The University of Edinburgh in the Late Eighteenth Century: Its Scientific Eminence and Academic Structure', *Isis*, vol. 62 (1971), pp. 158–71; J. B. Morrell, 'The Patronage of Mid-Victorian Science in the University of Edinburgh', *Science Studies*, vol. 3 (1973), pp. 353–88, also printed in Turner (ed.), *Patronage of Science*, pp. 53–93.
59. Sanderson, *Universities*, p. 148.
60. From 1831 the Scottish universities received £5,000 annually, from 1869, £42,000 (*Nature*, 77, 1907–8, p. 275; James Mountford, *British Universities*, London, New York, Toronto, 1966, p. 150).
61. Kenneth O. Morgan, *Rebirth of a Nation: Wales 1880–1980* (Oxford, 1981), p. 110.
62. Initially this grant amounted to £4,000 (1883–4); later it was raised to £31,000 (1910–11); *Nature*, 94 (1914–15), p. 551; Sanderson, *Universities*, pp. 121–145 (ch. 5,

Ireland had one university, founded in 1592 by Elizabeth I. It consisted of only one college, Trinity College, whose social standing equalled that of Oxford and Cambridge. Trinity College Dublin was Anglican, and until late in the eighteenth century Catholics, who made up three-quarters of Ireland's population, were excluded from it. Since the Emancipation Act of 1829, the government in London, urged by the Catholic Church in Ireland, had made many attempts to establish a Catholic university which would offer the Catholic Irish a university education independently of Trinity College, but all solutions proved to be only temporary. Thus in the nineteenth century, central government took an active part in establishing universities only in Wales and Ireland. In 1845 interdenominational Queen's Colleges were founded in Belfast, Cork and Galway; in 1850 they combined to form the Queen's University of Ireland. The Catholic bishops of Ireland, however, opposed this arrangement as much as the short-lived Catholic University of Ireland (1851–4). In 1879–80 the Royal University of Ireland was created as an examining body along the lines of the University of London. But a long-term and generally acceptable solution was not found until 1908, when Queen's College in Belfast became a separate university for the Protestant Irish, based on the existing colleges, and the National University of Ireland was founded for the Catholics. It consisted of three colleges, in Dublin, Cork and Galway, and was modelled on the University of Wales. Thus at the beginning of the twentieth century there were three *de facto* denominational universities in Ireland, mirroring the country's difficult political situation: Trinity College in Dublin for the Anglicans, the Catholic National University of Ireland with three colleges, and Protestant Queen's College in Belfast.[63]

Like the Scottish universities and the Universities of Wales and London at this time, the two new Irish universities in Dublin and Belfast were dependent on public funding. From 1889 the new university colleges also received a larger public subsidy. Unlike Oxford and Cambridge, they were therefore from the start subject to stringent state supervision exercised by government-appointed Visitors. On the other hand, this also guaranteed them recognition

---

'The Welsh Universities Movement 1850–1914'); D. Emrys Evans, *The University of Wales: A Historical Sketch* (Cardiff, 1953); Morgan, *Rebirth of a Nation*, pp. 107–11.
63. R. B. McDowell and D. A. Webb, *Trinity College Dublin 1592–1952: An Academic History* (Cambridge, 1982); T. W. Moody, 'The Irish University Question of the Nineteenth Century', *History*, vol. 43 (1958), pp. 90–109; T. W. Moody and J. C. Beckett, *Queen's, Belfast 1845–1949: The History of a University*, 2 vols. (London, 1959).

as academic institutions. By 1897 central government was subsidising tertiary institutions in England (three in London, ten in the provinces) to the tune of £26,000 per annum. This represented considerably less than the corresponding amount spent in Germany. Cardwell estimates that expenditure on English universities in 1897/8 was £26,000, while expenditure in Prussia was £476,000 in 1900/1.[64] According to another estimate, Prussia spent 17.3 million marks on tertiary institutions (universities and *Technische Hochschulen*) in 1901, which at contemporary exchange rates was equivalent to £865,000.[65] J. Norman Lockyer, editor of *Nature* and a well-known scientist, calculated that at the turn of the century the states of the German Reich spent a total of £1 million on their universities.[66]

All the new English universities founded shortly after 1900 were dependent on state subsidies because endowments and student fees, the other main sources of income, did not bring in enough to cover running costs. Especially after the turn of the century the significance and amount of government grants increased steadily. In 1903/4 the allocation for English universities totalled £27,000; in 1904/5, after the prime minister had received a deputation from the universities and larger scientific societies, it rose to £54,000, and by 1905/6 it had reached £100,000.[67] It has been calculated that in 1910 total state expenditure on tertiary education in the United Kingdom was £200,000.[68] In the academic year 1908/9 the cost of running English universities was met from the following sources: 26.6 per cent came from the Treasury, 16.3 per cent from local education authorities, 32.2 per cent from student fees, 14.9 per cent from donations and the proceeds of endowments, and 10 per cent from other sources.[69] This breakdown does not include Oxford and Cambridge. They declined financial support from the state

64. Cardwell, *Organisation of Science*, p. 202.
65. Pfetsch, *Wissenschaftspolitik in Deutschland*, p. 72.
66. J. Norman Lockyer, 'The National Need of the State Endowment of Universities', in idem, *Education and National Progress: Essays and Addresses, 1870–1905* (London, 1906), p. 219.
67. *Reports from those Universities and University Colleges in Great Britain which Participated in the Parliamentary Grant for University Colleges in the Year 1908–09*, Cd. 5246 (London, 1910), p. XX.
68. Harold Perkin, *Key Profession: The History of the Association of University Teachers* (London, 1969), p. 33. According to A. J. Meadows, the sum in 1914 was £250,000 (*Science and Controversy: A Biography of Sir Norman Lockyer*, Cambridge, Mass., 1972, p. 269).
69. *Reports from those Universities*, p. XI. The president of the Board of Education, Sir Walter Runicman, gave slightly different percentages: Treasury, 27 per cent; local education authorities, 15 per cent; student fees, 32 per cent; donations, 14 per cent; other sources, 12 per cent (Hansard, Parl. Deb., H. C., 5th Series, vol. 19, col. 406, 13 July 1910).

until 1919 because the proceeds of endowments covered their costs until the end of the war.[70]

The Committee on Grants to University Colleges, founded in 1904, was responsible for distributing government funds. It developed into the University Grants Committee, which took over the financial administration of the universities for the Treasury in 1919. It was made up of representatives of both science and industry, and is still similarly constituted today. This limits, but does not preclude, the possibility of the state exercising direct influence on the universities.[71]

The Era of Patronage

Until well past the turn of the century, government in London supported the view that the public purse could provide only a limited amount of money for universities. It believed that, as in the USA, large private endowments would still be desirable, indeed necessary, in future, and favoured a mixed system of financing universities from various sources, as had evolved in practice since the turn of the century. Austen Chamberlain, chancellor of the exchequer, pointed out to the Committee on Grants to University Colleges that state aid 'should be given with the object of stimulating private benevolence in the locality'.[72] In similar vein, Walter Runciman, president of the Board of Education in Asquith's cabinet, explained in a bluebook in 1910: 'State-aid to university teaching would, however, be of doubtful advantage if it did not stimulate private effort and induce benefactors to contribute in the present day as they did in the olden times, to give of their wealth for the support of that higher learning upon which now, more than ever, the "prosperity, even the very safety and existence of our country depend".'[73]

Runciman's remark about private support for universities was a reference to the university colleges outside London, Oxford and Cambridge, whose history illustrates the importance of private patronage in supporting science at institutional level in the nine-

---

70. State support was accepted most reluctantly in 1919. The Master of Trinity College in Cambridge, however, believed 'that the only alternative was to lose the efficiency of the University, and much as he disliked the receipt of money from the Government he disliked still more the idea of an inefficient University' (quoted in Owen, *English Philanthropy*, p. 355).
71. *Education in Britain* (London, 1974), pp. 6 and 36–7.
72. Treasury Minute, 30 March 1904, T 1/10215 B/3211/1904, PRO.
73. *Reports from those Universities*, p. IV.

teenth century. While scientific societies as a rule received donations only from members or their families,[74] the new colleges, especially in the industrial Midlands, were supported, often generously and on a long-term basis, by local patrons: individual families, industries or large sections of the bourgeoisie. Without this private financial aid most of the colleges, which later became provincial universities, would neither have been founded, nor have developed into centres of scientific research as quickly as did Manchester, for example. Extensive involvement by private patrons is certainly a characteristic feature of the history of British universities in the nineteenth and twentieth centuries. It has parallels in the USA, but not on the Continent. The only example which can be cited for the German-language area is Frankfurt. The University of Frankfurt was founded in 1914 purely on private initiative as an endowed university. It was largely dependent on the patronage of an industrialist, Wilhelm Merton.[75]

No detailed studies exist of the scope and forms of private patronage of the sciences, or of the humanities and the arts, for that matter, in Victorian Britain, and it is unlikely that it will ever be possible to reconstruct them fully.[76] Useful statistics on aid given to science, whether indirectly by founding and equipping scientific institutions, or directly by financing specific research projects, are available only for certain areas in which private patronage was of particular importance, such as the establishment of provincial universities and scientific research institutes. Thus Michael Sanderson, author of the most detailed study to date of industry's role in the founding of university colleges after 1850,[77] points both to the extent of private endowments and to the connection between economic boom periods and patronage. Generous private grants were made at times of economic prosperity, such as the period between 1868 and 1873, and from the mid 1890s to the First World War, that is, after the end of the Great Depression, now often more

74. For the Royal Society of London, for example, see the list in *Record of the Royal Society*, pp. 107–30 and 142–4; for the Royal Astronomical Society, the list in J. L. E. Dreyer and H. H. Turner, *History of the Royal Astronomical Society, 1820-1920* (London, 1923), p. 246.
75. Paul Kluke, *Die Stiftungsuniversität Frankfurt am Main 1914–1932* (Frankfurt, 1972).
76. Owen, *English Philanthropy*, provides an overview of philanthropic activity in England since 1660. Ch. 13, 'Philanthropy in Academe', pp. 346–71, discusses endowments for scientific purposes. Brian Harrison, 'Philanthropy and the Victorians', *Victorian Studies*, vol. 9 (1965–6), pp. 353–74, is also important. References to endowments for the arts can be found in Janet Minihan, *The Nationalization of Culture: The Development of State Subsidies to the Arts in Great Britain* (London, 1977).
77. Sanderson, *Universities*, particularly pp. 61–94.

accurately described as a long period of deflation.[78] Many cases could be cited to illustrate this thesis. In 1846, for example, John Owens (1790–1846) bequeathed more than £100,00 to the newly established institution in Manchester.[79] Owens's example was not immediately followed. Not until twenty years later, when the scientific and technical departments of Owens College were to be extended during a period of economic prosperity, was a public collection held, raising more than £200,000.[80] It was initiated by Thomas G. Ashton, a cotton-mill owner, who is therefore often considered the 'second founder' of Owens College.[81] A group of leading Manchester industrialists established a chair of mechanical engineering and endowed it with £10,000. They included Charles Frederick Beyer, a manufacturer of locomotives. German by birth, he was educated at the *Technische Hochschule* in Dresden and subsequently went to Manchester, where in 1854 he became one of the founders of Beyer, Peacock & Co.[82] He donated a total of £108,000 to Owens College and attempted to encourage it to develop along the lines of the German technical colleges. In the 1870s Owens College received another £131,000 in donations of £10,000 or more. The 1880s were years of economic depression and falling profits, but towards the end of the century private grants began to increase again. Between 1903 and 1914 the University of Manchester, formerly Owens College, received an average of £30,000 annually. It has been calculated that in the sixty years between the foundation of Owens College and the outbreak of the First World War, Manchester received a total of almost £700,000 in private

---

78. H. L. Beales, 'The Great Depression in Industry and Trade', *Economic History Review*, vol. 5 (1934), pp. 65–75; A. E. Musson, 'British Industrial Growth, 1873–96: A Balanced View', *Economic History Review*, vol. 17 (1964–5), pp. 397–403; S. B. Saul, *The Myth of the Great Depression, 1873–1896* (London, 1969); David S. Landes, *The Unbound Prometheus: Technological Change and Industrial Development in Western Europe from 1750 to the Present* (Cambridge, 1969), pp. 231–7.
79. B. W. Clapp, *John Owens: Manchester Merchant* (Manchester, 1965); Briggs, *Victorian Cities*, pp. 135–6.
80. See Sanderson, *Universities*, pp. 62–3; Kargon, *Science in Victorian Manchester*, pp. 191–4; Edward Fiddes, *Chapters in the History of Owens College and Manchester University 1851–1914* (Manchester, 1937); H. B. Charlton, *Portrait of a University 1851–1951 to Commemorate the Centenary of Manchester University* (Manchester, 1951); Joseph Thompson, *The Owens College: Its Foundation and Growth* (Manchester, 1886).
81. Owen, *English Philanthropy*, p. 364. Thomas G. Ashton (1855–1933) came from a family of Nonconformists and attended a German university. He was a Liberal Member of Parliament in 1885–6 and again from 1895 to 1911, and knighted in 1911.
82. Owen, *English Philanthropy*, p. 364; D. S. L. Cardwell, *Technology, Science and History* (London, 1972), pp. 188–9; D. S. L. Cardwell, 'Patronage of Science', pp. 102 and 106–7. Beyer (1813–76) was born in Plauen (Saxony).
83. Sanderson, *Universities*, p. 63; Fiddes, *Manchester University*, pp. 186–8;

grants.[83] Thus Manchester, one of the most flourishing communities in Victorian England, with strong industries and an export trade, had one of the most richly endowed of the new universities.

The initiative for the foundation of University College in Liverpool, by contrast, came from the municipal administration, and not from a rich individual donor. In this case, too, after a hesitant beginning, private money began to flow in. In 1881, when the College was established, £130,000 had been found, and by the time Liverpool joined Victoria University in 1884, a further £30,000 was available. Individual chairs were established and funded by particular branches of industry. Ship-owners, for example, endowed a chair of mathematics with £10,000, and later (1909) a chair of shipbuilding with £20,000. Sir John T. Brunner, a chemical manufacturer, endowed chairs of economic science, physical chemistry and Egyptology.[84] The Brunner Chair of Economic Science alone was endowed with £10,000.[85] Considerable sums were also given between 1885 and 1900, when Liverpool was a member of Victoria University: by 1886 the public had donated a total of £150,000, and the municipal administration had given £20,000; in 1888 the sugar producer Sir Henry Tate gave £16,000 for the library; in 1889 the brewer Sir Andrew B. Walker donated £15,000 for technical laboratories. These are just a few examples selected from a long list of private patrons. Patronage continued undiminished until the beginning of the war, and particularly after Liverpool left Victoria University in 1903 individual donations were sometimes remarkably large.[86] In 1914 Liverpool's income from its endowments was £15,000; among the new English universities only Manchester had a higher income (£22,500) from this source.[87]

Owen, *English Philanthropy*, p. 364; P. J. Hartog (ed.), *The Owens College, Manchester: A Brief History of the College and Description of its Various Departments* (Manchester, 1900).

84. Stephen E. Koss, *Sir John Brunner: Radical Plutocrat, 1842–1919* (Cambridge, 1970), pp. 169–70. John Tomlinson Brunner (1842–1919), son of a Swiss teacher resident in England, founded Brunner, Mond & Co. with Ludwig Mond in 1873. He was a Liberal Member of Parliament between 1885 and 1910 and was knighted in 1895. 'In private as well as public life, Brunner was the quintessence of the nonconformist Radical' (Koss, p. 3).

85. Deed of Trust Establishing the Brunner Chair of Economic Science, 7 May 1891 (The University Archives, Liverpool, 6/1/5).

86. Sanderson, *Universities*, pp. 63–5; Oliver Lodge, *Past Years: An Autobiography* (London, 1931), pp. 162–4; Stanley Dumbell, *The University of Liverpool 1903–1953: A Jubilee Book* (Liverpool, [1953]), pp. 10–12.

87. *Nature*, 94 (1914–15), p. 326. Of the University of Manchester's costs, 27 per cent were paid for from the proceeds of endowments. In the case of the University of Liverpool the figure was 18 per cent (Sanderson, 'The University of London and Industrial Progress 1880–1914', *Journal of Contemporary History*, vol. 7 nos. 2 and 3, 1972, p. 258).

Since 1887 Yorkshire College of Science in Leeds had been the third partner making up Victoria University. Yorkshire College was founded in 1874 'to provide instruction in such sciences and arts as are applicable or ancillary to the manufacturing, mining, engineering and agricultural industries of the county of Yorkshire'.[88] The suggestion had come from local industrialists who had visited the Paris International Exhibition in 1867 and had observed that European industry was becoming increasingly competitive.[89] None of them, however, particularly distinguished himself as a benefactor in the early years of the College's existence. By 1874 only £20,000 had been raised.[90] This is a relatively modest sum when the economic prosperity of those years is taken into account, and Sanderson has suggested that this situation may have been a result of the city's relatively weak economic structure.[91] Subsequently, the Clothworkers' Company in London became the College's most important benefactor. By 1912 it had raised over £160,000, compensating Leeds for the lack of substantial individual donations.

This form of private funding by a non-local association made Leeds exceptional among the new colleges; the development of Birmingham and Bristol resembled that of Manchester and Liverpool. In Birmingham, as in Manchester, a local industrialist made a large initial donation. Josiah Mason (1795–1881), among other things the largest manufacturer of pens in the world, gave £60,000, which enabled a piece of land to be bought and a building to be erected. By the time Mason College opened in 1880, Mason, who like John Owens had never gone to university himself, had donated a total of £200,000. This was patronage 'without parallel in the annals of modern education',[92] not only by contemporary standards. Other local industrialists and companies followed suit with large donations. Joseph Chamberlain was the driving force behind developments in science organisation in Birmingham. Mason Col-

---

88. Quoted in Sanderson, *Universities*, p. 66.
89. A. N. Shimmin, *The University of Leeds: The First Half Century* (Cambridge, 1954), p. 11.
90. E. J. Brown, *The Private Donor in the History of the University of Leeds* (Leeds, 1953), p. 3.
91. Sanderson, *Universities*, pp. 66–7. See also Briggs, *Victorian Cities*, ch. 4, 'Leeds, a Study in Civic Pride', pp. 139–83.
92. H. F. W. Burstall and C. G. Burton, *Souvenir History of the Foundation and Development of the Mason Science College and of the University of Birmingham, 1880–1930* (Birmingham, 1930), p. 11; Sanderson, *Universities*, p. 69; Eric W. Vincent and Percival Hinton, *The University of Birmingham: Its History and Significance* (Birmingham, 1947). Biographical sketches of Josiah Mason, who was knighted in 1872, can be found in Owen, *English Philanthropy*, pp. 408–13, and in the *Dictionary of National Biography*, vol. 36, pp. 434–5.

lege was founded during his period of office, and especially after 1897 he devoted a great deal of energy to helping it attain full university status. Within three years his appeals for support for the new university had raised the astonishing sum of £330,000, including individual donations of over £50,000.[93] Thanks to Chamberlain's efforts, a total of almost £500,000 was raised for Birmingham University around the turn of the century.[94]

The renewed prosperity experienced by British industry after the end of the Great Depression in the mid 1890s was an important factor in producing this situation in Birmingham. The evident connection between economic prosperity and patronage is also illustrated by developments in Bristol. After its foundation in 1876, University College in Bristol had received little financial support, and appeals for aid had evoked a poor response.[95] Contemporaries attributed this primarily to the unfavourable economic climate, the 'depression of trade', but also to the general economic stagnation which the city had been experiencing since the middle of the nineteenth century.[96] Around the turn of the twentieth century Bristol was financially one of the weakest colleges. Not until 1905, when new industries had been established in Bristol and the city had entered on a new period of growth, did the College receive more generous support from a local family in the tobacco industry. By the time the College became a full university in 1909, this family had donated more than £200,000. Thus, like Reading,[97] Nottingham and Dundee, Bristol was for decades supported by a

93. See Sanderson, *Universities*, pp. 69–70; Vincent and Hinton, *The University of Birmingham*, pp. 23–38. G. R. Searle writes of Chamberlain's motives: 'Birmingham: a university which Chamberlain intended to energize the commercial and industrial life of the entire West Midlands region' (*The Quest for National Efficiency: A Study in British Politics and British Political Thought 1899–1914*, Berkeley and Los Angeles, 1971, p. 121).
94. An internal Treasury minute of 1901 comments on the University of Birmingham: 'In asking for State aid to the extent of £2,000 a year the Representatives of the University of Birmingham were able to show that that City and its neighbourhood had already contributed, or promised, nearly £600,000 and that there was good reason to believe that still further subscriptions would be forthcoming from the Councils of the surrounding Counties' (Treasury Minute, 21 February 1901, T 1/9653 B/3328/1901, PRO).
95. An unusual feature that should be mentioned is that two Oxford Colleges (Balliol and New College) provided financial support for the College in Bristol. See Owen, *English Philanthropy*, p. 366; Basil Cottle and J. W. Sherborne, *The Life of a University*, 2nd edn (Bristol, 1959), pp. 8, 19.
96. B. W. E. Alford, *W. D. & H. O. Wills and the Development of the U.K. Tobacco Industry 1786–1965* (London, 1973), pp. 328–9; Sanderson, *Universities*, pp. 70–1, 73; Cottle and Sherborne, *The Life of a University* pp. 31–41.
97. Sanderson, *Universities*, p. 75. In 1914 proceeds of endowments represented 31 per cent of the income of University College in Reading. This was the highest proportion among all the new English universities at the time. See *Nature*, 94 (1914–15), p. 326.

single family — unusual among the new universities. This form of industrial patronage cannot be attributed to economic expediency alone.

We may conclude that private patronage of science in the second half of the nineteenth century focused on the new university colleges in the large provincial cities in addition to the old universities at Oxford and Cambridge.[98] The University of London, however, seems to have been conspicuously neglected around the turn of the century. A Treasury memorandum of 21 February 1901, written after the chancellor of the exchequer had received a deputation from the University of London, makes the following observation: 'So far as yet appears, the University of London has received practically nothing either from local or private sources. In these circumstances the Chancellor of the Exchequer feels that, in the amount of the Grant which he now proposes, the University of London is being treated with exceptional liberality'.[99] In this case, 'exceptional liberality' was represented by an increase from £8,000 to £13,000 in the annual grant. The University of London did not receive more generous support until the period between 1902 and 1912 when University College, for instance, received a total of £420,000, £100,000 of which was given by the ship-owner Donald Currie.[100] H. A. L. Fisher, president of the Board of Education from 1916 to 1922 in Lloyd George's cabinet, estimated that between 1906 and 1917 private endowments to English universities

---

98. On the basis of several case studies of grants made to colleges in Oxford and Cambridge, however, David Owen suggests 'that, until well towards the end of the nineteenth century, the flow of benevolence to the older universities was not much more than a trickle' (*English Philanthropy*, p. 347).
99. T 1/9653 B/3328/1901, PRO. It is paradoxical, but typical of contemporary attitudes towards supporting science, that Sir Francis Mowatt, permanent secretary to the Treasury, attributed the lack of private endowments in London to the granting of a state subsidy: 'Until the University ceases to be a State-paid Department, and has to make itself self-supporting . . . it will not be in a position to appeal to the City or to the Public for grants or bequests, and I am convinced that the longer we defer the change the higher will be the amount at which we shall have to fix the Grant.' For the rest, Mowatt was of the opinion that the university 'of the richest City of the World ought to depend solely on its own resource, and on municipal and private donations and bequests, that it should be altogether independent of State aid' (Memorandum to the chancellor of the exchequer, 5 November 1900, ibid.). According to Sanderson, the fact that the University of London never inspired 'that cohesive civic enthusiasm and local patriotism' from which the provincial universities profited so greatly was due to London's size ('The University of London', p. 261).
100. Owen, *English Philanthropy*, p. 362. London companies still provided little support. *The Daily Telegraph* commented: 'It is almost impossible to secure money in the metropolis for the applied science section of those colleges which are closely allied with London University . . . The London University does not seem to have been able to keep so closely in touch with the large works of the metropolis' (25 February 1907).

amounted to £200,000 annually. Taken by itself, this is an impressive sum. But as Fisher added, it was very small compared with private expenditure on universities in the USA, where during the same period it amounted to more than £4 million per year, or twenty times as much.[101] A survey of the preceding period, 1871 to 1901, estimates that private endowments for higher education in the USA were eight times as high as in the United Kingdom.[102]

Other scientific institutions, outside the university system, also owe their existence at this time to private endowments. Like the university colleges, several of the research laboratories founded at the end of the nineteenth century could not have been established and maintained without massive private support. Examples are the London School of Tropical Medicine (the foundation of which, it is true, was assisted by grants from the Colonial Office), and the Jenner Institute for Bacteriological Research. The Davy–Faraday Laboratory of the Royal Institution was able to start work in physical chemistry in 1896, after the industrialist Ludwig Mond had donated a total of £125,000 for its foundation in gratitude for his admission to the Royal Society. There had been calls for the establishment of this sort of laboratory since 1843.[103] Under the leadership of the physicist and Nobel Prize winner Sir William Bragg, who had previously taught at Leeds (from 1909) and London (from 1915), the Davy–Faraday Laboratory developed in the 1920s into one of Britain's leading scientific research institutes.[104]

National organisations for science and the humanities which were of a more representative nature also depended for their existence on private funds, but their activities were relatively inexpensive. A well-known example of this sort of organisation is the British Academy for the Promotion of Historical, Philosophical and Philological Studies. It was founded in London in December 1901 because while the Royal Society was a national academy for

---

101. *Nature*, 100 (1917–18), p. 452.
102. *Nature*, 68 (1903), p. 28. In the academic year 1906/7, the University of Chicago alone received almost £1,200,000 in endowments, while Yale received £200,000, Cornell £156,000 and Harvard £139,000 ('Higher Education in the United States', *Nature*, 80 [1909], p. 112).
103. *The Times*, 3 July 1894 ('A Magnificent Endowment'); MacLeod, 'Resources of Science', pp. 160–1; F. G. Donnan, *Ludwig Mond, F. R. S.: 1839–1909* (London, 1939), p. 11; John Michael Cohen, *The Life of Ludwig Mond* (London, 1956), p. 219; Edwin Edser, 'The Davy–Faraday Laboratory', *Nature*, 83 (1910), p. 40. The British Institute of Preventive Medicine in London, founded in 1891, was renamed Jenner Institute of Preventive Medicine in 1898. In 1903, after it had received a grant of £250,000 from Lord Iveagh, it was renamed again to become the Lister Institute of Preventive Medicine. During the First World War Chaim Weizmann worked there.
104. G. M. Caroe, *William Henry Bragg 1862–1942: Man and Scientist* (Cambridge, 1978).

the sciences, Britain had no academy for the humanities comparable with those on the Continent. The British Academy did not receive a state grant, although it would seem to have been an obvious candidate for one; since the seventeenth century almost all scientific academies in Europe had been financed by noble patronage or state subsidy. Six months before the new academy received its royal charter, the chancellor of the exchequer, Sir Michael E. Hicks Beach, was asked in the House of Commons about state aid for the British Academy. His answer was negative: 'This is a purely hypothetical question. I am not aware that the foundation of the proposed institution has yet been considered by His Majesty in Council; but, even if it should be agreed to, I am afraid that, personally, my feeling would be against a grant'.[105]

Unlike the Royal Society and some of the larger specialist scientific societies which received public money to pursue their scholarly aims, the British Academy thus survived its first two decades with the aid only of membership subscriptions and private grants, as did most scientific societies. Sir Israel Gollancz, one of the British Academy's original founders and subsequently a first secretary of long standing, was personally responsible for procuring many of the private grants made to the Academy. The family of Ludwig Mond, a chemist and industrialist of German descent, was one of the British Academy's more important benefactors. In 1912 Henriette Hertz, a friend of the Mond family, bequeathed it capital assets of £5,400, the proceeds of which have financed three lectures on general topics every year since 1914. This was one of the most substantial grants made to the British Academy before the First World War.

It would not be difficult to list more cases of patronage of science in Britain since the second half of the nineteenth century. But while the existence of widespread patronage is undisputed, many of its

---

105. Hansard, Parl. Deb., H.C., 4th series, vol. 104, col. 21 (28 February 1902). On the history of the foundation of the British Academy, see *Proceedings of the British Academy 1905–1906* (London, no date), pp. VII–IX; Kenyon, *British Academy*; Alter, 'The Royal Society'. On the attitude displayed by the state, see criticisms made by Sir Frederic G. Kenyon, historian of the British Academy and its president between 1917 and 1921: 'The Royal Society . . . had long been in receipt of a substantial subsidy from public funds; but the Treasury steadily refused to consider a corresponding grant for the service of the humanities. Whether its doctrine was that humane learning is of no public interest, or that a new institution must show its ability to stand and run alone before it receives public recognition, or merely the ingrained official tendency to say "No" to any application for money, may be uncertain, but the fact remains' (Kenyon, *British Academy*, p. 15). In 1913 the Treasury granted £400 for the publication of sources relating to British economic and social history. This money was stopped in 1915. From 1924 onwards, the British Academy received a government subsidy of £2,000 per annum.

features remain obscure. We still lack exhaustive answers to a number of questions. Who were the patrons? What were their motives in supporting science? David Owen's observation that British philanthropists in the late nineteenth century prove to be 'remarkably resistant to generalization' is undoubtedly correct.[106] The present state of research does not allow this issue to be tackled satisfactorily, but we can delimit with some degree of certainty the circle from which patrons came. And in spite of Owen's scepticism, we can make some — albeit hypothetical — generalisations. More regional and local studies will complete our picture of private patronage in Britain, the social changes it underwent, and the areas in which it concentrated its support.

At this stage we can suggest that patrons of British science in the second half of the nineteenth century were members of the prosperous urban industrial and commercial bourgeoisie, or, in some instances, members of the aristocracy. In social terms, therefore, they appear at first glance to be a remarkably homogeneous group, but each group can be further differentiated. Among the industrial bourgeoisie, for instance, patrons of science came preponderantly from the chemical, engineering, textile and food industries. In all these branches of industry, scientific research provided the basis for expansion. What remains to be explained is why heavy industrialists rarely supported science. The social structure briefly sketched here can best be illustrated by reference to the large grants made to the new provincial universities. In Manchester, local industrialists who followed John Owens's pioneering example and contributed to science between 1860 and 1890 included textile manufacturers, manufacturers in the engineering industries, wire manufacturers, bankers and a tea merchant.[107] Michael Sanderson writes of a 'regular flow of financial support from the local industrial class'.[108] In Liverpool, patrons of science came from important commercial and industrial families, especially those in the chemical, engineering and food industries, but also from ship and shipyard owners. Mason College in Birmingham received financial backing from the whole range of local industry: metal and engineering industries, food, glass and electrical industries, as well as breweries and the ceramics industry. As already noted, industry and commerce did not make such a large contribution in Leeds, and the large individ-

---

106. Owen, *English Philanthropy*, p. 472.
107. Fiddes, *Manchester University*, pp. 6, 63–5, 182; Sanderson, *Universities*, p. 62. See in general also W. D. Rubinstein, 'Wealth, Elites and the Class Structure of Modern Britain', *Past and Present*, vol. 76 (1977), pp. 99–126.
108. Sanderson, *Universities*, p. 63.

ual sums given in other cities were not received there. Prominent patrons in Leeds included the Duke of Devonshire (who had donated the Cavendish Laboratory for experimental physics to the University of Cambridge in 1874),[109] a banking family, an industrialist (Sir Andrew Fairbairn), and especially the Clothworkers' Company based in London.[110] Leeds received its first large individual donation (£10,000) in 1912. The founders and early benefactors of University College in Bristol in 1873 were a corn merchant, a chocolate and a soap manufacturer. After 1905 generous gifts from two resident industrialists enabled further developments to take place in Bristol. Substantial amounts were donated by several members of the Wills family, which owned one of the most important British tobacco concerns. By the 1920s this family had given a total of more than £1.3 million.[111]

The foundation of the London School of Economics and Political Science in 1895 was made possible by gifts from Sidney and Beatrice Webb, and from other intellectuals who belonged to their circle, including the philosopher Bertrand Russell and Charlotte Payne-Townshend, later to become the wife of George Bernard Shaw. A larger sum was given indirectly by Henry Hunt Hutchinson, a member of the Fabian Society and otherwise practically unknown as a patron. He died in 1894 and bequeathed £10,000 for 'propaganda and other purposes of the said Society and its Socialism', assigning Sidney Webb power of disposition over it.[112] In the following years large amounts of money were given by London County Council's Technical Education Board, set up in 1893 and chaired for many years by Sidney Webb, as well as by the Clothworkers' Company and Lord Nathaniel Rothschild of the well-known banking family, who donated £5,000. A public appeal for

---

109. The buildings of the Cavendish Laboratory cost a total of £6,300. Added to this was the cost of apparatus and interior furnishings (Alexander Wood, *The Cavendish Laboratory*, Cambridge, 1946, pp. 8–11). See also Owen, *English Philanthropy*, p. 354: 'Rarely has an academic investment returned more princely dividends.'
110. Brown, *The Private Donor*; Sanderson, *Universities*, pp. 66–7. See also *The Morning Post*, 22 February 1907: 'The textile and dyeing schools in the University of Leeds have been organised largely with the money given by manufacturers in Yorkshire, and the brewing school at Birmingham owes something to the brewers of the Midlands'.
111. Cottle and Sherborne, *The Life of a University*, pp. 1–19, 31–41.
112. Sydney Caine, *The History of the Foundation of the London School of Economics and Political Science* (London, 1963), pp. 1, 15–20, 56–7, 90–2. In the same book (pp. 10–14), a letter written by Sidney Webb on 3 January 1903 contains information about benefactors of the London School of Economics. Janet Beveridge, *An Epic of Clare Market: Birth and Early Days of the London School of Economics* (London, 1960), pp. 19, 31–2 and 70; Beatrice Webb, *Our Partnership* (London, 1948, reprinted 1975), pp. 84–95.

£2,500 for the library was supported by leading politicians of liberal imperialist persuasion, or at least sympathies: gifts were received from Lord Rosebery, after Richard B. Haldane probably the most influential supporter of the London School of Economics, the Duke of Devonshire, Arthur J. Balfour, Joseph Chamberlain, Lord Ripon, former viceroy of India (1880–4) and colonial secretary (1892–5), and Lord Alfred Milner, German-born governor of the Cape Colony (1897–1901) and later colonial secretary (1919–21). In the context of university foundations outside London, it is striking that no large London industrial firms or City banks made substantial contributions to the London School of Economics. This unusual caution was evidently due to the strong political motives of its founders, especially Sidney Webb, which gave the university its original social reformist and anti-capitalist character. The only exception was the £11,000 given by J. Passmore Edwards, a publisher from Cornwall, to pay for the erection of buildings in Clare Market.[113] Not until just before the First World War, when the London School of Economics was developing into the Economics Faculty of the University of London, newly constituted in 1899–1900, did support from industry increase, coming especially from the London Chamber of Commerce.

The Imperial College of Science and Technology was much more generously endowed by industry right from the start than were either the London School of Economics or the Colleges of the University of London. Founded in London in 1907, Imperial College received substantial donations, arranged by Richard B. Haldane, from Alfred Beit, Julius C. Wernher and Ernest Cassel, financiers and bankers resident in London. Beit gave £135,000, Wernher's contribution amounted to the enormous sum of £250,000, and Cassel's to £10,000. The firm of Wernher, Beit & Co., whose business activities were concentrated mainly in South Africa, donated a further £100,000, while the Goldsmiths' Company provided £137,000 for engineering subjects.[114] It is remarkable that Imperial College, which was intended to help British industry become competitive with German and American industry, did not receive any substantial grants from industry until the eve of the First World War. We can do no more than establish here that this was the case; why industry, which was, after all, dependent on

113. On J. Passmore Edwards's numerous endowments, which are said to have totalled between £200,000 and £250,000, see the biographical sketch in Owen, *English Philanthropy*, pp. 428–34. See also Webb, *Our Partnership*, p. 89; Jamie Camplin, *The Rise of the Plutocrats: Wealth and Power in Edwardian England* (London, 1978), pp. 81 and 231–3.
114. *Reports from those Universities*, p. V; Sanderson, *Universities*, pp. 116–17.

technically and scientifically trained experts, displayed this lack of interest and support, can only be a matter for speculation. Whatever the case, the contrast with the situation in Germany, and even with that in the Midlands and the north of England, is striking.

The British Academy for the Promotion of Historical, Philosophical and Philological Studies, founded in 1901, was supported by the chemical industrialist Ludwig Mond. Together with Alfred Beit, Julius C. Wernher and Ernest Cassel, Mond belonged to a small group of patrons, the extent and range of whose philanthropic patronage between the 1880s and the First World War, as well as its geographical distribution, distinguished them from the benefactors of universities in the Midlands and the north of England, who often had their own future interests in mind. The first group represented a nationally or internationally orientated type of patronage, in contrast to the more locally orientated patrons who concentrated their philanthropy in the relatively narrow circle of their home towns or regions and were, of course, more common in Britain during the nineteenth century. The biographies of patrons in the first group in many cases display a number of striking similarities. They were often naturalised Britons of German-Jewish families who, like Mond and Cassel, occasionally had played an important part in Anglo-German relations at the beginning of this century. They had international business connections, especially within the empire and frequently concentrated in southern Africa. They tended to achieve power and prestige after the turn of the century, during the reign of Edward VII, and their position was often expressed in an elevation to the peerage. Like many benefactors in the provinces, they generally did not inherit their wealth, but made their own fortunes. In some cases they kept up close personal and political ties with Cecil Rhodes, Joseph Chamberlain, Lord Rosebery and Richard B. Haldane — all enthusiastic imperialists.

Ludwig Mond (1839–1909), the son of a Jewish silk merchant, was born in Kassel. He did not come to England until 1862, after studying chemistry in Marburg and Heidelberg, and became a British citizen in 1880. He gained industrial experience while still in Germany, where he took out his first chemico-technical patents. With John T. Brunner, Mond built up the British ammonia-soda industry near Liverpool, using the Solvay process. Brunner, Mond & Co. became the largest producer of alkali in the world and in 1926, together with three other large chemical concerns, amalgamated with the newly established Imperial Chemical Industries (ICI). Mond was not only an extremely productive scientist with a large number of patents to his credit, but also 'a genius for divining the industrial possibilities of discoveries in pure science'.[115] His exceptional business success brought him a fortune estimated by contemporaries as worth more than £1 million, an enormous sum by the standards of those days. Mond was active in scientific societies, being a founder member of the Society of Chemical Industry in 1881, and its president in 1888. He became a Fellow of the Royal Society in 1891, and shortly before his death, a corresponding member of the Preußische Akademie der Wissenschaften. In addition to the large grant, mentioned above, which Mond made to the Davy–Faraday Laboratory, he also supported the Royal Society and the British Academy. The obituary notice in *Nature* described him as 'keenly interested in the progress of science in all its branches', and 'always ready to help experimenters in carrying out costly researches'.[116] Another obituary called him a 'princely benefactor of Science, great patron and lover of Art'.[117] He bequeathed his important collection of early Italian art to the National Gallery in London, and left large capital sums to various foreign institutions, including the University of Heidelberg, the Akademie der Schönen Künste in Munich and the Reale Accademia dei Lincei in Rome.

The *Dictionary of National Biography* describes Alfred Beit (1853–1906) as a 'financier and benefactor'. His career reveals several parallels with that of Ludwig Mond. Born into a family of baptised Jews from Hamburg, Beit worked in the diamond trade in Amsterdam and South Africa from the mid 1870s. In South Africa he acquired gold and diamond mines, and in 1890 he and Julius C. Wernher founded the commercial firm and financial holding Wernher, Beit & Co. in London. He and Cecil Rhodes became close personal friends as well as business partners in De Beers Consolidated Mines, founded in 1888. As director of the British South Africa Company, Beit gave strong financial support to both the expansion into what was later to become Rhodesia, and the Jameson Raid in the Transvaal in 1895–6. At the end of the 1880s Beit, who had become a British subject, made London his headquarters. It has been estimated that after the turn of the century his benefactions in Britain and Germany were worth a total of £2 million.[118] They included a chair of colonial history at Oxford; funds for the

115. *Dictionary of National Biography*, supplementary vol. 1901–1911, p. 633.
116. *Nature*, 83 (1910), p. 40.
117. Quoted in Donnan, *Ludwig Mond*, p. 11.
118. *Dictionary of National Biography*, supplementary vol. 1901–11, p. 129; G. Seymour Fort, *Alfred Beit: A Study of the Man and his Work* (London, 1932); Camplin, *Rise of the Plutocrats*, pp. 166–9.

Bodleian Library in Oxford to increase its holdings in colonial history; £135,000 for Imperial College of Science and Technology; £200,000 for charitable purposes in Rhodesia, the Transvaal, the Cape Colony, London and Hamburg; support for hospitals and South African universities; and finally, a bequest of paintings to museums in London and Berlin. In 1909 Sir Otto John Beit, Alfred's younger brother, set up the Beit Trust in memory of his brother, and endowed it with capital of £215,000. It was primarily intended to promote medical research.

Sir Julius C. Wernher (1850–1912), who became Alfred Beit's business partner and Cecil Rhodes's friend, acquired his fortune, like Beit, in the South African gold and diamond industries. He suffered a good deal of hostility because of his wealth and foreign origins. Bertrand Russell, in his autobiography, calls Wernher 'the chief of all the South African millionaires'.[119] The attempted irony of this description does not do justice to Wernher as a benefactor. As with Beit, little is known about Wernher's life and his business dealings; similarly, the reasons for his benefactions are largely unknown. Wernher was born in Darmstadt, the son of a Protestant railway engineer. From 1880 he directed his firm from London, although he did not become a British subject until 1898. He supported numerous charitable causes, especially London hospitals, as well as South African universities and Imperial College of Science and Technology in London. Richard B. Haldane had interested him in this project. In 1910 he gave £10,000 to the National Physical Laboratory near London. While this was a relatively small amount, it represented the largest individual grant so far made to the Laboratory, which had been working since 1900.[120] In 1905 Wernher was knighted. Like Mond and Beit, he bequeathed his collection of paintings to the National Gallery in London. It seems that in Britain, as in Europe and the USA,

---

119. *The Autobiography of Bertrand Russell*, vol. 1, 1872–1914 (London, 1967), p. 176. See also Beatrice Webb's diary entry of 2 July 1906: 'We both respect and like the man [Wernher]. He is a German giant, not unduly self-indulgent, and a real drudger at his business. But he is better than that. He is noted for generosity inside his own circle; regarding the South African commercial world as something for which he is responsible, perpetually carrying the weaker man on his back . . . He is also public-spirited in his desire for the efficiency of all industry, and the advancement of its technique' (*Our Partnership*, p. 346). Alfred Beit is said to have left £8 million at his death, Julius C. Wernher £10 million. This put them among the 'top wealth-leavers' in Britain before 1914 (W. D. Rubinstein, 'Modern Britain' in W. D. Rubinstein, ed., *Wealth and the Wealthy in the Modern World*, London, 1980, pp. 46–89).

120. *The National Physical Laboratory: Report for the Year 1910* (London, 1911), p. 13. On the National Physical Laboratory and Imperial College of Science and Technology, see below pp. 138–49 and 149–72 in this volume.

collections of this sort were an obligatory status symbol for all millionaires with aspirations towards becoming benefactors.

At the turn of the century Sir Ernest J. Cassel (1852–1921) was one of the most influential bankers in the City of London, although he had only been in England since 1869.[121] He became a British subject in 1878. According to his biographer, Cassel's life was 'one of the most extraordinary success stories of his or any other era'.[122] Born in Cologne as the son of a Jewish money-lender, Cassel acquired his enormous fortune via extensive financial transactions in Europe, America and in the Near East, as well as through industrial and railway interests in Sweden and England. After 1896 he was a close friend of the Prince of Wales, becoming his financial adviser and private banker.[123] Cassel's patronage, which began, characteristically, with Edward VII's accession to the throne, was directed in the main towards philanthropic organisations in England, Sweden, Germany and Egypt, all countries with which he had close personal or commercial ties. He made large grants to scientific and medical research institutions: £200,000 towards researching and finding a cure for tuberculosis (1902), which had killed his wife and daughter; £10,000 to the Imperial College of Science and Technology (1907); and £46,000 to the new Radium Institute (1909). In addition, Cassel financed university scholarships, and chairs in foreign languages, political economy and business law. In 1919 he provided £500,000 for further developing the London School of Economics into the Economics Faculty of the University of London. In 1911 he established a foundation, endowed with £200,000, to help British subjects residing in Germany and German subjects in Britain, who required assistance.[124] From

121. *Dictionary of National Biography*, supplementary vol. 1912–21, pp. 97–100; Brian Connell, *Manifest Destiny: A Study in Five Profiles of the Rise and Influence of the Mountbatten Family* (London, 1953), pp. 53–87. On Cassel's part in Anglo-German relations before the First World War, see Connell, pp. 75–8 and Alfred Vagts, 'Die Juden im englisch-deutschen imperialistischen Konflikt vor 1914' in Joachim Radkau and Imanuel Geiss (eds.), *Imperialismus im 20. Jahrhundert. Gedenkschrift für George W. F. Hallgarten* (Munich, 1976), pp. 113–43. Cyrus Adler, *Jacob H. Schiff. His Life and Letters*, 2 vols. (London, 1929) gives an insight into Cassel's business transactions. A small number of Ernest Cassel's private papers are kept at Broadlands, Romsey, the country seat of the Mountbatten family. They are largely Cassel's private correspondence and contain practically no references to his activities as a benefactor.

122. Connell, *Manifest Destiny*, p. 54.

123. Giles St Aubyn, *Edward VII: Prince and King* (London, 1979), p. 367; Connell, *Manifest Destiny*, pp. 65–6.

124. The Foundation's second annual report (1913, p. 4) reads: 'The Foundation has . . . been established by Cassel in memory of the late King Edward VII, with headquarters in Berlin and London . . . . The British section is intended to assist Germans residing in the United Kingdom, whilst the German section is for the assistance of British subjects in Germany. For each section the income from a capital

Science Patronage in Nineteenth-Century Britain

1913 the King Edward VII British–German Foundation also provided scholarships for British students in Germany. Cassel's numerous benefactions are said to have totalled more than £2 million; the largest of these were made in the last years of his life. He had been knighted in 1899.

The extensive patronage exercised by Ludwig Mond, Alfred Beit, Julius C. Wernher and Ernest J. Cassel raises the question of what motivated benefactors of science in the last half of the nineteenth century. As expected, no generally valid answer emerges. Only in very few cases can motives be established with certainty. When John Owens and Josiah Mason, for example, made the endowments which started the development of the British university system, did they do so on their own initiative, or did they require encouragement — random or directed — from a third party, that is, from scientists with an interest in gaining financial support? In other words: did patrons themselves determine the purposes and aims of their benefactions on the basis of their own judgement, or did they follow advice, or perhaps even allow themselves to be 'guided' towards one particular field rather than another? It is unlikely that unequivocal answers will ever be found to these questions. According to W. K. Jordan, author of the most important study of English philanthropy in the early modern period, the motives of acts of patronage remain 'buried deep in the recesses of our nature, immune, perhaps happily, from the fumbling probing of the historian and, certainly happily, from the too arrogant enquiry of the psychoanalyst'.[125]

Although Jordan is correct in pointing out that an exact historical analysis of the motives of patrons will never be made, the historian can sometimes make inferences about patrons' motives and intentions. Thus, for example, the lack of a direct heir to take over a family fortune, or the early death of a son, are often important factors. Frequently, too, a benefactor simply wanted his foundation to be a memorial to himself: the majority — apparently over 90 per cent — of all foundations which are still operative today bear the name of their donor.[126] It could be argued that the purpose of

---

sum of £100,000 is available.' This report and several other pamphlets about the Foundation can be found in Cassel's papers, Broadlands Archives.
125. W. K. Jordan, *Philanthropy in England, 1480–1660* (London, 1959), p. 144.
126. Ben Whitaker, *The Foundations: An Anatomy of Philanthropy and Society* (London, 1974), p. 49. Leopold von Wiese writes: 'If we seek the motives which govern patrons, we find ourselves in the middle of the rich, wonderful world of human desires. And if an explanation is based on only one wish or expectation, it will rarely capture the whole motive' (translated from *Die Funktion des Mäzens im gesellschaftlichen Leben*, Cologne, 1929, p. 18). On the motives of American patrons

benefactions represents no more than a secondary problem. Certain trends, areas which happen to become fashionable at the time, can influence patrons and determine who will benefit from a particular benefaction. This is clearly illustrated by endowments for science, which in Victorian England increased in number and size at the cost of purely humanitarian endowments, but without ever approaching their level.

During the last third of the nineteenth century and until the eve of the First World War, public discussion increasingly stressed the benefits, indeed the necessity, of applied science in order to achieve economic and industrial progress. Consequently, behind a large number of endowments made for science in the widest sense in the second half of the nineteenth century was a more or less firm expectation that intensifying scientific research and founding universities which tailored their research and teaching programmes to meet the needs of local industry would, in the middle or long term, produce results that could be put to economic use. This expectation goes a long way towards explaining both the creation of individual colleges outside London and the two old English universities, and the numerous endowments made by the chemical, engineering and food industries to fund chemical, physical or technical laboratories and appropriate chairs.

In 1909, for example, Sir John Brunner, a chemical manufacturer and MP of long standing, explained in a House of Commons debate on state aid for science: 'I have often said that every penny I have has come from the application of science to industry'.[127] In 1910 the ship-owners T. Fenwick Harrison, J. W. Hughes and Heath Harrison donated just under £40,000 to the University of Liverpool for new technical laboratories. At the opening ceremony they clearly emphasised the commercial aspect of their gift and pointed to the area of research they wished to encourage: 'As shipowners who use three hundred thousand tons of coal a year they see the advantages to be derived from the internal combustion engine so far as ships are concerned'.[128]

Yorkshire College in Leeds and other new colleges attached no value to the teaching of the humanities, adding them to their syllabuses only later, and then reluctantly. When Josiah Mason made his donation to Birmingham in 1875 he was apparently

---

in the early twentieth century, see Frederic Cople Jaher, 'The Gilded Elite: American Multimillionaires, 1865 to the Present' in W. D. Rubinstein (ed.), *Wealth and the Wealthy in the Modern World* (London, 1980), pp. 210–14.
127. Hansard, Parl. Deb., H.C., 5th Series, vol. 11, col. 2250.
128. Quoted in Sanderson, *Universities*, p. 65.

'deeply convinced from his long and varied experience . . . in different branches of manufacture of the necessity and benefit through systematic scientific instruction specially adapted to the practical mechanical and industrial pursuits of the Midland district'.[129] Mason, too, who had enjoyed only a rudimentary education, originally wanted to see the humanities excluded from Birmingham. 'The provincial business classes', writes David Owen, 'tended to think of the older universities as not for them and to dismiss the classical-mathematical training as irrelevant to their special needs. They felt themselves excluded, spiritually or in fact, by their social background, their industrial orientation, and their Nonconformist religion.'[130] Only the sciences and technical questions captured the full interest of these men.[131] In the eyes of many nineteenth-century benefactors, therefore, endowing science was, in the final analysis, investing in local industry and the local economy. The provincial colleges, too, repeatedly stressed their willingness to comply with the wishes of patrons and the needs of industry, and as Michael Sanderson has shown, some achieved this most successfully.

The expectation that investment of this nature would prove profitable was not always present, or at least it was overlaid or weakened by other motives. In the case of John Owens in Manchester, and in that of the Wills family's support of Bristol University, philanthropic motives dominated — a desire, often rooted in religious conviction, to improve the educational facilities of a region or city for the general good.[132] At the end of 1886, for example, Sir Andrew Walker, a brewer, wrote to Lord Derby, chancellor of University College Liverpool:

> It has come to my knowledge that there is urgent need for a building with efficient equipment for the Engineering School of your College. I feel the importance of Liverpool having facilities second to none in the Kingdom for the proper training and technical education of persons intending to enter the Engineering profession. I am also very desirous of

129. Quoted in ibid., p. 69. See also Owen, *English Philanthropy*, p. 412.
130. Owen, *English Philanthropy*, p. 361.
131. Joseph Chamberlain's comment is typical: 'We desire to systematize and develop the special training which is required by men in business and those who either as principals or as managers and foremen will be called upon to conduct the great industrial undertakings' (letter by Chamberlain, 11 December 1899, quoted in Sanderson, *Universities*, p. 82).
132. John Owens (1790–1846) stipulated that his endowment was to be used to educate young men 'in such branches of learning and sciences as are now and may be hereafter taught in the English universities' (quoted in Clapp, *John Owens*, p. 173). On the Wills family's endowments, see Alford, *W. D. & H. O. Wills*, pp. 328–9.

doing anything I can to promote the interests of the artisans of Liverpool and the neighbourhood, and I understand that one of the features of the Engineering Department is to give an opportunity for working mechanics to avail themselves of such technical instruction as the building in question will give facilities for. Such a building would also become the natural centre of the Technical Schools which are necessary to complete our educational system.

He then went on to say:

I shall esteem it a privilege to be permitted to defray the cost of such building, with the necessary machinery and appliances, to the extent of £15,000. This sum . . . is ample to make the Institution one of the most efficient in the Kingdom. I propose the acceptance of this building as my gift to Liverpool to commemorate the Jubilee of Her Majesty the Queen.[133]

The example of other cities in the vicinity which had already established universities or were planning to do so was also a powerful stimulus, as Walker's letter shows. Asa Briggs and Michael Sanderson have pointed out that a university was a status symbol for Victorian cities. Competition between prospering Victorian cities and growing local patriotism meant that the wealthy bourgeoisie and municipal administrations sometimes provided generously for their developing universities, both financially and in terms of departments. In the early 1880s Gerald Rendall, first vice-chancellor of University College Liverpool, ended his public appeal for funds with the words: 'A great city without a university is but an uncrowned queen.'[134] None of the frequent public appeals of this time fail to refer to the existence of a university college in a neighbouring city, to the generous financial support provided in other cities, to London's real or claimed superiority, to the modernity of other university laboratories or libraries: in short, to the generally exemplary nature of other universities. Liverpool in competition with Manchester, and especially Birmingham, the rising industrial city of the Midlands, probably provide the most impressive examples of this contest, with its appeal to the loyalty, solidarity and pride of the citizens.

In many benefactions made at this time, no commercial motives at all can be recognised. Examples are the endowing of humanities

---

133. Quoted in H. A. Ormerod, 'History of University College, Liverpool', typescript, n.d., Library of the University of Liverpool, ch. 6, p. 21.
134. Quoted in Lodge, *Past Years*, p. 162.

departments in university colleges and, after the turn of the century, in the new universities, as well as many of the research laboratories and chairs in the medical and biological sciences, both within the university system and outside it. In these cases, private interests and preferences were decisive, or perhaps experience of specific medical problems into which the patron's business provided some sort of insight. Thus the ship-owner and industrialist Alfred L. Jones, who had helped to finance the School of Tropical Medicine at University College Liverpool when it was founded in 1898–9, on his death in 1909 bequeathed £500,000 for charitable and educational purposes in England and the British colonies in West Africa, as well as for research into the causes of tropical diseases on the West Coast of Africa.[135] Henry Solomon Wellcome, a wealthy manufacturer of pharmaceutical products, founded the Wellcome Physiological Research Laboratories in 1894, and two years later, the Wellcome Chemical Research Laboratories. The Wellcome Trust, which grew out of Wellcome's endowments, is the largest private trust in Britain today. When Sir Julius C. Wernher donated £10,000 to the National Physical Laboratory in 1910, he expressed the hope that this would encourage the state to support the laboratory more generously. Informing the director of the National Physical Laboratory of his donation, Wernher wrote: 'You told me that the requirements for the proposed new Metallurgical Building were £10,000, and I have much pleasure in enclosing that amount, and hope it will lead to a good grant from the Government to complete the scheme'. He continued: 'When the building is completed you will perhaps place a tablet in some obscure corner stating that it had been erected at my expense'.[136]

An analysis of legacies published in contemporary newspapers shows that at the end of the nineteenth century endowments for medical research, hospitals and sanatoria increased sharply. By the turn of the century more than half of all endowments were apparently designated for these purposes.[137] Three of the most im-

---

135. *The Times*, 14 December 1909 (Obituary for Alfred L. Jones); P. N. Davies, *Sir Alfred Jones: Shipping Entrepreneur par Excellence* (London, 1978); *Reports from those Universities*, pp. IV–V. Further examples of foundations for medical research can be found in Owen, *English Philanthropy*, p. 490.
136. Printed in *National Physical Laboratory: Minutes of the Executive Committee*, vol. 3, p. 180 (Wernher's letter dated 24 July 1910). See also below, p. 147 in this volume. In 1908 Beatrice Webb noted: 'Wernher . . . is a big man – big in body and big in mind and even big in his aims. To make wealth was his first aim; to carry on great enterprises because he delights in industrial construction was his second aim, and now to advance technology and applied science has been his latest aim' (Passfield Papers: Diary of Beatrice Webb, vol. 26, 27 July 1908. A slight variation on this text appears in Webb, *Our Partnership*, p. 412).
137. Harrison, 'Philanthropy and the Victorians', p. 354; Owen, *English Phil-*

portant benefactors of Owens College in the nineteenth century — Sir Joseph Whitworth, John Rylands and R.C. Christie — apparently had only philanthropic, religious and educational motives for their widespread support of hospitals, homes for old people, orphanages and almshouses, university libraries and educational institutions.[138] The lack of direct commercial motives for many endowments suggests that in late-nineteenth-century Britain spectacular support for the arts and the sciences, in addition to that for charities, which as a whole was of incomparably greater significance, in many cases was characteristic of successful self-made businessmen or industrialists. Nevertheless, many endowments were also made by second- or third-generation owners of wealth. Much supports David Owen's argument that 'industrial wealth was developing a quasi-aristocratic sense of obligation, that the plutocracy was doing its best to attain the status of aristocracy'.[139]

Generous philanthropy in the widest sense was traditionally expected by society of wealthy industrialists, who were often happy to fulfil this expectation.[140] They were repaid, as a rule, by an enhanced social position expressed in a knighthood or elevation to the peerage. Some contemporary observers suggested, therefore, that the real motive behind the large grants made by industry and commerce, especially by 'self-made men', was not a 'sense of obligation', as David Owen supposes, but simply recognition that patronage was the key to advancement in Britain's social hierarchy.[141] A modern economic historian such as D. C. Coleman,

---

anthropy, pp. 474, 479.

138. Sir Joseph Whitworth (1803–87), engineer and owner of a machine-tool factory in Manchester, one of the largest in the world, was a Fellow of the Royal Society of London from 1857 and donated more than £150,000 to Owens College. John Rylands (1801–88) owned Britain's largest textile concern. He and his widow donated £100,000 to the John Rylands Library, which opened in 1899 in Manchester.

Richard C. Christie (1830–1901), the son of a wealthy spinning mill owner, was Professor of History, Law and Political Economy at Owens College, and donated the Christie Library (75,000 volumes).

In addition to the *Dictionary of National Biography*, see Owen, *English Philanthropy*, p. 364; Sanderson, *Universities*, pp. 63, 90; Cardwell, 'Patronage of Science', pp. 106–107; Henry Guppy, *The John Rylands Library Manchester* (Manchester, 1935), pp. 7–10.

139. Owen, *English Philanthropy*, p. 470. On this see also Camplin, *Rise of the Plutocrats*.

140. Social expectations of this kind already existed in the early eighteenth century in England. See Whitaker, *Foundations*, p. 47: 'In early 18th-century England . . . the failure of a merchant to settle some substantial and conspicuous charitable trust or gift was generally regarded as little short of shocking unless his estate had suffered some exceptional misfortune.'

141. How far this also applied to Jews in Britain is discussed by Vagts, 'Die Juden im englisch-deutschen imperialistischen Konflikt vor 1914'; U. R. Q. Henriques,

too, examining the biographies of the more important nineteenth-century entrepreneurs, concludes that 'social advancement was one of the most prized possessions to be bought by an English business fortune'.[142] But here too there were exceptions, such as Ludwig Mond, who refused the knighthood offered him, pointing to his German origins and his religious affiliations.[143] One motive which is important today was of practically no significance in the nineteenth century: the attempt to avoid high income and property taxes by establishing philanthropic endowments.

Given the present state of research in this field, it is difficult to make any generalisations about private patronage in Britain since the mid nineteenth century on the basis of this rather cursory survey. Quantification is practically impossible. There is no question that widespread and well-endowed patronage for science and the arts existed.[144] Benefactors came from the prosperous industrial, financial and commercial bourgeoisie, and had a variety of motives and ideas about what they wanted to achieve. In the foreground were philanthropic and religious motives, and the desire to make science useful for medicine, industry and the economy. As late as the beginning of this century the president of the British Academy, Lord Reay, considered that it was the 'private benefactor on whose munificence we depend to a large extent in this country for the advancement of scientific knowledge'.[145] In many cases, new institutions could only be established because of endowments, which thus provided an important stimulus for the

---

'The Jewish Emancipation Controversy in Nineteenth-Century Britain', *Past and Present*, vol. 40 (1968), pp. 126–46; and Stephen Aris, *The Jews in Business* (London, 1970), pp. 50, 236, 267–8. See also Vivian D. Lipman, *Social History of the Jews in England 1850–1950* (London, 1954), p. 162: 'Socially, the late Victorian and Edwardian age was perhaps the most brilliant Anglo-Jewry had experienced. The friendship of Edward VII, as Prince of Wales and King, had given the *entrée* to the Court of many talented, cultured and wealthy Jews'.
142. Coleman, 'Gentlemen and Players', p. 95. See also Wiener, *English Culture*, pp. 13–15 and 127–54; Camplin, *Rise of the Plutocrats*.
143. Cohen, *Ludwig Mond*, p. 221. On the situation in Germany, which was in some respects similar, see Friedrich Zunkel, *Der Rheinisch-Westfälische Unternehmer 1834–1879. Ein Beitrag zur Geschichte des deutschen Bürgertums im 19. Jahrhundert* (Cologne and Opladen, 1962), pp. 99–127; Lamar Cecil, 'The Creation of Nobles in Prussia, 1871–1918', *American Historical Review*, vol. 75 (1970), pp. 757–95. Recently also Fritz Stern, *Gold and Iron: Bismarck, Bleichröder, and the Building of the German Empire* (London, 1980), pp. 112–13, 167.
144. David Owen, however, concludes that 'in the hierarchy of eighteenth- and nineteenth-century philanthropic giving, higher education held a curious and uncertain position' (*English Philanthropy* p. 346). If a *Times* editorial is to be believed, the income from private donations of charitable institutions in London alone was higher than the national budget of countries like Sweden, Denmark, Portugal or Switzerland (*The Times*, 9 January 1885).
145. Speech held at the annual meeting of the British Academy, 26 June 1903, printed in *Proceedings of the British Academy 1905–1906* (London, n.d.), p. 15.

development of research and university education. Private patronage for science in Britain can therefore be described as decidedly innovative and risk-taking in character, all the more so as private benefactions could be granted relatively quickly and with a minimum of bureaucracy.

This represents a remarkable contrast with the majority of charitable foundations in the late nineteenth century. David Owen sums up his analysis of late-Victorian bequests as follows: 'The chief impression left by late-century bequests is not that of eccentricity but of conventionality. Few wills contained anything particularly venturesome or imaginative. Money went, on the whole, to maintain established institutions or to create new ones of the same sort'.[146] In addition, it is clear that private patronage for science concentrated on establishing and developing new scientific institutions such as university colleges in the provinces after 1850, as well as research laboratories and institutes. This indirect support for science, channelled through institutions, by far exceeded direct aid for research projects of individual scientists or groups of scientists. Until 1914 scientific societies, too, derived very little income from endowments. The relatively modest amounts which they received from this source were generally donated by members or their heirs. Good examples are not only the Royal Society, with its universal orientation, but also specialist societies such as the Entomological Society of London, or the Linnean Society.[147]

It has not yet been adequately explained how private endowments came about in specific individual cases — how and under what conditions patrons made money available for particular projects. So far, almost nothing is known about the processes governing these decisions, from the vague first conception through to the final definition of goals. Who, for example, provided the decisive impetus for a private endowment which established or supported a research institute? Did the initiative come from the patron himself, who perhaps wanted a particular scientific problem solved, or wanted to support specific sciences, or did it come from scientists? Or did an endowment develop out of a large number of suggestions and initiatives, which had to combine before the desired result was achieved?

---

146. Owen, *English Philanthropy*, p. 474.
147. On the Royal Society, see above, p. 20 in this volume. On the Entomological Society, see S. A. Neave and F. J. Griffin, *The History of the Entomological Society of London, 1833–1933* (London, 1933), pp. 103–4 and 112–14. Referring to the Linnean Society between 1800 and 1914, Gage Writes: 'Nearly half of the total bequests was left by three Fellows' (Gage, *Linnean Society*, p. 149).

Available evidence suggests that the interplay between patrons and science took place in very different ways. We have already pointed out that in the course of their professional or commercial activities, 'potential' patrons were often confronted with specific scientific problems, or gained some sort of insight into the work of particular disciplines. Sometimes they owed their fortunes to the industrial application of the results of scientific research. From this could develop a 'will to exercise patronage', which provided the necessary impetus to innovate in the area of science organisation. Examples are benefactions made by Ludwig Mond, Charles F. Beyer, John T. Brunner, the ship-owner Alfred L. Jones and Henry S. Wellcome, manufacturer of pharmaceutical goods. In other cases, one or more scientists had to approach a potential patron with a concrete proposal, thus directing his attention towards a specific research project. Once the support of such a person had been gained, the benefaction could develop in close contact with scientists, taking one of several forms. It might be limited to one or more substantial payments, provide continuous financial support or be anchored in an independent legal corporation for the administration of endowed capital.

The extremely important function of mediating between scientists and patrons, between the needs of science and the will to exercise patronage, which inevitably involved an element of chance, was assumed by scientists, and frequently also by politicians. Sir Israel Gollancz, scholar and science organiser, fulfilled it with extraordinary virtuosity for the British Academy, which faced severe financial problems because it did not receive state aid. Lord Rosebery, former Liberal prime minister, took on the role of mediator for both the London School of Economics and Imperial College of Science and Technology. In numerous other cases, Richard B. Haldane, who had extensive political and social connections, was an active mediator — his function could almost be described as that of a catalyst. The new university colleges created after the middle of the nineteenth century also obtained private funds by public appeals launched by university planners and spokesmen for science. These examples often give the impression that it was less a case of a lack of money for scientific purposes than of a lack of objects worth supporting. If this was so, a modification of John D. Rockefeller's dictum would apply: that for benefactors, too, giving is not an easy art. Benefactors need expert knowledge as well as the advice of spokesmen for science or scientists with a talent for administration.

Indifference and Improvisation: The Role of the State

In spite of their proven flexibility, private endowments for science had disadvantages: they were made unsystematically, and their purpose was usually determined by their donors and could not be changed. Since the sixteenth century Britain had a well-developed charity law which recognised and encouraged, under the concept of 'charitable purposes', contributions made to social welfare and hence also to science. Nevertheless, by the nineteenth century, incorporated foundations had not yet been set up to administer large company or family fortunes in trust and make their proceeds available over the years for science and research in general, or for more precisely defined projects. Large industrial foundations, such as have existed in the USA since the early twentieth century, admittedly involving incomparably larger sums of money, had neither precursors nor contemporary parallels in Britain.[148]

Private patronage stimulated science in Britain in many significant ways, and provided large amounts of money, especially in the second half of the nineteenth century. Nevertheless, it was not a suitable permanent basis for the promotion of science. This required continuity and rapidly growing expenditure, two factors which were made more pressing by the enormous development of science during the second half of the nineteenth century. Private endowments for scientific purposes were not only established in an unsystematic and uncoordinated manner, and usually tied to a specific purpose, but they could not be counted on in the long term, and experience showed that they were very sensitive to fluctuations in the economy. It has not yet been established, for example, whether the amount of money made available in private grants kept pace with increases in Gross National Product. Ultimately, too, private patronage could not supply enough money to finance large research undertakings and institutes in the long term, or to build and maintain the new universities. Even in Manchester, where the prosperous industrial bourgeoisie of the city and its

---

148. In Britain, on the other hand, there were numerous charities; it has been estimated that there were almost 29,000 in 1837 (Whitaker, *Foundations*, p. 39). Foundations comparable in conception to American foundations were not established in Britain until during the Second World War. The best known of these are the Nuffield Foundation and the Isaac Wolfson Foundation. The Atlantic was bridged by foundations established in Britain by American millionaires, such as the Scottish Universities Trust (1901) and the Carnegie United Kingdom Trust (1913), both financed by Andrew Carnegie, or the Pilgrim Trust (1930). In the USA, there were only five general-purpose foundations before 1900. In 1914 there were 30, in 1935, 200, in 1964 almost 15,000 and in 1976, 26,000 (Whitaker, *Foundations*, p. 12; Jaher, 'Gilded Elite', pp. 206–10).

environs formed a solid core of patrons, private patronage could not provide a stable basis. The conviction that the promotion of science went beyond the means of private patrons and was basically the responsibility of the state, therefore gained momentum towards the end of the century. It was accepted not only by scientists but also in some quarters of the public.

The next chapter will show in more detail that since the early nineteenth century the scientific community in Britain had been demanding more state involvement in promoting science, and advocating that the state assume more and greater responsibilities in this area. The scientific community was calling for a redefinition, and ultimately an expansion, of the state's role as a patron of science and the arts. The state, it was argued, was evading a responsibility which fell to it as a result of the development of science. It was the state's duty to defray the rising costs of scientific research, especially in disciplines which had expanded rapidly since the middle of the century, such as, for example, meteorology, astronomy, experimental physics and chemistry, at least in cases where private patronage was inadequate. By the end of the nineteenth century it was no longer possible to set up and maintain a physics institute on the basis of private funding alone.

In the early 1870s the government had instructed the Devonshire Commission to investigate existing facilities for the teaching of science and for scientific research, and in 1875 the Commission pointed to the general discrepancy in science between rising costs and availability of money: 'Whatever may be the disposition of individuals to conduct researches at their own cost, the Advancement of Modern Science requires Investigation and Observation extending over areas so large and periods so long that the means and lives of nations are alone commensurate with them'.[149] Public statements by scientists, many of whom uncritically took as their model the reputedly superior conditions in Germany, produced the image of a passive state, which did practically nothing to eliminate this discrepancy, indeed, abandoned science almost entirely to a situation governed by *laissez-faire* attitudes. This has been accepted, largely untested, by many modern historians of science. Hilary and Steven Rose, for example, write that for science in nineteenth-century Britain, 'self-help was the order of the day'.[150] And Gerard Turner has emphasised that it was perfectly normal for scientists to finance their own research in the nineteenth century, even if this

149. *Eighth Report of the Royal Commission on Scientific Instruction and the Advancement of Science*, C. 1298 (London, 1875), p. 24.
150. Rose and Rose, *Science and Society*, p. 32.

was becoming less common.[151] The thesis that the state adopted a *laissez-faire* attitude is supported by Cardwell's survey of the development of scientific organisations and institutions between 1800 and 1914.[152] It confirms the view that as late as the second half of the nineteenth century the state had no systematic and long-term policy for promoting science.

On closer examination, however, Hilary and Steven Rose's general assessment needs to be differentiated, as in the final analysis it does not accurately describe the role of the state in supporting science in Britain during the nineteenth century. In addition, it is based on the unspoken assumption that state support is the ideal method of promoting science, and that a lack of it disastrously retards the scientific and economic development of a country. From at least the middle of the century it is clear that the state was increasingly ready to offer financial aid to specific areas of science, and to exercise patronage over the most diverse scientific activities, primarily for utilitarian reasons. But as so many government departments and ministries were involved, it is extremely difficult to find exact figures for expenditure, even for the late nineteenth century, and so the extent of this form of patronage can only be estimated. As a rule, government agencies did not coordinate with each other the aid they offered. Government patronage, like private patronage, therefore, took place in an unsystematic manner, although specific projects were sometimes supported for years or even decades. Admittedly, in almost all of these cases it was not the original intention to provide regular financial support. The history of the Geological Survey of Great Britain is a good example. This survey, the first of its kind undertaken by a state anywhere in the world, was begun in 1835.[153] Initially, it was thought that it would be completed in a few years. But no end was in sight by the beginning of the twentieth century, and it continued to require considerable sums of money.[154]

A rough categorisation of state patronage of science in Britain in the nineteenth century shows that financial aid was offered in four main areas: universities, establishments which provided scientific expertise for the general administration of the country, scientific

151. See examples in Turner (ed.), *Patronage of Science*, pp. 5–6.
152. Cardwell sums up: 'The denial of State aid during the crucial period 1850–80 was the final reason why applied science was later in making its appearance in England than in Germany' (*Organisation of Science*, p. 244).
153. Edward Bailey, *The Geological Survey of Great Britain* (London, 1952); John Smith Flett, *The First Hundred Years of the Geological Survey of Great Britain* (London, 1937).
154. Morrell, 'Patronage of Mid-Victorian Science', p. 359; Archibald Geikie, *A Long Life's Work: An Autobiography* (London, 1924), p. 304.

expeditions and learned and scientific societies. In individual cases, these categories cannot always be clearly separated.[155]

(1) Reference has already been made to the Treasury's gradually increasing expenditure on the new universities in England. After 1839, when central government first granted the University of London a subsidy of £5,000, specific conditions governed the use to which allocations of public money could be put. The sums involved were relatively small. The university colleges in Wales received £4,000 for the year 1883/4, but by 1888/9 the government grant had risen to £14,000.[156] In 1889, when the government made a total of £15,000 available for all the new university colleges in England, Manchester's share was £1,500. This was little enough, but as has already been shown, the general trend was for government subsidies to rise. Together with student fees and regular payments from local administrations responsible for education, government subsidies were a constant factor in college budgets. Their significance increased steadily up to the eve of the First World War.

(2) During the nineteenth century certain characteristic developments took place in Britain as well as in other European countries: while state expenditure was generally rising, and the state's activities expanding, administration was becoming increasingly 'scientific' and differentiated.[157] This covers the entry of 'scientific experts' into the civil service, and the creation of establishments to develop and implement administrative measures in an industrialising and urbanising society. The National Vaccine Establishment, founded in 1808 to combat smallpox, made pioneering achievements in preventive medicine. The setting up of the Geological Survey of Great Britain in 1835 has already been mentioned. On

---

155. See MacLeod's basic study, 'Science and the Treasury', esp. pp. 120–2. Also Roy M. MacLeod, 'The Support of Victorian Science: The Endowment of Research Movement in Great Britain, 1868–1900', *Minerva: A Review of Science, Learning and Policy*, vol. 4 (1971), pp. 197–230.
156. *Nature*, 94 (1914–15), p. 551.
157. Of the extensive literature, see especially Oliver MacDonagh, 'The Nineteenth-Century Revolution in Government: A Reappraisal', *Historical Journal*, vol. 1 (1958), pp. 52–67; Henry Parris, 'The Nineteenth-Century Revolution in Government: A Reappraisal Reappraised', *Historical Journal*, vol. 3 (1960), pp. 17–37; Henry Parris, *Constitutional Bureaucracy: The Development of British Central Administration since the Eighteenth Century* (London, 1969); Gillian Sutherland (ed.), *Studies in the Growth of Nineteenth-Century Government* (London, 1972); Norman Chester, *The English Administrative System 1780–1870* (Oxford, 1981); Roy M. MacLeod, 'The Alkali Acts Administration, 1863–1884: The Emergence of the Civil Scientist', *Victorian Studies*, vol. 9 (1965), pp. 85–112; David Roberts, *Victorian Origins of the Welfare State* (New Haven, Conn., 1960). Roy M. MacLeod, 'Science and Government in Victorian England: Lighthouse Illumination and the Board of Trade, 1866–1886', *Isis*, vol. 60 (1969), pp. 5–38, gives an example of the conflicts which frequently occurred between administration and scientific experts.

the basis of existing initiatives, special departments were introduced in the Treasury, Board of Trade, Home Office, Colonial Office, Royal Mint, Privy Council, Local Government Board, Admiralty, War Office and, after 1904, in the Post Office, to set up, carry out and monitor routine scientific tasks. These included, for example, trigonometric and hydrographic measurements, testing materials, chemical analyses, collecting medical data, public health measures, maintaining observatories and meteorological stations, oceanographic surveys and collecting data on problems in agriculture, fisheries and mining. By the turn of the century it had become clear that as a reaction to new tasks and the needs of society, 'scientific and specialist expertise, within limits, was becoming part of the accepted orthodoxy'.[158] The state had also taken on responsibility for an increasing number of public museums since the second half of the nineteenth century. The most important and costly of these, after the British Museum, was the complex in South Kensington, established in 1851 after the Great Exhibition. Added to this was the maintenance of botanical gardens in London (Kew Gardens), Edinburgh and other parts of the country. Almost all of these institutions, which employed an increasing number of scientists, saw themselves as research establishments. But they were primarily involved in applied science and were intended to be of direct use to government, administration and public education. By 1870 the government was spending £140,000 annually to maintain these various institutions.[159]

Institutes which were partly financed by the government because the work they did was of interest to British imperial policies, also to some extent belong to this category. Other research was also done in these institutes, independently of government agencies. The Imperial Institute in London, founded in 1887 on the occasion of Queen Victoria's Golden Jubilee and opened in 1893, provides one example. As 'a lasting emblem of the unity and loyalty of the Empire',[160] it was to work on aspects of the culture, history, society and economy of individual parts of the empire. But its real function was to publicise the idea of empire, and any scientific

---

158. Roy M. MacLeod, 'Statesmen Undisguised', *American Historical Review*, vol. 78 (1973), p. 1,404. Arthur J. Taylor points to the difficulty of speaking of the nineteenth century as a century of *laissez-faire* in view of this development (*Laissez-faire and State Intervention in Nineteenth-Century Britain*, London and Basingstoke, 1972). No data exist about the number of scientists employed by central government in London during the nineteenth century. At the turn of the century it is said to have been 200 (Rose and Rose, *Science and Society*, p. 35).
159. MacLeod, 'Support of Victorian Science', p. 204.
160. Queen Victoria at the opening of the Institute in 1893 (Humberstone, *University Reform*, p. 77).

work was in fact secondary to this. The foundation of the Imperial Institute was made possible by a collection which, under the auspices of the Prince of Wales and the lord mayor of the City of London, raised the considerable sum of £413,000 by 1892. No exact data exist about the donors. In general terms, it is known that £236,862, or over half of the sum raised, came from private donors, and that the government of India gave £101,000, and Canada £20,000.[161] This secured the necessary buildings, but it did not cover the cost of maintenance and routine work. Right from the start the day-to-day running of the Institute was plagued by financial difficulties, because in spite of the Institute's political function, the government refused to provide the full cost of running it. When the University of London, at the government's suggestion, took over a large section of the ostentatious Imperial Institute building in 1890 and set up its own administration there, the significance of the Institute, founded with so much imperial pomp and such high-flown political expectations, rapidly declined. In 1902 it became the responsibility of the Board of Education, and in 1916, of the Colonial Office.

Like the Imperial Institute, the London School of Tropical Medicine was also financed from a mixture of sources, funds in this case coming from private endowments and the Colonial Office. This Office, which had supported the work of the School and that of its sister institute in Liverpool, especially during Joseph Chamberlain's period of office, granted the School an annual subsidy. By 1910 it amounted to £1,300 and almost covered routine running costs.[162] Similarly, the Sleeping Sickness Bureau and the African Entomological Research Committee, which studied the transmission of tropical diseases by insects, were almost totally funded from the Colonial Office's budget. A similar function was fulfilled by specific chairs which the government established and financed at various universities (e.g. in London and Scotland) between 1850 and 1890.

(3) Transitions are clearly visible here to the common nineteenth-century practice of the state funding learned and scientific societies to carry out specific projects or to organise international conferences. Support could take the form of regular or one-off

---

161. Ibid., p. 75. See also Wyndham R. Dunstan (ed.), *Imperial Institute: Technical Reports and Scientific Papers* (London, 1903).
162. *Nature*, 85 (1910–11), p. 28; MacLeod, 'Support of Victorian Science', pp. 226–7. By 1914 the subsidy had risen to £1,500. To this was added £2,245 contributed by the dominions and colonial governments (*Report of the Advisory Committee for the Tropical Diseases Research Fund for the Year 1914*, Cd. 7796, London, 1915, p. 1).

payments. The best-known cases are the financial aid given to the Royal Geographical Society, the Royal Astronomical Society and Royal Society of London for expeditions all over the world in the nineteenth and early twentieth centuries under their leadership and responsibility. From time to time relatively large sums were involved. The Royal Geographical Society, for example, received £3,000 in 1876 for expeditions, and £5,000 in 1883. The Marine Biological Association of the United Kingdom received similar support in the 1880s and 1890s for specific projects and for building and maintaining its marine biology research station at Plymouth.[163] Enterprises which were thought to be connected with national prestige or which promised to be of economic and strategic value received particularly generous support. The Board of Trade, the Admiralty and the War and Colonial Offices, therefore, were frequently benefactors of science. This was especially true of expensive, prestigious projects such as the research ship *Challenger*'s world trip in the 1870s. During the voyage, extensive oceanographical, meteorological and zoological research was carried out. The results were published in a fifty-volume report, subsidised by the Treasury to the extent of almost £30,000. Imperial prestige was also involved in the Arctic Expedition of 1875, expeditions to observe solar eclipses between 1869 and 1905, and expeditions to observe the transit of Venus in 1874 and 1882. The 1882 expedition alone received £15,000, which by contemporary standards represented a considerable sum in the context of aid granted for scientific projects.

Much higher sums were made available for Robert F. Scott's and Ernest H. Shackleton's Antarctic Expeditions of 1901–4, 1907–9 and 1911–12, which were supported largely by the Admiralty. Special claims were made on the state's generosity in this case. Reaching the South Pole was obviously not only of scientific interest. It was also seen as a test of the country's technical and industrial efficiency in the face of international competition, and as both a national and a sporting challenge. Shackleton's expedition of 1907–9 therefore received 'substantial help', amounting to £45,000, from the government. 'These are', *Nature* commented, 'genuine

163. MacLeod, 'Science and the Treasury', pp. 121, 156. The foundation of the Marine Biological Association in 1884 under the leadership of T. H. Huxley goes back to a suggestion made by the German zoologist Anton Dohrn. He visited Britain in 1867, 1870 and again in 1873. In 1870 he expounded before the British Association in Liverpool his plans for a network of marine biology research stations spanning the world. The Marine Laboratory established at Plymouth in 1888 was modelled on the Zoological Station which he had founded at Naples in 1872 (officially opened in 1874).

examples of the aid given by the state for science purposes, but they are from their very nature variable in amount, and not assigned as part of a definite policy of encouraging systematic scientific research as a continuous and necessary part of national organisation. They owe their existence to external forces, and if these were withdrawn or lessened they would disappear.'[164]

(4) The regular support received by scientific societies referred to above seems extremely modest in comparison with the sums made available for a project like the exploration of the South Pole, which was certainly not motivated by purely scientific interests. Occasionally, however, even these small allocations met with resistance. For example, Robert Lowe, chancellor of the exchequer between 1868 and 1873, rejected in principle public aid for learned and scientific societies and their research projects. The arguments he used to justify his attitude are revealing: 'I think it a very bad plan for any government to select societies as their agents, and to give them large sums of money, because the tendency is to give large salaries, and they give rise to a suspicion of jobbery'. He continued: 'We are called upon for economy. Now, the first maxim of economy is that Government should not be called upon to do that which there is a reasonable probability people will do for themselves'.[165] Nevertheless, scientific societies did receive government support, and it should be mentioned that there were no conditions attached to its use. Thus it could be channelled into 'pure' research, as in the case of the £4,000 granted annually to the Royal Society. For this reason, the Royal Society's subsidy has rightly been described as 'until about 1890 the major continuous source of direct government finance earmarked solely for the support of original scientific investigation'.[166]

It is very difficult to gain any general overview of total state expenditure on science and research in Britain between 1850 and 1900. Existing statistical data are inconsistent, and for long periods they are not comparable because they are based on different criteria. The sums granted annually by Parliament for universities, learned and scientific societies, 'civil research' and 'scientific investigations' did not always comprise the same areas, and applied to different expenditures by various government agencies. Often the figures

---

164. *Nature*, 94 (1914–15), p. 548. In the financial year 1910/11 Scott received £20,000 for his Antarctic Expedition (Hansard, Parl. Deb., H.C., 5th Series, vol. 29, col. 1487, 10 August 1911). The support given to expeditions of this kind was occasionally criticised in the House of Commons. For example, there was talk of 'these silly objects' on 16 August 1911 (see ibid., cols. 2025–7).
165. Quoted in MacLeod, 'Science and the Treasury', p. 137.
166. MacLeod, 'The Royal Society', p. 324.

Table 1.3. State expenditure on science, 1850–1900

Year	Sum in absolute terms(£s)	Percentage of national budget
1850/1	34,328	0.9
1859/60	77,764	0.9
1869/70	271,138	2.8
1879/80	405,889	2.7
1889/90	652,371	4.1
1899/1900	617,787	2.6

Source: Roy M. Macleod, 'Science and the Treasury: Principles, Personalities and Policies, 1870–1855' in G. L'E. Turner (ed.), *The Patronage of Science in the Nineteenth Century* (Leyden, 1976), p. 122.

include sums spent on secondary education or on the arts. Over the years certain items of expenditure appear in different departments. Roy MacLeod has calculated the amounts of state expenditure on science and on activities which promoted science (Table 1.3), as discussed under the categories listed above.[167]

The trend towards increased expenditure on science after the middle of the century revealed by these figures is also reflected in budget figures for 'scientific investigations'. Under this heading were subsumed: (a) subsidies to scientific societies in Britain and Ireland, (b) the running of scientific institutes, (c) expeditions and (d) academies, public libraries and music academies. Table 1.4 shows the expenditure on these categories between 1869 and 1914. At first glance, expenditure seems to vary widely from year to year. A trend emerges more clearly, however, if total expenditure is compared in ten-year cycles:

1869–79	1879–89	1889–99	1899–1909
£120,720	£215,510	£267,769	568,609

The periods from 1879 to 1889 and from 1899 to 1909 in particular show sharp increases in expenditure on 'scientific investigations'. Expansion continued in the five years between 1909 and 1914, when it totalled £419,026. But as a proportion of Gross National Product, the total expenditure on science as set out in these figures

---

167. See also the following contemporary reference, which fits in well with MacLeod's figures: 'At present [1876], about £300,000 are expended annually by the nation on various scientific establishments' (Richard A. Proctor, *The Wages and Wants of Science Workers*, London, 1876, reprinted 1970, p. 45).

Table 1.4. State expenditure on science, 1869–1914

Year	Amount(£s)	Year	Amount(£s)	Year	Amount(£s)
1869/70	12,300	1884/5	23,400	1899/1900	36,724
1870/1	12,370	1885/6	21,400	1900/1	50,724
1871/2	12,450	1886/7	24,400	1901/2	53,154
1872/3	12,450	1887/8	23,900	1902/3	68,396
1873/4	12,450	1888/9	21,900	1903/4	90,780
1874/5	13,300	1889/90	27,003	1904/5	46,407
1875/6	12,550	1890/1	25,253	1905/6	53,900
1876/7	12,550	1891/2	25,790	1906/7	57,650
1877/8	15,550	1892/3	25,896	1907/8	54,479
1878/9	17,050	1893/4	26,163	1908/9	56,295
1879/80	17,050	1894/5	26,247	1909/10	57,964
1880/1	17,050	1895/6	26,827	1910/11	74,228
1881/2	21,600	1896/7	28,154	1911/12	61,603
1882/3	20,900	1897/8	27,984	1912/13	125,523
1883/4	23,650	1898/9	28,452	1913/14	99,708

Source: *Nature*, 94 (1914–15), pp. 552–3.

was extremely small. Indeed, as late as 1934 the figure for Britain was only 0.1 per cent, compared with 0.6 per cent in the USA.[168]

The exact amount of public money spent on science in the second half of the nineteenth century is as difficult to ascertain as is the extent of private patronage. Nevertheless, the figures above do show that in absolute terms state expenditure on 'science' increased by leaps and bounds. The increase is particularly noticeable after the turn of the century. But expenditure on 'science' was not the same thing as expenditure on 'research'. 'The aid given by the State towards the advancement of science', wrote C. A. Buckmaster in 1915, in his article 'State Aid for Science',

> has increased in amount and variety so far as applied science is concerned. Questions dealing with matters of immediate utility connected with physics, biology, pathology, or agricultural practice have secured the attention of the State with a fair measure of success. But the support and encouragement formally given to pure science have dwindled down till they are now almost lost among the grants given for the study of Shakespeare and tailors' cutting, the practice of cooking and Morris dancing, or the cultivation of languages from ancient Greek to modern Esperanto. All excellent objects, and all deserving of the nation's support, but in their rapid growth they bid fair to choke the tender plant of

168. Bernal, *Social Function of Science*, pp. 64–5.

pure science, to the prior existence of which they owe their own flourishing condition, and the early struggles of which insured their own success.[169]

Buckmaster may be deliberately overstating his case, and his assessment undoubtedly reflects the resentment felt by scientists at the — in their opinion — inadequate provision made for science by the state, and at the improvised nature of the provision that was made. But essentially it points to several important circumstances, which we will briefly recapitulate.

(1) During the nineteenth century the state in Britain took on more and more scientific functions which were able to assist it in its growing and increasingly differentiated administrative tasks. In some cases the state itself organised this scientific 'auxiliary' aid in response to obvious societal needs. The state's involvement in promoting science increased automatically, as it were, as a consequence of its expanding functions. But these measures were not based on a long-term or coordinated policy, or on any specific criteria by which scientific projects and institutions could be judged to be worth supporting. Whether the institutions we have described as providing 'auxiliary' aid received sufficient financial support, remained a disputed question. The Meteorological Office, for example, which was set up by the Board of Trade in 1854 but remained administratively independent, was kept in a state of 'chronic poverty', according to the physicist Arthur Schuster in 1906. He wrote that it had 'to restrict itself to work of the most pressing necessity'.[170] Roy MacLeod has suggested that the state's increasing participation in the promotion of science after about 1850 was 'one of the most significant characteristics of nineteenth-century science'.[171] It remains to be seen whether this claim is justified.

(2) In spite of all justified reservations about the distinction between theoretical, or 'pure', science and practically orientated, or 'applied', science, this remains an important division. During the nineteenth century, state support was mainly directed at what can roughly be called the applied sector. While senior civil servants and

---

169. *Nature*, 94 (1914–15), p. 553. A few months later *Nature* called state aid for basic research 'a sham supplemented by a few doles' (*Nature*, 96, 1915–16, p. 336).

170. Arthur Schuster, 'International Science', *Nature*, 74 (1906), p. 259. Schuster was a professor at Owens College in Manchester. From 1912 to 1919 he was one of the two secretaries of the Royal Society.

171. MacLeod, 'The Royal Society', p. 323. Brock's assessment is more restrained ('Science Patronage', p. 173). He asks whether private patronage of science was not in fact of much greater significance in the nineteenth century.

scientists used the same terms, they only too often meant different things by the word 'science'. Where science was at least potentially useful for government, the state soon displayed an interest in encouraging it, without motives being precisely identifiable in each case. Subsequently, this resulted in an expansion in the state's 'scientific services', with more and more being spent on staffing and equipment. The Treasury could point to this, in answer to criticism made by scientists, as proof of Britain's growing expenditure on science. Only if discussion is restricted to 'applied' science is it legitimate to speak of a 'dramatic' increase in the amount spent by central government on research after 1850.[172]

(3) Basic research, by contrast, was largely neglected by the state during the nineteenth century in Britain. *Nature* commented on this in 1873:

> The lack of pecuniary means can be the main difficulty which has hitherto, in the richest country in the world, hindered original investigation in the sciences. The natural harvest of scientific discoveries which England might annually reap has . . . been checked by the irregularity with which the labourers have been rewarded and the comparative indignity with which they have been treated. For a certain class of scientific investigators of a strikingly practical character, the public will always be willing to sanction large Parliamentary grants; but for the permanent Endowment of Research and the continuous support in a worthy position of the researchers, not only the aid of the nation at large, but the wealth and prestige of our ancient universities are required.[173]

Even if it is not correct to say that the state gave no support for basic research at this time, it was in general granted hesitantly and unwillingly — according to *Nature*, with 'comparative indignity' and 'in a grudging way'[174] — and was limited to an absolute minimum. The £4,000 granted annually to the Royal Society represented for a long time the largest amount made available for this purpose as 'continuous support'. But as it was not increased after 1876, its significance decreased proportionately towards the end of the century, and by 1900 it apparently made up only 5 per cent of the state's total expenditure on 'science' in the widest sense.[175]

172. As does MacLeod, 'Support of Victorian Science', p. 205.
173. *Nature*, 8 (1873), p. 297. Edward Shils surveys the situation in the USA in 'The Order of Learning in the United States from 1865 to 1920: The Ascendancy of the Universities', *Minerva*, vol. 16 (1978), pp. 159–95.
174. *Nature*, 85 (1910–11), p. 31.
175. MacLeod, 'The Royal Society', p. 356. Since 1849 the Royal Society had

As late as the second half of the nineteenth century the state took no initiatives in the area of basic research, by, for example, creating and improving the institutional prerequisites for scientific research. It seldom willingly took up initiatives which came from scientists. Their complaints about the state's lack of interest in science, or more precisely, in basic scientific research, had dominated public discussion since the first third of the nineteenth century.[176] If scientific research did not have obvious economic or military value, state interest was minimal. The Treasury in particular, the 'department of departments',[177] was committed, as several recent studies have shown,[178] not only to a strict cost – benefit analysis in forming its attitude to science but also to a principle of negation, the logical consequence of which was that all increases in expenditure were basically rejected as undesirable, and money made available for 'pure' science only in exceptional cases. In 1878 the permanent secretary to the Treasury wrote:

> Science as such is no part of the provision of political government and nothing can be more out of place than direct discussions about it in official correspondence. My Lords select the most promising *savants* and give them their heads, so long as they show no signs of running wild, or of spending too much or of producing no result.

Three years later he expressed the Treasury's attitude even more precisely:

> My Lords consider generally that scientific inquiry is not the direct business of government, but rather of Societies and bodies, which . . . government may occasionally assist, e.g., by such grants as those of £4,000 and £1,000 now made to the Department of Science and Art and to the Royal Society. If any more special grant is ever in question, the object of it should be definite, of great importance, beyond any public views, and promising of success.[179]

---

received an additional £1,000 to subsidise the cost of scientific publications. See the bitter criticism made in 1916: 'Several times this amount is expended each year upon stationery alone used by members of the House of Commons' (Richard A. Gregory, *Discovery or the Spirit and Service of Science*, London, 1916, p. 35).
176. On this topic, see Chapter 2 in this volume.
177. Maurice Wright, *Treasury Control of the Civil Service, 1854–1874* (London, 1969), p. 1.
178. MacLeod, 'Science and the Treasury'; Henry Roseveare, *The Treasury 1660–1870: The Foundations of Control* (London and New York, 1973), esp. pp. 186–92; Maurice Wright, 'Treasury Control 1854–1914' in Sutherland (ed.), *Nineteenth-Century Government*, pp. 195–226; Chester, *Administrative System*, pp. 222–81.
179. Quoted in MacLeod, 'Science and the Treasury', p. 154. In 1904 the

With these views, the permanent secretary was in total accord with the liberal theory of the state and fiscal policy of his time, and in particular, with the political convictions of the prime minister, Gladstone. Gladstone had headed the Treasury himself from 1852 to 1855, and again between 1859 and 1866, and in the 1870s frequently opposed both state support for science and the creation of a Ministry of Science. He explained in 1879 that 'the Chancellor of the Exchequer should boldy uphold economy in detail; and it is the mark of a chicken-hearted Chancellor when he shrinks from upholding economy in detail, when, because it is a question of only two or three thousand pounds, he says that is no matter'.[180] And a memorandum on Treasury expenditure policy submitted to the newly appointed chancellor of the exchequer, George J. Goschen, in 1887 read: 'The first object of the Treasury must be to throw the departments on their defence, and to compel them to give strong reasons for any increased expenditure, and to explain how they have come to have to demand it'.[181]

Considering this restrictive and inflexible attitude in the conduct of public finances in Britain, it is not surprising that generous financial gestures, far-sighted cooperation and spectacular long-term support of science for its own sake were lacking on the part of the government. As late as the turn of the century there was no question of the state, its officials and politicians placing any trust in the work and ideas of scientists. Extreme economy and the drive to avoid any increase in expenditure were the iron principles governing fiscal policy in Britain. Therefore, private initiative and private patronage had to step in and take over functions which, because of the accelerated development of science, had long been assumed by the state in Germany, for example.[182] This situation in Britain

---

vice-president of the Society of Arts wrote: 'There seems to be a dread at the Treasury of any of the present departments having more to do with science than is absolutely forced upon them' (William Abney, 'Science and the State', *Nature*, 71, 1904–5, p. 90).

180. Quoted in Henry Roseveare, *The Treasury. The Evolution of a British Institution* (London, 1969), p. 190. Sir Michael E. Hicks Beach, chancellor of the exchequer from 1885 to 1886, and again from 1895 to 1902, is still described by R. C. K. Ensor as 'a strong conservative in party matters and an equally strong Gladstonian in the field of finance' (*England 1870–1914*, Oxford, 1936, reprinted 1968, p. 349).

181. Printed in Wright, 'Treasury Control', p. 223. On the occasion of *Nature*'s fiftieth anniversary in 1919, Richard A. Gregory reviewed the situation fifty years ago: 'When the publication of *Nature* was begun fifty years ago, experimental research received little or no support from the State. Astronomical work was carried on at the Royal Observatory, Greenwich, and natural history objects were displayed at the British Museum, but there was absolutely no provision in this country for the support of experimental investigation of a modern type' ('The Promotion of Research', *Nature*, 104, 1919–20, p. 259).

182. See Pfetsch, *Wissenschaftspolitik in Deutschland*. See also Theodor Schieder,

provides the essential background to the thesis of the state's *laissez-faire* attitude, and the reproach arising from it that science and the system of university education were being neglected to the nation's loss. To what extent this thesis holds true for actual conditions in Britain in the second half of the nineteenth century and at the beginning of this century, remains to be investigated.

'Kultur, Wissenschaft und Wissenschaftspolitik im Deutschen Kaiserreich' in Gunter Mann and Rolf Winau (eds.), *Medizin, Naturwissenschaft, Technik und das Zweite Kaiserreich* (Göttingen, 1977), pp. 9–34; Bernhard vom Brocke, 'Hochschul- und Wissenschaftspolitik in Preußen und im Deutschen Kaiserreich 1882–1907: das "System Althoff"' in Peter Baumgart (ed.), *Bildungspolitik in Preußen zur Zeit des Kaiserreichs* (Stuttgart, 1980), pp. 9–118; Ernst Rudolf Huber, *Deutsche Verfassungsgeschichte seit 1789*, vol. 4, *Struktur und Krisen des Kaiserreichs* (Stuttgart, 1969), ch. 13, 'Der Staat und die Hochschulverfassung', pp. 925–70.

2
Science and the Awareness of Crisis

Throughout the reign of Queen Victoria scientists in Britain complained that the government was not interested in science and did not understand its needs and problems. This aspect of the relationship between science and the state has already been touched upon. But the dialogue which scientists wanted did not eventuate. Almost without exception, attempts by scientists to influence political decision-making processes and to obtain some form of long-term financial commitment to science from the government were unsuccessful, although scientists increased their efforts throughout the nineteenth century.

In retrospect, therefore, we can see that the debate about science was a very one-sided affair. It was a debate from which one of the partners consistently withdrew, and for this reason it remained for the most part restricted to the scientific community,[1] a small group which did, however, grow in size during the nineteenth century. Only occasionally, under exceptionally favourable political circumstances, did these issues impinge upon a wider public. 'We are not strong enough', a scientist could write in *Nature* as late as 1914, 'in making our demands heard; and, in my opinion, this is not a virtue, but a neglect of duty.'[2] But for scientists, the debate, in which basic issues of science policy were aired, was not without benefits. It fulfilled three main functions: it allowed stock to be taken of the relationship between science and the state in Britain; it provided a forum in which to express concern and dissatisfaction about the state of support for science in Britain; and it provided an opportunity to list desiderata and to develop aid programmes for the future. This debate began in the early nineteenth century and was particularly intense in the periods around 1830, 1870 and between

---

1. The term 'scientific community' refers to a closed social and communication system of scientists, or (perhaps more appropriately) of groups of scientists, segmented by discipline. See Warren O. Hagstrom, *The Scientific Community* (New York and London, 1965).
2. Ronald Ross, 'Organisation of Science', *Nature*, 94 (1914–15), p. 367.

1900 and the outbreak of the First World War.

This chapter does not attempt to follow in detail the course and the ramifications of the discussions among scientists about the inadequacies of the British education system, especially in the university sector, and about the problems of aid for science and the role of science in modern society. Its aim is more modest. It attempts to describe the protagonists in this debate. The 'science lobby'[3] in Victorian England is presented briefly and necessarily incompletely, and this is followed by a description of its organisational forms, the levels at which it was active and the forms of activity in which it engaged. The fundamental question addressed is: did the groups of people involved, the methods they used and the levels at which they worked, change in the period up to the First World War, or was there a large degree of continuity? Any discussion of methods and levels of agitation also involves the arguments used by British scientists to justify their demands for state aid for science. There are so many variations, however, that they cannot all be reported in detail. The approach taken here is to concentrate on certain representative examples in order to allow the essential arguments to stand out more clearly.

The British Science Lobby in the Nineteenth and Early Twentieth Centuries

As Theodor Schieder has pointed out,[4] one of the contradictions characteristic of the nineteenth century is that while the natural sciences experienced a tremendous boom and the results of scientific research were having a greater impact on people's lives than ever before, in public awareness science did not achieve the status due to it. This was true of almost all European countries and North America. Against the background of this incongruity, already noted by contemporaries, there developed, and not only in Britain,

---

3. On this modern historiographical term, see Peter Weingart, *Die amerikanische Wissenschaftslobby. Zum sozialen und politischen Wandel des Wissenschaftssystems im Prozeß der Forschungsplanung* (Düsseldorf, 1970). Following Weingart (pp. 16–18), who correctly points to the link between 'lobbying' and the specific conditions of the American system of government, the term 'science lobby' here refers to the more or less organised representation of the interests of science, which since the nineteenth century has attempted to influence the British Parliament and government in line with its own objectives. For pragmatic reasons alone, therefore, it is used here as a collective term.
4. Theodor Schieder, 'Europa im Zeitalter der Nationalstaaten und europäische Weltpolitik bis zum I. Weltkrieg (1870–1918)' in idem (ed.), *Handbuch der Europäischen Geschichte*, vol. 6 (Stuttgart, 1968), pp. 4–5.

a practically continuous campaign for the recognition of science as a decisively important element in modern society. Connected with this campaign was the demand for a new assessment of the role of scientists who, it was becoming clear, were of crucial importance in an expanding industrial economy. The British science lobby, which had developed in various forms since early in the nineteenth century, was aware that it had an educative task going beyond the listing of demands. Society's ignorance about the work, aims and the technical and economic benefits of science seemed to preclude any intelligent response to mere catalogues of demands.

While those who spoke for the interests of science at this time formed a group which was relatively homogeneous in social terms, their forms of organisation and their goals and methods differed widely. As it is almost impossible to separate them neatly from each other, we shall concentrate on some typical cases. To simplify matters, the representatives of scientific interests in Britain can be divided into four groups. The earliest spokesmen for the interests of science, those who were constantly in the public eye from the beginning of the nineteenth century until well into the First World War, were scientists and journalists, and sometimes politicians with an interest in science organisation, in research and in promoting it. Three 'spokesmen for science', Charles Babbage, Sir Alexander Strange and Sir J. Norman Lockyer, their biographies and public activities will be discussed briefly here as examples typical of the whole of the British science lobby in the nineteenth century.

Apart from individual scientists and informal groups of scientists, the specialist scientific societies regarded it as one of their most important tasks to draw public attention to the concerns of science, to create understanding for its demands and to advocate state aid for basic research. Until the end of the nineteenth century and even beyond, however, there were no clear and generally accepted conceptions about what forms state aid should take, and under what conditions it should be given. At the time, the main concern was to achieve recognition for a principle which was directly opposed to the dominant liberal theory of state and society. Under the circumstances, putting this principle into practice was of secondary significance. Given the lack of precision in working out models for the organisation and support of science in Britain, it was almost inevitable that the state reacted cautiously, and intervened only after a relatively long delay.

Since at least the second half of the century the Royal Society of London had carried more weight than any of the other specialist societies because it was recognised as a sort of unofficial British 'national academy of sciences'. But the specialist societies as a whole had only a limited impact — many of them avoided or totally rejected the idea of exercising any direct political influence. They rarely reached a broad public audience. As a rule, their journalistic forum was the specialist journals, whose number had increased sharply during the nineteenth century, reflecting the increase in the number of scientific and technical societies. Journals such as *The Chemical News* (founded in 1859), *The Engineer* (founded in 1861), *The Quarterly Journal of Science* (founded in 1864), *The English Mechanic* (founded in 1865), *Engineering* (founded in 1866), *The Academy* (founded in 1869) and, of course, *Nature* (founded in 1869), achieved particular prominence. But because the learned and scientific societies had proved to be relatively ineffectual in representing the interests of science, professional associations of scientists were founded very early. Although they also contributed to the discussion of specialist problems, they went beyond this and expressly defined their goal as promoting the interests of science in public. Their goals, organisational forms and methods make them the precursors of modern interest groups.

Charles Babbage (1792–1871) was an impressive individual lobbyist for science as early as the first half of the nineteenth century. The methods he developed to mobilise popular interest and his arguments in favour of more state aid for science set the tone for the discussion which has continued ever since. Babbage, one of the founders of the Royal Astronomical Society in 1820, was one of the most widely known mathematicians of his time, and held a chair at Cambridge from 1828 to 1839. While teaching there he published a controversial book in which he compared the state of scientific research in England, France and Prussia, and came to negative conclusions about conditions in England.[5] Personal bitterness about inadequate financial backing for an adding machine he had designed, and the feeling that there was little understanding of the benefits science could offer the general public and the national economy, may have helped to turn Babbage into a spokesman for science.

As L. Pearce Williams has suggested, Babbage's polemical book on the decline of science in England must also be seen in connection with the controversy which had long been smouldering about the constitution of the Royal Society in the early nineteenth century.[6] Like several other publications, Babbage's book severely criticised

---

5. Charles Babbage, *Reflections on the Decline of Science in England and some of its Causes* (London, 1830, reprinted Farnborough, 1969).
6. See Williams, 'The Royal Society', and above, pp. 18–19 in this volume.

the Royal Society, its allegedly ossified administrative structure and its outdated admissions practice, which gave more weight to social status than to scientific achievement. The election of the Duke of Sussex, a son of George III, to the presidency of the Royal Society was for the minority of scientists among its Fellows (including Babbage) merely another confirmation that it had become no more than a social club like many others in London in which non-scientists had the say — a visible expression of the relative decline of science in England. Going beyond this topical issue, however, Babbage's book precipitated a long overdue debate on matters of principle: the lack of public aid for science in England, the limited career opportunities for scientists, the inadequate institutional preconditions for research, and the general lack of scientific instruction in schools. This debate, occasionally involving *The Times*, raised for the first time issues that were to be discussed again and again, in different forms, to the point of weariness: complaints about the neglect of science and its lack of social recognition in Britain, comparisons between conditions in Britain and other countries, the economic benefits which scientific research could have for the general public, and the contribution which it could make to solving practical industrial problems.

In contrast to Babbage, who after releasing a flood of discussion with his controversial book in 1830 rarely took part in the subsequent debate on science policy, Alexander Strange (1818–76) remained a spokesman for science for many years. According to J. G. Crowther, he was 'one of the most, perhaps the most outstanding British Statesman of Science in the second half of the nineteenth century'.[7] Like Babbage, Strange belonged to several scientific societies. From 1863 he held leading positions in the Royal Astronomical Society, whose ranks included some of the most active spokesmen for science in the nineteenth century, and from 1867 to 1869 he was on the Committee of the Royal Society. Initially Strange followed his scientific interests while pursuing a military career. Until 1861 he was involved in taking basic trigonometric and geodetic measurements for the British Army in India. After his return from India, Strange actively advocated that basic research — that is, research which has little direct economic use and therefore depends to a large extent on support from a third party — should be the responsibility of the state. He expressed his ideas within the Royal Society and the British Association for the Ad-

---

7. J. G. Crowther, *Statemen of Science* (London, 1965), p. 233. See also MacLeod, 'Support of Victorian Science', pp 202–3, and Cardwell, *Organisation of Science*, pp. 119–21.

vancement of Science, as well as through journalistic channels. The title of a speech Strange gave in 1868 at the annual meeting of the British Association in Norwich, 'On the Necessity for State Intervention to Secure the Progress of Physical Science', articulated a demand common among the scientific community. Strange drew conclusions from the observations he had made a year earlier at the Paris International Exhibition, where, as one of the British experts present, he had seen first-hand evidence of the rapidly growing competition which the British economy faced from other countries.

Within a few years Strange, who possessed remarkable administrative skills, had achieved great prestige as a lobbyist for science. The royal commission chaired by the Duke of Devonshire was set up in 1870 largely as a result of his efforts. Its appointment was the most important success achieved by the British science lobby in the second half of the nineteenth century. The remarkably detailed report which the commission submitted after almost six years of work,[8] confirmed all the essential points made by Strange and described state aid for science in Britain as inadequate. One of the commission's most important recommendations followed up Strange's suggestion that a Ministry of Science be established in view of the enormously enhanced significance of science in a modern industrial society. But this recommendation did not have the undivided support of all scientists.[9] Many feared that a ministry would interfere in scientific questions and in research. They therefore rejected a central authority for science on principle, and also opposed state aid in any form as posing a potential threat to the independence of science.

The debate about whether a Ministry of Science should be set up in Britain made clear for the first time that different opinions existed, not only among large sections of the public but also within the scientific community, about what form the relationship between science and the state should take. While the majority of scientists who expressed opinions in public favoured more state involvement in the science sector and advocated accepting public money, a minority firmly rejected this position until well into the 1920s, when certain forms of state intervention had long been the norm.[10] For them, state involvement in science — and here they

8. *Eighth Report of the Royal Commission on Scientific Instruction and the Advancement of Science*, C. 1298 (London, 1875).
9. Ibid., pp. 45–6; MacLeod, 'Support of Victorian Science', pp. 207–8, 220, 222; Jean-Jacques Salomon, *Science and Politics* (London, 1973), p. 22.
10. In April of 1881 the Astronomer Royal, Sir George B. Airy, wrote: 'I think

followed Gladstone's argument — was necessarily unwanted state interference in science. It is difficult to judge how common this attitude was among British scientists at the time.

The controversy of the mid 1870s, however, made apparent the full extent of the contradictory situation in which science constantly found itself: while freedom to determine research goals and independence from state and social control rank highly in scientists' perceptions of the place of science in society, accepting public aid almost inevitably involves state regulation and intervention. But most scientists reject the right of state authorities to have a say in the allocation of public money for scientific research, because this restricts the autonomy of science. The search for a compromise between the demand for the relative independence of science from the state and the fear of financial dependence on the state and of the conditions which could be imposed as a result, remains a general problem of science policy. The lack of agreement about the state's role in supporting science, first apparent after the publication of the Devonshire Report, resulted in scientists making contrary demands. This often provided the politicians to whom demands for initiatives in science policy were addressed with a pretext for inactivity or postponing necessary decisions. The innovative but costly recommendations for promoting science made by the Devonshire Commission, which with few exceptions (such as an increase in the grant for the Royal Society) were not implemented until decades later, provide an early and well-known illustration of this situation. Richard Proctor, a science journalist, wrote of the Commission's recommendations that the time had not yet come 'when the nation would look with satisfaction on any wide scheme of scientific endowment'.[11]

J. Norman Lockyer (1836–1920) earned his first laurels in the area of science organisation as the secretary of the Devonshire Commission, a post he held for the full six years of its work.[12] From the end of the 1860s until his death in 1920, he was one of the most

---

that successful researches have in nearly every instance originated with private persons or with persons whose positions were so nearly private that the investigators acted under private influence, without incurring the danger attending connection with the State' (*The English Mechanic*, 831, 1881, pp. 586–7). Further examples of this attitude are given by MacLeod, 'Science and the Treasury', pp. 131–2, and George Haines, *Essays on German Influence upon English Education and Science, 1850–1919* (Hamden, Conn., 1969), pp. 67–8. Reservations about any expansion in state activity were still widespread in Britain in the early twentieth century. When the Royal Society celebrated its 250th anniversary, *The Times* wrote: 'Like many useful things in this country the Royal Society owes its origin to private effort' (*The Times*, 17 July 1912).

11. Richard A. Proctor, *The Wages and Wants of Science-Workers* (London, 1876, reprinted 1970), pp. 84–5.

12. A. J. Meadows, *Science and Controversy: A Biography of Sir Norman Lockyer*

energetic and influential spokesmen for British scientists. Lockyer was a scientist and science journalist as well as a science organiser. As an astronomer, he had discovered the gas helium in the emission spectrum of the sun (1868), thus securing his reputation as one of the most brilliant scientists of his time. A Fellow of the Royal Society at thirty-three, Lockyer became professor of astrophysics at the Royal College of Science in London in 1890. As founder, editor and leader-writer of *Nature*, he tried to keep his readers informed about important new findings and current research problems. Beyond this, *Nature* was to 'urge the claims of Science to a more general recognition in Education and in Daily Life' — according to the aims defined in the first issue.[13]

Lockyer was probably the first science journalist in late Victorian times, apart from T. H. Huxley, who also had very broad scientific interests, to recognise the potential of the expanding British market for newspapers and journals for publicising scientific issues. Lockyer used the media with remarkable skill, as testified by *Nature*'s continuing success after some initial difficulties. This success was achieved partly at the expense of similar periodicals,[14] and in the 1870s and 1880s inspired the founding of similar journals on the Continent in direct imitation. It should not, however, obscure the fact that even by contemporary standards its circulation was small. Exact figures do not exist for the period before the beginning of the First World War, but it has been estimated that *Nature* had between 100 and 200 subscribers in its first years, and between 300 and 500 by 1895. Material held by the London publisher Macmillan, who managed the journal, shows that *Nature* had 970 subscribers in 1913. Only one-quarter of them lived in the United Kingdom; the rest lived in the British colonies and dominions, in the USA or in Europe. Even by 1920 the number of subscribers was still just under 1,100.[15] Lockyer's ability as a science organiser is also illustrated by his work for the British Association for the Advancement of Science, of which he was president in 1903, and by the founding on his initiative of the British Science Guild in 1905, of which more later.

(Cambridge, Mass., 1972); T. Mary Lockyer and Winifred L. Lockyer, *Life and Work of Sir Norman Lockyer* (London, 1928).

13. On the foundation of *Nature*, see Meadows, *Science and Controversy*, pp. 25–31.

14. On the history of *Nature*, see the issue published on 1 November 1969, on the occasion of the journal's 100th anniversary (*Nature*, 224, 1969, pp. 417–81), particularly the brief outline of scientific periodicals in the nineteenth century on pp. 423–7.

15. *Nature*, 224 (1969), p. 443.

The influence of individual scientists who voiced the demands of science in specialist journals and at lectures, at scientific conferences and in letters to the daily newspapers,[16] was limited. But local and national specialist societies, too — the Royal Society at their head — were rarely able to reach beyond the narrow circle of specialists and address a wider public in their role as spokesmen for science. In an article published in 1906 *Nature* criticised the Royal Society, the most venerable institution of British science, as an 'exclusive and retired body, known to few', and attributed this situation to the Society's peculiar aversion to publicity.[17] This simple explanation for the Royal Society's lack of public impact must of course be questioned. It is, however, true that despite its universal orientation, the Royal Society made no attempt to coordinate the representation of the interests of science. The activities of the specialist societies were also limited in scope: their resolutions, appeals, declarations, conferences, submissions and deputations to authorities and Parliament generally did not take place in a public forum. Their journalism, published in specialist journals and widely distributed literary weeklies, as a rule evoked little response from the general public. The major issues of the time, such as Britain's economic development since the mid 1870s, the social problems created by unemployment, strikes and the squalor of urban slums, imperial rivalry among the European powers and the question of Irish Home Rule all eclipsed the concerns of science, which affected only a very small minority of the population and could expect to arouse very little interest. In the nineteenth century, specialist societies were not able to create a social climate in which promoting science came to be seen as a political necessity. 'Science' was very low on the list of contemporary political priorities.

In the final analysis, prominent scientists working together informally were as ineffective as individual scientists and the specialist societies. They attempted to do something for the problems of science by making private arrangements to intervene personally with influential politicians and officials, and by working on royal commissions or advisory bodies of various sorts. Well-known examples of such informal associations, whose minimal organisational structure allowed them great flexibility of action, are the X-Club and the Endowment of Research Movement. From 1864

16. The impact of the biologist T. H. Huxley's (1825–95) journalism was described as follows in 1894: 'In England when people say "science" they commonly mean an article by Prof. Huxley in the "Nineteenth Century", but Huxley was busy with scientific journalism long before then' (quoted in Cyril Bibby, *T. H. Huxley: Scientist, Humanist and Educator*, London, 1959, p. 101).
17. *Nature*, 74 (1906), p. 466.

the X-Club — its name was purely fortuitous and had no significance – consisted of nine influential scientists, including the famous biologist and later president of the Royal Society, T. H. Huxley.[18] Three members had taken their doctoral degrees in Germany: the chemists Edward Frankland (1825–99) and John Tyndall (1820–93) and the mathematician Thomas Archer Hirst (1830–92). All members of the Club held the view, no longer original at the time, that the foundations for the teaching of science and for research were inadequate in Britain, and that any improvement would require state aid.[19] They advocated a radical change in social attitudes towards science, as well as greater security for researchers and the creation of more jobs for scientists. The argument they used to justify these demands has become sufficiently familiar: that the benefits which science offers society oblige society to give science financial support. Members of the X-Club, which existed until the beginning of the 1890s, played a decisive part in setting up the Devonshire Commission. Two of them sat on the Commission, five held the office of president of the British Association for the Advancement of Science, and three became presidents of the Royal Society: Joseph D. Hooker 1873–8, William Spottiswoode 1878–83 and T. H. Huxley 1883–5.

Like the X-Club, the Endowment of Research Movement also pleaded for a fundamentally new attitude towards scientific research on the part of the state and the universities. Above all, it advocated state research scholarships to finance long-term scientific projects.[20] The term 'research scholarship' must be understood in this context as a euphemism for a salary for scientific work undertaken without pressure to produce economically useful results within a short time. The core of the movement consisted of a group of twenty scholars from Oxford and Cambridge. Later this group briefly numbered seventy. Influenced by scientific developments in Europe, they had since 1872 advocated reforming the two old English universities to make them more strongly research-orientated. They saw the solution for their problems in the universities defining their role differently and not in more state intervention. Instead of

18. Bibby suggests that the 'X' refers to three things: to the name still to be found for the club, to the original number of members planned, and to the still unknown identity of the tenth member (*T. H. Huxley*, p. 249).
19. Roy M. MacLeod, 'The X-Club: A Social Network of Science in Late-Victorian England', *Notes and Records of the Royal Society of London*, vol. 24 (1969–70), pp. 305–22.
20. MacLeod, 'Support of Victorian Science', pp. 207–9; idem, 'Resources of Science', pp. 113–15; Haines, *German Influence*, pp. 99–100. See also Charles E. Appleton (ed.), *Essays on the Endowment of Research by Various Writers* (London, 1876).

state interference and control, they advocated financial aid for an independent science, and their criticism of the status quo was combined with an optimistic belief in the universities' ability to reform themselves. In the case of this movement, too, scientists were divided about the advisability of representing the interests of science and about the demands to be made. This movement's attempt to claim state aid and at the same time to keep the state at a distance in order to preserve the autonomy of science, was recognised. Nevertheless, some members of the scientific community accused it of seeking its own material advantage under a thin disguise, reproaching the movement with being less concerned with the 'endowment of research' than with 'research for endowment'.

Informal associations of scientists could not represent science on a permanent basis any more adequately than the organised specialist societies. In many cases, like that of the Endowment of Research Movement, internal disagreements limited their effectiveness. Improvisation, a rudimentary institutional structure and a temporary existence made them unsuitable vehicles for representing science. The necessity of developing other means to influence and mobilise the public, Parliament and government in accordance with the wishes of science was, therefore, recognised early. In the first half of the nineteenth century, organisations were formed which claimed to represent, at least in theory, the whole of science. The best known and longest surviving of these was the British Association for the Advancement of Science (BAAS). In addition to functioning as a learned society where scientific topics and recent research results could be discussed, it aimed to provide a forum in which the interests of science could be presented.[21]

The growing controversy in the Royal Society between professional scientists and 'amateurs' with scientific interests provided the immediate incentive for the foundation of the BAAS. Scientists had long been uneasy about the decline of the Royal Society, which, despite its adherence to an out-dated view of science, was still regarded as 'the natural guardian of English science' in the early nineteenth century.[22] This uneasiness turned into open opposition when in 1830 the Duke of Sussex, who had no scientific qualifi-

21. O. J. R. Howarth, *The British Association for the Advancement of Science: A Retrospect, 1831–1931* (London, 1931); Cardwell, *Organisation of Science*, pp. 59–61. Arnold Thackray and Jack Morrell, *Gentlemen of Science: The Origins and Early Victorian Years of the British Association for the Advancement of Science* (London, 1981), and Roy M. MacLeod and Peter Collins (eds.), *The Parliament of Science: The British Association for the Advancement of Science, 1831–1981* (Northwood, 1981) were published on the occasion of the British Association's 150th anniversary.
22. Sir David Brewster to Charles Babbage on 12 February 1830; quoted in Williams, 'The Royal Society', p. 223.

cations whatsoever, was elected president.[23] The Royal Society's orientation towards activities of a social nature, reflected in the disproportionate number of non-scientists among its Fellows, had prevented it participating in the 'professionalisation' of science which had taken place in France and Prussia since the end of the eighteenth century. The failure of the Royal Society in its role as a central agency for encouraging research and representing the interests of scientists *vis-à-vis* the public and the state in Britain was obvious. In the first half of the nineteenth century the 'gentleman tradition' continued unbroken in the Royal Society.[24] The presidential elections of 1830 sparked off Babbage's attack on the Royal Society. Its unwillingness to reform, reflected in this election, gave rise to the idea of founding a purely scientific rival organisation. Babbage attended a meeting in Berlin of the Gesellschaft Deutscher Naturforscher und Ärzte (GDNÄ), chaired by Alexander von Humboldt, the celebrated geographer, and reported on it in Britain. In a circle around Charles Babbage and Sir David Brewster, physicist and editor of the *Edinburgh Journal of Science*, the idea of a rival organisation crystallised into the BAAS. It was 'the direct reaction of the Royal Society reformers to their defeat by the amateurs' in 1830.[25]

Of some significance for this initiative was the fact that scientists in other European countries were making similar efforts to represent scientific interests more adequately. In France these efforts resulted in the foundation of the Congrès Scientifique de France in 1833. Eleven years earlier the GDNÄ had been established in Leipzig as a central society for the sciences and medicine. As the first organisation of its kind, it provided a model for the BAAS and for the Congrès Scientifique de France in terms of both organisation and goals. Founded in 1831, the BAAS differed from the GDNÄ in two respects: it defined its goals more widely, and it excluded medicine. The fact that medical training in Britain was much more practically orientated than in Germany may have been a factor in arriving at this decision.

23. See above, p. 18 in this volume.
24. See Chapter 5 of this volume, pp. 214–45.
25. Williams, 'The Royal Society', pp. 230–1. See also *Record of the Royal Society*, p. 56; W. H. G. Armytage, *The Rise of the Technocrats: A Social History* (London, 1965, reprinted 1969), pp. 94–5; and the chapter 'The Scientific Revolt: 1820–1860' in Lyons, *The Royal Society*, pp. 228–71. In her wide-ranging study, *Science in Culture: The Early Victorian Period* (Folkestone and New York, 1978), Susan Faye Cannon correctly points out that, independently of the controversies surrounding the Royal Society, the foundation of the British Association goes back above all to the suggestion of scientists outside London. Their main aim was to have a society which was not limited to London, and which was not dominated by 'amateurs' (pp. 170–5, 179–80, 213–14).

Several other points in the Association's statutes are also worth noting.[26] Elitist criteria were dropped, members being drawn from a wide circle. Membership was open to anyone who belonged to a 'philosophical' society in Britain or its overseas possessions and was prepared to pay a small membership fee. The aim of these fairly non-specific conditions was to recruit all scientists in Britain for the BAAS and achieve as broad a membership as possible, while keeping at a distance lay people with an active interest in science. While these conditions were generally adhered to in the early years of the British Association's existence, later they were interpreted more and more liberally. In 1883 the BAAS had 4,000 members, the majority of whom were not scientists in the strict sense of the word. The BAAS's main offices, however, and especially that of president, were reserved for those 'who appear to have been actually employed in working for science'.[27] The criterion was to have had at least one scientific work published. But this rule was not followed strictly either. In 1859, for example, Prince Albert, lauded for his scientific and artistic interests, took over the presidency for the annual meeting in Aberdeen. London was selected as the British Association's permanent seat. The inaugural meeting defined the Association's aims as: 'to give a stronger impulse and a more systematic direction to scientific inquiry, to obtain a greater degree of national attention to the objects of science, and a removal of those disadvantages which impede its progress, and to promote the intercourse of the cultivators of science with one another, and with foreign philosophers'.[28]

In the nineteenth and early twentieth centuries the BAAS developed into the 'closest approach . . . to a parliament of scientific workers' in Britain.[29] It attempted to fulfil its aim of mediating between science and the public as well as between specialist scientific disciplines, which were moving further and further apart, in four ways:

(1) Modelling itself on the GDNÄ, the British Association, which catered for all the sciences, held a conference of several days' duration in a different British city every year. It was divided into a central session and sections for specialist disciplines. This important annual event was prepared and organised by local scientific societies, whose involvement in the BAAS's scientific and social

---

26. For details, see Howarth, *The British Association*; Williams, 'The Royal Society'; and A. D. Orange, 'The British Association for the Advancement of Science: the Provincial Background', *Science Studies*, vol. 1 (1971), pp. 315–30.
27. Howarth, *The British Association*, p. 24.
28. Ibid., pp. 16–17.
29. Bernal, *Social Function of Science*, p. 41.

activities allowed them at least temporarily to overcome their oft-lamented isolation in the 'provinces'. Scientific societies in the provinces, especially in the north of England and in Scotland, therefore urged that a principle of rotation be followed in selecting the city to host the conference each year.[30] Characteristically, the BAAS's constituent assembly took place in York in 1831, at the invitation of the Yorkshire Philosophical Society based there. It is therefore often called the BAAS's parent society.[31]

Another advantage of the principle of rotation was that it meant that the normal concentration of scientific activities in London, Oxford and Cambridge was avoided. London was expressly excluded as a meeting place. Annual meetings usually had a greater impact in provincial cities, both on the general public and on scientists. This was reflected most obviously in the foundation of local scientific societies, many of which established a loose connection with the BAAS after 1880 as 'corresponding societies'.[32] In anticipation of this effect, speakers at the inaugural meeting in York had spoken of their intention 'to animate the spirit of philosophy in all the places through which the meetings may move',[33] 'philosophy' being understood by contemporaries to mean the natural sciences. A speaker at the 1848 annual meeting attributed 'the awakening of many important places to scientific activity' to the Association's practice of holding its conference in a different place each year.[34] This conclusion appears to have been correct: in 1834 a Geological Society was established in Edinburgh immediately after the British Association had held its meeting there. One year after the meeting took place in Norwich in 1868, the Norfolk and Norwich Naturalists' Society was founded. The 'naturalists' and geologists always made a large impact because these disciplines were flourishing.[35] The number of new learned and scientific

30. Orange, 'The British Association'; Cannon, *Science in Culture*, pp. 201–24.
31. Howarth, *The British Association*, p. 15; Orange, 'The British Association', pp. 315–17. The Yorkshire Philosophical Society was founded in 1821. Its aim was 'to promote science in the district by establishing a scientific library, scientific lectures, and by providing scientific apparatus for original research' (quoted in Howarth, p. 14).
32. In 1903 the British Association was associated with seventy 'corresponding societies' with a total of 25,000 members (according to figures given by Norman Lockyer in *Nature*, 68, 1903, p. 441).
33. Quoted in Howarth, *The British Association*, p. 24.
34. Quoted in ibid., p. 93.
35. 'The meeting at York', wrote Justus von Liebig to Michael Faraday after the BAAS's annual meeting there, which he had attended, 'did not satisfy me in a scientific point of view. It was properly a feast given to the geologists, the other Sciences serving only to decorate the table' (L. Pearce Williams, ed., *The Selected Correspondence of Michael Faraday*, vol. 1, *1812–1848*, Cambridge, 1971, p. 430, letter dated 19 December 1844).

societies in the provinces increased sharply after 1840; the wave of new foundations reached its peak between 1870 and 1880.[36] After 1884 the British Association held some of its meetings in cities of the empire in order to promote, as it liked to emphasise on these occasions, solidarity between English-speaking scientists. Meetings held in Montreal (1884), Toronto (1897), Capetown and Johannesburg (1905), Winnipeg (1909) and various Australian cities (1914) showed that, in the words of the president at the Winnipeg meeting, the British Association had successfully taken the step 'from an Insular into an Imperial Association'.[37]

(2) As the setting for the British Association's annual meeting changed each year, so did its president. This was an honorary office, which conferred on the holder the obligation to give a public lecture at the annual meeting on the general implications of a current research problem. Normally this office was reserved for a well-known scientist, but for the meeting in Cambridge in 1904 the Conservative Prime Minister Arthur J. Balfour was elected. His lecture, 'Reflections suggested by the New Theory of Matter', testified to an unusual scientific interest.[38] Since 1888 Balfour had been a Fellow of the Royal Society; in 1920 and again between 1925 and 1929, as lord president of the Privy Council, he was responsible for the newly created Department of Scientific and Industrial Research; and from 1921 to 1928 he was president of the British Academy. The annual rotation of the presidency and the setting for the annual meeting was an ostentatious demonstration of the difference between the British Association and the ossified organisational structure of the Royal Society in the first half of the nineteenth century. The Royal Society's failure as the leading scientific organ-

36. See Appendix 1 in this volume, p. 256.
37. *Nature*, 81 (1909), p. 248. In his presidential address to the 1907 annual meeting in Leicester, Sir David Gill, astronomer and later president of the Royal Astronomical Society (1909–11), referred to the meeting which had been held in South Africa two years earlier: 'Having myself only recently come from the Cape, I wish to take this opportunity of saying that this southern visit of the Association has, in my opinion, been productive of much good: wider interest in science has been created amongst colonists, juster estimates of the country and its problems have been formed on the parts of the visitors, and personal friendships and interchanges of ideas between thinking men in South Africa and at home have arisen which cannot fail to have a beneficial influence on the social, political, and scientific relations between these colonies and the mother country' (*Nature*, 76, 1907, p. 319).
38. Printed in *Nature*, 70 (1904), pp. 368–70. See also Robert J. Strutt, *Lord Balfour in his Relations to Science* (Cambridge, 1930); Kenneth Young, *Arthur James Balfour: The Happy Life of the Politician, Prime Minister, Statesman and Philosopher 1848–1930* (London, 1963), p. 208; Almeric Fitzroy, *Memoirs*, vol. 2 (London, 1925), p. 491. Also Max Egremont, *Balfour: A Life of Arthur James Balfour* (London, 1980); Sydney H. Zebel, *Balfour: A Political Biography* (London, 1973); Blanche E. C. Dugdale, *Arthur James Balfour*, 2 vols. (London, 1936, reprinted Westport, Conn., 1970).

isation in the United Kingdom was attributed to the fact that it was tied to London and to the rigidity of its leadership structure.

(3) A large part of the British Association's agitation consisted of deputations and petitions making specific requests of the government. The earliest deputation sent by the BAAS was one to the Treasury in 1834, asking for more rapid publication of British ordnance maps.[39] In 1868 the annual meeting in Norwich decided that the state of science teaching and research in Britain should be investigated. In this case the government's reaction was extraordinarily positive, resulting in the Royal Commission on Scientific Instruction and the Advancement of Science being set up under the Duke of Devonshire in 1870.[40] Its work, however, largely came to nothing, because very few of its recommendations were implemented in the years that followed. Deputations from the BAAS repeatedly pointed out the importance of immediate action, but to no avail.

(4) The British Association for the Advancement of Science could allot only relatively small sums for scientific purposes, but providing direct financial support for research and the publication of research results had not been among its original aims. For the first twenty years of its existence the Association spent £15,000 from membership subscriptions on supporting 300 research projects. According to figures in its official history, it spent a total of £92,000 in its first 100 years on almost 700 scientific undertakings, subsidising both projects and publications. During this time the main emphasis in terms of money spent was on mathematics and physics, followed by zoology, anthropology and geology.[41] The details of projects financed which O. J. R. Howarth lists in his history of the BAAS show that in most cases it was a matter of small, sometimes trivial, amounts.[42] The most expensive single commitments at this time were maintaining the observatory at Kew near London, on which £12,300 was spent between its establishment in 1843 and 1872, when it was taken over by the Royal Society, and financing a table at a cost of £4,640 at Anton Dohrn's Zoological Station in Naples after 1876.[43] The British Association

---

39. Howarth, *The British Association*, p. 212.
40. Meadows, *Science and Controversy*, p. 82. On the Devonshire Commission, see also Cardwell, *Organisation of Science*, pp. 119–26; Rose and Rose, *Science and Society*, pp. 31–2; Haines, *German Influence*, pp. 51–2. The Commission and its work have not yet been the subject of a scholarly publication.
41. Howarth, *The British Association*, p. 147.
42. Ibid., pp. 266–92.
43. Ibid., pp. 267 and 280. Tables at Naples had been rented by the University of Cambridge (since 1873) and the University of Oxford (since 1890). British (and

tried to avoid supporting large projects which tied up large amounts of money for long periods of time, and this principle was followed well into the twentieth century.

At the 1921 annual meeting the president, with some justification, considered the BAAS responsible for the fact that 'the Government was induced to extend a direct national encouragement to science and to aid in its organisation'.[44] In spite of all it had achieved since its inception, however, the Association was not immune to criticism of its effectiveness as a mediator between science and the public and as a representative of the interests of science. Its annual meetings, the Association's most important method of agitation, continued to attract large numbers.[45] More and more often, however, they were regarded as 'excuses for holiday-making'.[46] Towards the end of the century the social functions of these meetings, undoubtedly an important feature, sometimes did come to dominate proceedings. Meetings assumed the character of specialist conferences at which scientific questions and developments were widely discussed, embedded in an extended social programme. As a forum for scientific communication, the British Association was certainly successful. But it did not achieve the aim of becoming an umbrella organisation representative of all scientific interests in the way that had been intended. Its primary goal, which in 1831 had been to educate the British public about the problems of science, was rather overshadowed towards the end of the century.

The increasing criticism of the British Association, which must be seen in the context of a general, politically orientated debate about national efficiency, therefore often referred to the inadequacy of the Association's efforts to publicise its work. In his presidential address at the annual meeting in Southport, Lockyer regretted that when it came to the big national issues, science in Britain had neither a voice of any weight nor a corporate identity with authority to express its expectations and wishes. Lockyer was overstating his case, but he saw this as the reason why science in Britain had

---

American) practice was unusual in as much as in all other cases tables at the Zoological Station were 'rented' by countries, not by scientific bodies.

44. T. Edward Thorpe, 'Some Aspects and Problems of Post-war Science, Pure and Applied', *Nature*, 108 (1921), p. 45. Thorpe (1845–1925) taught chemistry at the Royal College of Science/Imperial College of Science and Technology in London. He was foreign secretary of the Royal Society from 1899 to 1903.

45. Between 1831 and 1914 the number of participants at the annual meetings fluctuated, on average, between 1,300 and 2,500. The highest numbers of participants ever registered were 3,838 at Manchester in 1887 and 3,335 at Newcastle-on-Tyne in 1863 (Howarth, *The British Association*, pp. 117–18).

46. *Spectator*, 2 September 1882, p. 1,127.

no backing in public or from the government. He said it was necessary 'to enlighten public opinion on the importance of science', without, however, explaining in detail what form this should take in specific cases. Lockyer concluded that 'our great crying need is to bring about an organisation of men of science and all interested in science, similar to those which prove so effective in other branches of human activity'.[47]

When Lockyer's rhetorical and dramatic appeal for the BAAS to undertake more effective publicity work evoked no appreciable response, he decided to form a rival organisation, which was to be primarily a mouthpiece for British science. Two years after Lockyer's speech, in October 1905, the British Science Guild came into being. Lockyer, its founder, did not become its president. With an eye towards publicity, Richard B. Haldane, who only a few weeks later became secretary of state for war in Campbell-Bannerman's Liberal cabinet, was named president. Joseph Chamberlain became deputy president, which conferred a form of honorary membership. He was one of many people to hold this office, including the lord mayor of the City of London, well-known scientists, MPs and members of the House of Lords.[48] There were no formal restrictions as to who could become a member of the British Science Guild. Theoretically, therefore, it could have developed into a mass organisation for science, as Lockyer had originally intended.[49] But the memorandum in which the conception behind the planned organisation was outlined early in 1904 was unmistakably addressed to a particular circle of people. Members and supporters were desired from both Houses of Parliament, from the legislative assemblies developing in the dominions, from local authorities in Britain, from learned and scientific societies, chambers of commerce, federations of industry, the trades union movement and, finally, from the universities and from among British academics.[50] This elitist membership structure was achieved in time, with varying success in the different groups addressed. The British Science Guild had attracted 600 members by the beginning of 1907; at the end of the

---

47. J. Norman Lockyer, 'The Influence of Brain-power on History', *Nature*, 68 (1903), p. 441.
48. On the British Science Guild, see Meadows, *Science and Controversy*, pp. 272–9; Searle, *National Efficiency*, pp. 84–5; W. H. G. Armytage, *Sir Richard Gregory: His Life and Work* (London, 1957), pp. 67–97. On the inaugural meeting, see the report in *The Times*, 31 October 1905; also printed in *Nature*, 73 (1905–6), pp. 10–13.
49. *Nature*, 68 (1903), p. 441. Lockyer was thinking of half a million members, and an annual income of £12,000 from members' contributions.
50. The memorandum, presumably written by Lockyer, is printed in *Nature*, 70 (1904), p. 343.

same year the number had risen to 793. At the end of 1910 it was 872, and in May of 1912 about 900.[51]

When he founded the British Science Guild, Lockyer, perhaps the most experienced science lobbyist of his time, cleverly sought to associate the interests of science with those of the whole nation, indeed, of the Empire. He concealed hefty demands for more expenditure on universities and research behind a veil of patriotic language, taking advantage of the campaign for 'national efficiency'[52] which had started after revelations about the health of British recruits and the weaknesses of British army organisation during the war in South Africa. At the inaugural meeting of the British Science Guild, Lockyer expressly pointed to the ostensibly close connection between science, economic strength and political power in a competitive international system, something which he thought was not fully appreciated in Britain at the beginning of the twentieth century:

> Mr Chamberlain, Lord Rosebery, and others have referred to the relative advance — I may say the *great* relative advance — of the commerce and industry of Germany and the United States. Let me again point out that these are *par excellence* the lands of complete and numerous State-aided universities. Surely it is more than a coincidence that we find in those lands the State service and all the national activities carried on in the full light of modern science, by men who have received a complete training both in science and the humanities in close touch with the Governments. If the guild helps in any way to improve our national position in this respect, it will not have been founded in vain, but there is certainly much for it to do along many lines.[53]

The British Science Guild, which reflected the social Darwinism then prevalent, held an essay competition on the topic: 'The best way of carrying on the struggle for existence and securing the survival of the fittest in national affairs'. One of the questions to be discussed was how science could best be of use to the nation in

---

51. *Nature*, 75 (1906–7), p. 328; ibid., 86 (1911), p. 218; ibid., 89 (1912), p. 296.
52. This phrase, which quickly became a slogan, seems to have been coined by Sidney Webb in the title of his book *A Policy of National Efficiency* (London, 1901). Lord Rosebery defined 'national efficiency' as 'a condition of national fitness equal to the demands of our Empire – administrative, parliamentary, commercial, educational, physical, moral, naval, and military fitness – so that we should make the best of our admirable raw material' (speech in Glasgow, 10 March 1902, quoted in Bernard Semmel, *Imperialism and Social Reform: English Social-Imperial Thought 1895–1914*, London, 1960, p. 63); E. J. T. Brennan (ed.), *Education for National Efficiency: The Contribution of Sidney and Beatrice Webb* (London, 1975).
53. *The Times*, 31 October 1905. Also in *Nature*, 73 (1905–6), p. 10.

peace and in war.[54]

Branches of the British Science Guild were established in Canada and Australia, 'to foster goodwill between the colonies and the Mother Country, thus helping to strengthen the fabric of the Empire'.[55] The imperial orientation of the British Science Guild's objectives remained equally strong after the First World War. In 1922, for example, it described itself as an association of citizens 'for the purpose of making the Empire strong and secure through science and the application of scientific method'.[56] This articulated the Guild's essential goals. At its foundation in 1905, these had been expressed purely in terms of politics and political spheres of interest, and promoting research in the true sense was not mentioned. The British Science Guild was intended to put into practice a resolution passed, at Lockyer's insistence, by the British Association's annual meeting in 1903; namely that

> scientific workers and persons interested in Science be so organised that they may exert permanent influence on public opinion in order more effectively to carry out the third object of this Association . . ., viz. 'to obtain a more general attention to the objects of Science, and a removal of any disadvantages of a public kind which impede its progress', and that the Council be recommended to take steps to promote such organisation.[57]

In addition to this publicity work, the British Science Guild was to encourage the application of scientific working methods and principles of organisation to all areas of life, especially to the solution of social problems and in government: it was to 'promote and extend the application of scientific principles to industrial and general purposes' and to 'bring home to all classes the necessity of making the scientific spirit a national characteristic which shall inspire progress and determine the policy in affairs of all kinds'.[58] This was a programme which *Nature* had been advocating since the 1880s,[59] and it reflected the growing self-confidence of science about its importance in industrial society. The fact that the British

---

54. *Nature*, 83 (1910), p. 100. See also Meadows, *Science and Controversy*, p. 279.
55. Report on the third annual meeting of the British Science Guild (*Nature*, 79, 1908–9, p. 382).
56. *Nature*, 110 (1922), p. 798.
57. Quoted in Lockyer and Lockyer (eds.), *Life and Work*, p. 444.
58. *Nature*, 72 (1905), p. 585 ('The British Science Guild').
59. See, for example, *Nature*, 21 (1880), p. 295: 'There is surely no reason why political action, the conduct of the state, should not be guided by scientific method quite as much as the conduct of a scientific exploring expedition.' Taking up this theme, Richard B. Haldane explained to the Chemical Society in 1905 that there was

Science Guild adopted this programme shows that it also saw itself as an 'educational movement', aiming 'to educate the people at large and the Government and political parties not to undervalue the great resources of science in the development of the kingdom'.[60]

Like other similar organisations within the British science lobby, the British Science Guild worked through specialist committees, the coordination of which was Lockyer's task, and after 1918, Richard A. Gregory's, Lockyer's successor as editor of *Nature*. Specialists from the university sector, industry and the higher ranks of the civil service sat on these committees, which concentrated on specific areas in which science was involved, or in which the British Science Guild thought science should be involved: agriculture, industry, patent law, technical training, schools and universities, the energy supply, the prevention of water pollution, and public administration. In the early years of the First World War these committees sometimes functioned in a semi-official capacity. In 1914–15, for example, the Institute of Chemistry commissioned the British Science Guild to investigate the supply situation of optical glass in Britain.[61] By and large, the methods used by the British Science Guild were the traditional ones of passing resolutions, holding meetings, pamphleting, organising deputations to the government, presenting petitions and holding exhibitions and lectures. But the Guild did not limit itself to publicising its major concerns in these various ways. Its committees also took an interest in practical and seemingly trivial details such as reducing postage rates for scientific publications, naming London streets after famous British scientists and synchronising all public clocks. The British Science Guild's journalistic forum was provided by the scientific and technical journals, the most important of these being *Nature*. Although it enjoyed easy access to *Nature*, the Guild founded its own Journal in the summer of 1915. The *Journal of the British Science Guild* was intended to promote communication between members of the Guild and to publicise its objectives.

The many activities undertaken by the British Science Guild under the leadership of Richard B. Haldane unquestionably made it

---

practically no government body 'which did not require science, if its policy was to be an effective policy' (*Nature*, 71, 1904–5, p. 589).
60. *The Times*, 31 October 1905.
61. See Roy M. MacLeod and E. Kay MacLeod, 'War and Economic Development: Government and the Optical Industry in Britain, 1914-1918' in J. M. Winter (ed.), *War and Economic Development. Essays in Memory of David Joslin* (Cambridge and London, 1975), p. 171.

one of the most influential interest groups in British science until the end of the First World War. Its initiatives had some public impact: they were not only mentioned in the specialist journals, but sometimes also found their way into the daily press. But it achieved no spectacular successes. Neither the government nor industry responded other than hesitantly to the British Science Guild's demands and suggestions for more support for science. 'Institutions had no reason to listen to a non-political, non-professional body representing no visible interest group in society, and political parties no obligation to respond.'[62] In addition, the Guild's precise objectives were largely misunderstood. What, for example, was meant by adopting scientific working methods in the conduct of state affairs: applying scientific thinking to the solution of political problems or simply thinking logically? 'I, as you well know', noted Arthur J. Balfour in a letter to Lockyer in 1906, 'am most anxious to promote the cause of science in every way in my power: but, frankly, I am not sure that I form a clear idea of the exact methods which your Guild proposes to adopt to attain that end. The words "science" and "scientific method" are . . . loosely used; and, especially when applied to such subjects as Government or taxation, often mean little more than "reasonably" or "rationally" conducted.'[63]

After the death of its founder in 1920, the British Science Guild quickly diminished in importance, despite various reform attempts. It was pushed into the background by new forms of representation for the scientific community's interests, such as the National Union of Scientific Workers (founded in 1917) and the Association of Scientific Workers (founded in 1927) which succeeded it. They adapted more successfully to changed conditions after the First World War.[64] In 1936 the British Science Guild, which by this stage had been reduced to just over 300 members, amalgamated with the revivified British Association for the Advancement of Science.

Other forms of interest representation, apart from the British Science Guild, had already emerged during the First World War. Although they all proved to be ephemeral, their close connections with scientific societies allowed them to achieve surprising results. Two of these organisations will be briefly discussed here. In Febru-

62. According to Roy M. MacLeod, 'Into the Twentieth Century', *Nature*, 224 (1969), p. 458.
63. Quoted in Meadows, *Science and Controversy*, p. 274.
64. Werskey, *The Visible College*, pp. 39–41 and 235–7. See also Harold Perkin, *Key Profession: The History of the Association of University Teachers* (London, 1969); Rose and Rose, *Science and Society*, pp. 52–3; MacLeod and MacLeod, 'Social Relations of Science', pp. 318–19.

ary of 1916 thirty-six university teachers and scientists published a memorandum entitled 'Neglect of Science'. From this initiative the Neglect of Science Committee developed in May 1916. It was chaired by the physicist Lord Rayleigh, and its chief activities were holding conferences and passing resolutions.[65] Germany's military successes in the early years of the First World War, and the financial difficulties experienced by the scientific societies as a result of a general rise in prices, provided another opportunity to point out, with increasing bitterness, how undervalued science and technology were in Britain. 'For more than 50 years', the February memorandum said, 'efforts have been made by those who are convinced of the value of training in experimental science to obtain its introduction into the schools and colleges of the country as an essential part of the education given therein.' But in practice, nothing had happened to secure for science its proper place in society, and in particular, within the British education system. 'It is clear that the old methods and old vested interests have retained their dominance, at least as far as our ancient universities and great schools are concerned.' The memorandum continued: 'The one and effective way of changing this attitude and of giving us both better educated Civil servants and a true and reasonable appreciation of science in all classes is in the hands of the Legislature, and of it alone'.[66]

The memorandum proposed, in addition to other reforms, transforming the Board of Trade into a Department of Science, Economy and Industry. This suggestion was readily accepted, because under the exceptional conditions created by the war, similar ideas were already circulating in the Asquith cabinet. The proposal to redefine the Board of Trade's areas of responsibility was, however, soon dropped. Instead, the Committee of the Privy Council for Scientific and Industrial Research, which had been created in the summer of 1915 with £1 million to spend on promoting research projects, was transformed into the Department of Scientific and Industrial Research.[67]

The Conjoint Board of Scientific Societies, which came into being in 1916, like the Neglect of Science Committee, with the support of the Royal Society, clearly identified with the objectives

---

65. Also called Committee on the Neglect of Science. *The Neglect of Science: Report of the Proceedings at a Conference at Burlington House, 3rd May 1916* (London, 1916); *Nature*, 97 (1916), pp. 230–1. The suggestion of writing the memorandum had come from the zoologist Sir E. Ray Lankester (1847–1929).
66. Printed in *The Times*, 2 February 1916 ('Neglect of Science: A Cause of Failures in War').
67. See below, p. 208–10 in this volume.

of the British Science Guild. The Board's main goals were 'promoting the co-operation of those interested in pure or applied science; supplying a means by which the scientific opinion of the country may, on matters relating to science, industry, and education, find effective expression'. A further aim was 'taking such action as may be necessary to promote the application of science to our industries and to the service of the nation'.[68] But while in theory at least membership of the British Science Guild was open to all, only representatives of scientific and technical societies could join the Board. Early in 1920 seventy-five societies belonged to the Board — almost all of the leading British organisations 'concerned with the advancement of science and technology'.[69] During the war the Conjoint Board of Scientific Societies saw itself as assisting the government in an advisory capacity, and in this role it played an important part in British economic planning. Its publicity work was therefore minimal. As a committee of experts working for the government its activities centred on writing reports of enquiries and submitting expert opinions. The Conjoint Board of Scientific Societies was dissolved in the spring of 1923.

'Competition between Nations': The Arguments and Interests of Science

This survey of the activities of British scientific organisations between 1820 and 1920 shows that agitation for science did not take place constantly and with equal intensity throughout the whole period. There were of course times when the discussion within the scientific community was livelier than at others, when it went beyond the conferences and specialist journals which were its normal forum and found its way into the daily papers. Seen as a whole, the campaigns waged by scientists from the early nineteenth century to the First World War proceeded in a kind of wave motion, with successive peaks and troughs.

As a rule, outside factors which clearly demonstrated the importance of science in the expanding industrial economy of the Victorian period were necessary to give greater force to the agitation for science and the discussion about promoting it. In the nineteenth

---

68. *Nature*, 105 (1920), p. 317; ibid., 97 (1916), p. 104. See also the minutes of the Royal Society's meeting on 24 February 1916 in *Minutes of Council*, vol. 11, 1914–20, p. 129, and the material in file CD 370–587, Royal Society Archives.

69. *Nature*, 105 (1920), p. 317.

century many different factors of this sort existed; common to all was the shock effect they had on the public. The longest-lasting effects were achieved by two factors in particular: alarming reports of Britain's economic backwardness in comparison with other countries, which were becoming more and more frequent in the second half of the nineteenth century; and wars, the Crimean War (1854–6) and the Boer War (1899–1902) being the clearest examples. The Crimean War precipitated reforms which were overdue at the universities of Oxford and Cambridge to strengthen the position of science there.[70] The debate about 'national efficiency' referred to above took place in the wake of the expensive war in South Africa, which brought dismaying revelations about failures in the British Army's organisation and supply situation.[71] The problem of supporting science adequately for the benefit of both state and society was central to this debate. From a military and economic point of view, therefore, the Boer War was more than merely a distant colonial episode for Britain. But wars in which it was not directly involved, such as the wars of German unification after 1864, could also stimulate the discussion taking place in Britain about the role of science because they raised basic issues to do with modern technology and science organisation.[72] Their increasing significance in modern warfare was always in the air after 1870, when the Devonshire Commission began its work.

The symptoms which, as factors fuelling the debate about science in Britain, seemed to indicate that the oldest industrial nation was losing its economic and industrial position and technological lead, had an effect which was no less shocking than that produced by wars. In this context the great international exhibitions, a new development in the second half of the nineteenth century, played a key part. As the liberal capitalist system's 'giant new rituals of self-congratulation',[73] these exhibitions put the growing number and diversity of industrial products on display at regular intervals after the middle of the century. At the same time they were 'milestones of progress', highly regarded at the time, for measuring

---

70. MacLeod, 'Support of Victorian Science', p. 200.
71. Searle, *National Efficiency*, pp. 35–53 (ch. 2, 'The Lessons of the Boer War'); Ensor, *England*, pp. 251–6, 344–8; Clive Trebilcock, 'War and the Failure of Industrial Mobilisation: 1899 and 1914' in J. M. Winter (ed.), *War and Economic Development: Essays in Memory of David Joslin* (Cambridge and London, 1975), pp. 143–4, 148.
72. See 'Science and War', *Nature*, 16 (1877), pp. 37–8, 57–8; Michael Howard, *The Franco-Prussian War: The German Invasion of France, 1870–1871* (London, 1961), pp. 1–18; Haines, *German Influence*, pp. 47–8, 94.
73. Eric J. Hobsbawm, *The Age of Capital 1848–1875* (London, 1975), p. 32.

national technical and industrial efficiency.[74]

The first of these exhibitions, the Great Exhibition of 1851, was held in London and had been, on the whole, a triumph for British industry. The majority of prizes awarded by an international jury had gone to British exhibits. Although warning voices had been raised in Britain, even on this early occasion, pointing to the increasing competitiveness of other countries, the Great Exhibition had demonstrated 'the zenith of Victorian technical achievement'.[75] The exhibition of 1862, again held in London, seemed to confirm the technical superiority of Britain's industrial production — long symbolised by Crystal Palace, scene of the Great Exhibition of 1851. But at the Paris Exhibition of 1867 it was already becoming apparent that a change was taking place. For the first time British products in many fields, from steelworking to the aniline dyes newly invented in England, could not compete with those produced in other countries. While industries on the Continent and in the USA had made rapid progress during the world-wide economic take-off after 1849,[76] technical and scientific innovation and industrial growth in Britain seemed to stagnate by comparison. In 1851 British industry had won the majority of awards for its exhibits; in 1867 it took only a disappointingly small fraction.

It was perceived clearly on the British side, by Alexander Strange, for example, and the chemist and science journalist Lyon Playfair, that the success of Britain's competitors was due to the application of scientific and technical knowledge to industrial production. It was perceived no less clearly that the prerequisite for this was a broadly based system of technical education. 'The one cause upon which there was most unanimity of conviction [at the International Exhibition of 1867] is that France, Prussia, Austria, Belgium, and Switzerland possess good systems of industrial education for the masters and managers of factories and workshops,

---

74. Utz Haltern, *Die Londoner Weltausstellung von 1851. Ein Beitrag zur Geschichte der bürgerlich-industriellen Gesellschaft im 19. Jahrhundert* (Münster, 1971); idem, 'Die "Welt als Schaustellung". Zur Funktion und Bedeutung der internationalen Industrieausstellung im 19. und 20. Jahrhundert', *Vierteljahrschrift für Sozial- und Wirtschaftsgeschichte*, 60(1973), pp. 1–40; Evelyn Kroker, *Die Weltausstellung im 19. Jahrhundert. Industrieller Leistungsnachweis, Konkurrenzverhalten und Kommunikationsfunktion unter Berücksichtigung der Montanindustrie des Ruhrgebiets zwischen 1851 und 1880* (Göttingen, 1975); Cardwell, *Organisation of Science*, pp. 76–7. The following International Exhibitions took place before the First World War: 1851 London, 1855 Paris, 1862 London, 1867 Paris, 1873 Vienna, 1876 Philadelphia, 1878 Paris, 1889 Paris, 1893 Chicago, 1900 Paris, 1904 St Louis.
75. Morrell, 'Patronage of Mid-Victorian Science', p. 356.
76. Out of the vast literature on this subject, see the basic studies by Walt W. Rostow, *The Stages of Economic Growth* (New York, 1962); idem (ed.), *The Economics of Take-Off into Sustained Growth*, 2nd edn (London, 1965); idem, *Politics and the Stages of Growth* (Cambridge, 1971); and Landes, *Unbound Prometheus* pp. 124–230.

and that England possesses none.'[77] Playfair therefore called for more science teaching in schools and sharply attacked the humanistic ideal dominating education. The striking discrepancy between Britain's position as the leading industrial nation and the state of its scientific and technical education system gave rise to pessimistic forecasts about the country's economic development. In 1867 the future foreign secretary, Lord Granville, wrote in *The Times* about the lessons which 'the late war in Germany and the present Exhibition in Paris afforded us, if we wish to hold our own with other nations in the arts of peace and war'.[78] British industry's relatively bad performance at the Paris Exhibition was also a factor, at least indirectly, in the founding of the Iron and Steel Institute in London in 1869, and the Yorkshire College of Science in Leeds in 1874.

In somewhat simplified form, four periods in which the debate on the role of science in modern society was carried on with particular intensity can be distinguished between the early nineteenth century and the years before the First World War. The earliest period covers the controversy about the decline of British science in the 1830s, which flared up when it was suggested that science was better organised and funded in France and Germany. This controversy had visible results which also indicated the direction of future developments: the foundation of the BAAS in 1831, the establishment of the University of London in 1836, the reform of the Royal Society, the granting of public funds to the Royal Society for the promotion of scientific research, and finally the foundation of the Royal College of Chemistry in London in 1845. The first director of this institute was the 27-year-old German chemist, August Wilhelm Hofmann, appointed on Justus von Liebig's recommendation. Hofmann worked at the Royal College — 'German-inspired and German-run'[79] — for twenty years, until he became a professor in Berlin in 1865.

The discussion about reforming Oxford and Cambridge and about the necessity of state aid for science began in the late 1850s against a background of two factors: the Crimean War, which had revealed faults in the British Army's armament technology and organisation, and accelerating industrialisation in Europe, which

---

77. Lyon Playfair (1818–98), a student of Justus von Liebig and British referee at the Great Exhibitions of 1851 and 1867, in an open letter to Lord Taunton in 1867 (quoted in Ashby, *Technology and the Academics*, p. 58). See also Cardwell, *Technology* p. 201; Mendelsohn, 'Emergence of Science', pp. 33–5.
78. *The Times*, 29 May 1867.
79. Cardwell, *Organisation of Science*, p. 105. See also W. H. G. Armytage, *A Social History of Engineering*, 4th edn (London, 1976), p. 151.

could be assessed by the developments displayed at the international exhibitions after 1851. The discussion of reform, which centred primarily on popularising technical and scientific knowledge and disseminating it more widely, was carried on by university staff, individual scientists, MPs, journalists and, after 1864, by members of the London X-Club who, like many other British scientists, had gained first-hand experience of the German university system by studying there, or in relatively long visits undertaken since the 1830s.

Towards the end of the 1860s efforts to reform the English universities merged into a fundamental debate about the faults of the British education and science system, primarily under the impression created by the Paris International Exhibition of 1867 and British industry's ostensibly bad performance there.[80]

> In order ... to make Oxford a seat of education [wrote Mark Pattison in 1868] it must first be made a seat of science and learning. All attempts to stimulate its *teaching* activity without adding to its solid possession of the field of science will only feed the unwholesome system of examination which is now undermining the educational value of the work we actually do.[81]

At that time in England, Pattison's ideas were progressive and symptomatic of the changing way in which universities saw their role. On the initiative of the BAAS and several individual scientists, the Devonshire Commission on scientific instruction and research in Britain was set up in the course of the programme of political and social reform which Gladstone introduced after the Liberal victory at the polls in 1868. Members of the X-Club were in a majority on the Commission, which, like others before and since, prepared a set of detailed suggestions for reform. Based on the Commission's report and its suggestions, the Endowment of Research Movement developed after 1876. It attempted to get the Commission's suggestions implemented and to bring about a fundamental change in the attitude of the government and the universities towards science. All in all, however, these attempts proved unsuccessful, although after 1874 Disraeli's Conservative government took a more positive attitude towards science than had its predecessor under Gladstone. One effect of this was an increase in the Royal Society's annual research fund from £1,000 to £5,000.

---

80. Michael Sanderson writes of a 'traumatic experience' (*Universities*, p. 9).
81. Mark Pattison, *Suggestions on Academical Organisation with Especial Reference to Oxford* (Edinburgh, 1868), p. 171.

Another positive factor was that during the 1870s and 1880s the British public was more favourably disposed towards issues to do with science and education. Between 1882 and 1884 a Royal Commission on Technical Instruction conducted an extensive survey in the United Kingdom and in other countries. 'There is a fashion in education, as in pictures and dress and china', wrote *The Times* in a leader in 1884: 'Technical education is the fashion at present'.[82] A concrete result of this 'fashion' was the foundation of several institutions for technical and industrial training in the secondary-education sector. But attempts to establish a nation-wide system of technical training did not go beyond this for the time being.

But the most intense debate about the position of science in Britain began towards the end of the nineteenth century, despite all the science lobby's efforts before that date. To what extent the period between the turn of the century and the First World War was of exceptional significance for emerging British science policy, in conceptual as well as institutional terms, will be discussed further below. Like the Crimean War of the 1850s, the Boer War (1899–1902) exposed organisational and logistic weaknesses in the British Army, and technical faults in its armaments, all of which were discussed with growing alarm in Britain. Together with other factors, the debate stimulated by the war revealed a profound crisis in the national consciousness,[83] the first symptoms and precursors of which were already visible in the last third of the nineteenth century. These factors included the impression that Britain was falling behind other comparable countries in scientific and technical fields as well as in economic efficiency; social pressures which were vented in long-drawn-out and sometimes remarkably harsh strike movements; political rivalry with an economically expanding Wilhelmine Germany; and finally, increasingly intense rivalry in world markets for industrial products. From this point of view, the reality of the Edwardian age in Britain did not match up to the cliché of a country at the height of its brilliant imperial career. Understandably enough, this image developed in an atmosphere of nostalgia after the First World War.

In view of this increasing awareness of crisis, British scientists

---

82. *The Times*, 16 May 1884.
83. See Hobsbawm, *Industry and Empire*, pp. 162–3: 'There was, especially in the last years before the First World War, an atmosphere of uneasiness, of disorientation, of tension, which contradicts the journalistic impression of a stable *belle-époque* . . . . They were the years when wisps of violence hung in the English air, symptoms of a crisis in economy and society which the self-confident opulence of the architecture of Ritz hotels, proconsular palaces, West End theatres, department stores and office-blocks could not quite conceal'.

continued their efforts at many levels until the early years of the First World War. Especially after the turn of the century, under the impact of an ideology of efficiency, these efforts were aided by a change in public attitudes towards science. At the beginning of the twentieth century the mass circulation *Daily Mail* commented pessimistically on Britain's shrinking role in the world economy: "If the position which this country is slowly losing is to be regained, there can be no doubt that England must improve her education and more richly endow research'.[84] 'Research', wrote *Nature*, commenting on the broad impact of scientists' agitation 'is a word much used in newspapers and in public discussions nowadays.'[85] Both quotations show that the intensification of the discussion about science is also one of the most striking indicators of internal social crisis in Britain at this time. The science lobby's success in obtaining improvements in the institutional conditions for science in Britain at this stage will be discussed in more detail below.[86]

Of all the factors which allowed the latent doubts and feelings of insecurity in British society to come to the surface after the Boer War, the realisation which dawned during the last decades of the nineteenth century that Britain's industry was stagnating by comparison with other countries was of special importance. Many contemporaries documented an awareness that Britain was falling behind economically, and that its industry was becoming less efficient. Ernest E. Williams's famous book, *Made in Germany*, begins: 'The Industrial Supremacy of Great Britain has been long an axiomatic commonplace; and it is fast turning into a myth . . . The industrial glory of England is departing, and England does not know it'.[87] 'There is now a general opinion', wrote Norman Lockyer in 1904, 'that Britain is in danger of falling behind in the industrial competition now going on between the most highly civilised States.'[88] In the context of the debate about protective tariffs caused to a large extent by foreign trade competition, the

---

84. *Daily Mail*, 8 June 1901. On criticisms made around the turn of the century, see Cardwell, *Organisation of Science*, pp. 193–5.
85. *Nature*, 85 (1910–11), p. 29. Similarly, Richard A. Gregory in 1919: 'The recognition of the value of scientific research as a determining factor of progressive development has been a common note of many public utterances in recent years' ('The Promotion of Research', *Nature*, 104, 1919–20, p. 259).
86. See below, pp. 138–77 in this volume.
87. Ernest E. Williams, *Made in Germany* (London, 1896), p. 1. Ross J. S. Hoffman, *Great Britain and the German Trade Rivalry, 1875–1914* (Philadelphia, Penn., and London, 1933), p. 246, called Williams's book 'the most famous and sensational piece of alarmist literature . . ., a trumpet blast to awaken his countrymen'.
88. 'The National Need of the State Endowment of Universities' in J. Norman Lockyer, *Education*, p. 216. In general, see also Searle, *National Efficiency*, pp. 73–5.

Treasury had pointed to the loss of new industries which depended heavily on research — an unmistakable sign that the problems of science development in Britain since the end of the nineteenth century were at last being perceived more clearly by the government too. A cabinet memorandum by the Treasury, listing past omissions in the promotion of science, reveals a perplexing similarity with the same sorts of documents emanating from the science lobby. It contains an observation which even at that time was already a journalistic commonplace: 'We are no longer the workshop of the world without a rival: we have formidable competitors. Germany and the United States have sprung into great manufacturing countries'.[89] At the same time *Nature* wrote:

> The country has felt more and more the competition of other nations. The colour industry has forsaken our shores, the finest electrical machinery is made abroad, we go to America for labour-saving appliances . . . . But when it was first suggested that our deficiency in scientific and technical education was at the root of the matter, those who dared to make the suggestion were, if not mocked at, at any rate treated with scant courtesy.[90]

The indicators of Britain's relative industrial decline[91] from about 1870 could no longer be ignored by the turn of the century. Two dramatic tendencies were becoming clear around the end of the century. First, economic data showed that after a long period of expansion and prosperity between 1849 and 1873, Britain's industrial growth rate was slowing down, while in other countries growth rates were being maintained or even increased. This factor, together with the larger populations of these countries, meant that the USA's and Imperial Germany's total industrial output outstripped Britain's towards the end of the nineteenth century (Table 2.1).[92] Around 1870 Britain's share of world industrial output had

---

89. Memorandum of 25 August, 1903, p. 1 (CAB 37/66/55, PRO).
90. *Nature*, 68 (1903), p. 203.
91. Economic historians refer to 'relative industrial decline', 'industrial retardation', 'industrial failure' or 'deceleration'. See Alexander Gerschenkron, *Economic Backwardness in Historical Perspective: A Book of Essays* (Cambridge, Mass., 1962), pp. 5–30.
92. Slightly different figures are given by Harry W. Richardson, 'Retardation in Britain's Industrial Growth, 1870–1913' in Derek H. Aldcroft and Harry W. Richardson (eds.), *The British Economy 1870–1939* (London, 1969), p. 105. Of the vast literature, see also Hoffman, *German Trade Rivalry*; A. L. Levine, *Industrial Retardation in Britain 1880–1914* (London, 1967); D. J. Coppock, 'British Industrial Growth during the "Great Depression" (1873–96): a Pessimist View', *Economic History Review*, vol. 17 (1964–5) pp. 389–96; A. E. Musson, 'British Industrial Growth, 1873–96: A Balanced View', *Economic History Review*, vol. 17 (1964–5), pp.

Table 2.1. Average annual growth in industrial output in the three largest industrial nations, 1860–1913 (%)

	United Kingdom	Germany	USA	World
1860–80	2.4	2.7	4.3	3.2
1880–1900	1.7	5.3	4.5	4.0
1900–13	2.2	4.4	5.2	4.2

Source: Michael Sanderson, *The Universities and British Industry 1850–1970* (London, 1972), p. 9.

been almost one-third (31.8 per cent), and that of its closest rival, the USA, had been less than one-quarter (23.3 per cent). In 1913 the figures were 14.1 per cent for Britain, 35.5 per cent for the USA and almost 16 per cent for Imperial Germany.[93] Between 1880 and 1913 Britain's share of the world export market for manufactured goods fell from 41.1 per cent to just under 30 per cent; during the same period Germany's rose from 19.3 per cent to 26.5 per cent, and that of the USA from just under 3 per cent to 12.6 per cent.[94]

Up to a point the development reflected in these figures was not surprising. Contemporary and later British observers saw that Britain would not be able to maintain for ever the large economic and technological lead which, as the pioneer of the Industrial Revolution, it had possessed until the beginning of the Great Depression in the early 1870s. Despite protectionist tendencies all over the world, Britain's share of world trade, which at the time of

---

397–403; J. W. Grove, *Government and Industry in Britain* (London, 1962), pp. 11–18; Landes, *Unbound Prometheus*, pp. 231–358; P. Deane and W. A. Cole, *British Economic Growth 1688–1959: Trends and Structure*, 2nd edn (Cambridge, 1967), pp. 296–7; Paul M. Kennedy, *The Rise of the Anglo-German Antagonism 1860–1914* (London, 1980), pp. 291–305; S. B. Saul, *Industrialisation and De-Industrialisation? The Interaction of the German and British Economies before the First World War* (London, no date [1980]). Reference should also be made to Charles P. Kindleberger's observation that per capita income continued to be higher in Britain than in Germany. It was not until the 1960s that West Germany's per capita income overtook Britain's ('Germany's Overtaking of England, 1806 to 1914' in idem, *Economic Response: Comparative Studies in Trade, Finance, and Growth*, Cambridge, Mass., and London, 1978, p. 186).

93. Aldcroft and Richardson, *British Economy*, p. 8; H. J. Habakkuk, *America and British Technology in the Nineteenth Century: The Search for Labour-Saving Inventions* (Cambridge, 1962), p. 204. See also W. G. Hoffmann, *British Industry, 1700–1950* (Oxford, 1955).

94. D. H. Aldcroft (ed.), *The Development of British Industry and Foreign Competition, 1875–1914* (London, 1968), p. 21. Figures for the changing structure of British imports after 1870 (increase in industrial finished goods): ibid., pp. 25–6. See also Sidney Pollard, *The Development of the British Economy, 1914–1967*, 2nd edn (London, 1969), pp. 3–10.

the Franco-Prussian war was bigger than that of France, Germany and Italy combined, naturally had to decrease to the extent that other countries were industrialising. And economic growth in these countries was bigger than Britain's, because they were starting from a much lower initial level of manufacturing output. Therefore it was not so much the industrialisation of other countries that was seen as disturbing in Britain; more alarming were the pace of its own economic growth, symptoms of industrial stagnation, and international competition, which British industry was less and less able to cope with both at home and abroad, as the boom in imports of finished products in the 1890s seemed to show. From now on Britain with its 'mature economy' was only one of many industrial nations. It was no longer setting the pace of industrialisation: 'Monopoly had given way to competition'.[95]

A second factor seemed to confirm dramatically the warnings and forecasts scientists had been issuing since the last third of the nineteenth century. Economic data revealed that during the Great Depression, from the beginning of the 1870s to the mid 1890s, Britain lost its leading position, both quantitatively and qualitatively, in traditional branches of production like iron and steel to the USA and Germany (Table 2.2).[96] But of greater import for the future of British industry was the fact that Britain was not taking a full part in the development of new research-based and growth-intensive industries, and could not compete with foreign rivals in this field. These industries expanded so rapidly and had such wide-ranging effects from the end of the nineteenth century that David S. Landes speaks of a second industrial revolution,[97] but their significance for Britain's national economy remained small until the war. This was true especially of the organic chemical industry and the closely related optical and pharmaceutical industries. It was also true of the mechanical engineering and electrical industries,[98] as suggested in the quotation from *Nature* above. These were all industries in which ready availability of raw materials

---

95. Landes, *Unbound Prometheus*, p. 239; Donald N. McCloskey (ed.), *Essays on a Mature Economy: Britain after 1840* (London, 1971).

96. Brian R. Mitchell, *European Historical Statistics 1750–1970* (London, 1975), pp. 393-4 and 399-401; *Historical Statistics of the United States: Colonial Times to 1970* (Washington, DC, 1975), pp. 599–600 and 693–4. On the eve of the First World War Britain was the largest importer of steel in the world, and Germany the largest exporter. Peter Temin, 'The Relative Decline of the British Steel Industry, 1880–1913' in Henry Rosovsky (ed.), *Industrialization in Two Systems: Essays in Honor of Alexander Gerschenkron* (New York, 1966), p. 143.

97. Landes, *Unbound Prometheus*, p. 235. Also Levine, *Industrial Retardation*, p. 6.

98. Chemical and electrical products made up 16 per cent of German exports in 1913, and 8 per cent of British exports (Coppock, 'Industrial Growth', p. 393).

Table 2.2. Iron and steel production in the leading industrial nations, 1870–1910 (000s tons)

	United Kingdom	France	Germany	USA
Iron				
1870	6,059	1,178	1,261	1,665
1880	7,873	1,725	2,468	3,835
1890	8,031	1,962	4,100	9,203
1900	9,104	2,714	7,550	13,789
1910	10,173	4,038	13,111	26,674
Steel				
1870	334	84	126	77
1880	1,316	389	690	1,397
1890	3,636	683	2,135	4,779
1900	4,980	1,565	6,461	11,227
1910	6,476	3,413	13,100	28,330

Source: Brian R. Mitchell, *European Historical Statistics 1750–1970* (London, 1975), pp. 393–4, 399–401; *Historical Statistics of the United States: Colonial Times to 1970* (Washington, DC, 1975), pp. 599–600, 693–4.

at home was less important than technological knowledge and know-how, and which consequently employed more and more scientists and engineers.

Scientists, as well as contemporary political economists, therefore saw Britain's general weakness in applying scientific findings to industrial production as an important factor in the relative decline of the British economy at the end of the nineteenth century. Of equally far-reaching consequence for the British economy was the fact that British industry did not systematically concentrate in the way characteristic of Germany and the USA, where trusts, cartels and syndicates had been developing since the 1880s.[99] While tendencies towards concentration and centralisation existed in the British political economy prior to 1914, they were weakly developed in the national economy as a whole. Only after several spectacular cases of American concerns buying up British firms had occurred did the general public become aware of structural changes

---

99. Hans Medick, 'Anfänge und Voraussetzungen des organisierten Kapitalismus in Großbritannien 1873–1914' in Heinrich August Winkler (ed.), *Organisierter Kapitalismus. Voraussetzungen und Anfänge* (Göttingen, 1974), pp. 58–83; J. Morgan Rees, *Trusts in British Industry 1914–1921: A Study in Recent Developments in Business Organisation* (London, 1922); Kindleberger, 'Germany's Overtaking', p. 214–15; Landes, *Unbound Prometheus*, pp. 246–7; H. W. Macrosty, *The Trust Movement in British Industry: A Study of Business Organisation* (London, 1907).

in the industrial economy.[100] But Britain was 'too deeply committed to the technology and business organization of the first phase of industrialization, which had served her so well, to advance enthusiastically into the field of the new and revolutionary technology and industrial management which came to the fore in the 1890s'.[101] What went wrong? This was the question asked by *The Daily Mail* in June of 1901 in a series of articles about the American export industry. Its answer was: 'Our economic system is based upon theories which facts are each day proving more delusive. Yet for nations as for men it is a terrible wrench to break from the beaten track, to leave the groove and to go forth into the unknown land of experiment and innovation'.[102] Given that British industry generally adhered to the technological status quo, only one traditional way out remained for Britain: 'the economic (and increasingly the political) conquest of hitherto unexploited areas of the world. In other words, imperialism'.[103] No more than brief mention will be made here of Hobsbawm's rather provocative thesis, as it requires more detailed investigation. According to this view, imperialist expansion represented an attempt, made under the pressure of foreign competition, to establish privileges overseas enabling Britain to import raw materials and to export manufactured products. The debate about preferential duties for British goods and protective duties for the empire as against other countries, which took place after the Empire Conference of 1897, provides abundant material in support of this thesis.

The aniline dye industry was often held up by contemporaries, as well as by later economic historians, as a perfect example of Britain's loss of its scientific and technical lead in the second half of the century. The first aniline dye was discovered in 1856 by

100. Henry Pelling, 'The American Economy and the Foundation of the British Labour Party', *Economic History Review*, vol. 8 (1955), pp. 3–5.
101. Hobsbawm, *Industry and Empire*, p. 107. See also Rose and Rose, *Science and Society*, p. 24: 'Colonies provided essentially soft, politically protected markets for British goods. British ships carried them there quickly and cheaply, and the huge demand that this created throughout the whole of the nineteenth century and the first quarter of the twentieth, provided British industry with a guaranteed sale for its goods, without the need for continued technological innovation.' In this context the colonies' share of British exports is interesting: in 1890 it was 34.6 per cent, and in 1902 it was 42.1 per cent (Semmel, *Imperialism*, p. 149).
102. *Daily Mail*, 6 June 1901 ('Why We Regress'). One letter to the editor, written in response to the series, claims: 'The British manufacturer regards the inventor as an enemy who is trying to upset his business; whereas the American manufacturer is always on the look-out for something to enable him to get ahead of his trade rival' (ibid., 20 June 1901).
103. Hobsbawm, *Industry and Empire*, p. 107. On this see also Karl Rohe, 'Ursachen und Bedingungen des modernen britischen Imperialismus von 1914' in Wolfgang J. Mommsen (ed.), *Der moderne Imperialismus* (Stuttgart, 1971), pp. 60–84.

Table 2.3. Number of patents granted for synthetic dyes in Britain, 1856–1913

	To German individuals and companies	To British individuals and companies
1856–60	8	20
1861–5	21	54
1866–70	17	23
1871–5	8	11
1876–80	47	13
1881–5	113	15
1886–90	201	39
1891–5	386	29
1896–1900	427	52
1901–5	447	38
1906–10	561	30
1911–13	252	11

Source: Ed. 24/1579, February 1915, PRO.

William Henry Perkin, just 18 years old at the time and a student of August Wilhelm Hofmann at the Royal College of Chemistry in London.[104] Britain retained its superiority in the potentially rich field of organic chemistry only for a few years. By 1862, at the Great International Exhibition in London, German and Swiss companies were already competitive. The extent to which development in this new branch of industry depended on chemical research undertaken in company laboratories or in cooperation with the universities is revealed by figures on the granting of patents for synthetic dyes and allied products in Britain, taken from Board of Education documents (Table 2.3).[105] Figures for the period 1856 to 1913, which distinguish between patents issued to British and German individuals and firms, clearly illustrate the different levels of activity in chemical research and in applying its results to industry in the two countries. In 1900 the world market for synthetic dyes was dominated almost completely by big German companies. On the eve of the First World War their factories, at home and abroad, produced a good 85 per cent of world requirements for dye (Table 2.4).[106] About 80 per cent of the British textile

---

104. See Cardwell, *Technology*, p. 191; W. J. Reader, *Imperial Chemical Industries: A History*, vol. 1, *The Forerunners* (London, 1970), p. 13.
105. Almost identical figures for the period 1886 to 1900 can be found in *Nature*, 65 (1901–2), p. 138.
106. In 1915 *Nature* pointed to the decline in Indian indigo production: the value

Table 2.4. World production and consumption of synthetic dyes, 1913 (000s tons)

	Production	Consumption
Germany	135	20
Switzerland	10	3
France	7	9
United Kingdom	5	23
USA	3	26
China	—	28
Others	2	53
Total	162	162

Source: William Joseph Reader, *Imperial Chemical Industries: A History*, 2 vols. (London, 1970 and 1975), vol. I, p. 258.

industry's high demand for dye had to be imported, almost all of it from Germany.[107] What was left of the domestic dye industry constantly lost out to dynamic German competition under the conditions of free trade, to which Britain alone of the large trading nations adhered, despite Joseph Chamberlain's protests. In addition, individual British companies were so small that they had little financial and institutional capacity for research.[108]

Future prospects for Britain's optical industry, usually regarded as including precision engineering, seemed to be no better in 1900. During the Boer War the high demand among the armed forces for binoculars, telescopes, range-finders and so on could only be met by increasing imports. German products were also considered to be of higher quality than British products. Between 1903 and 1910 the

---

of Indian indigo exports fell from a good £3,570,000 in 1896 to £225,000 in 1911 (*Nature*, 95, 1915, p. 49). During the same period German indigo exports rose from 3,520 to 22,970 tons (Eduard Farber, *The Evolution of Chemistry*, New York, 1952, p. 293).

107. *Nature*, 114 (1924), p. 113 ('Applied Chemistry in Peace and War'). The author of this article points out that in 1924, 80 per cent of the demand for dye in Britain had been met from domestic production.

108. Lutz F. Haber, 'Government Intervention at the Frontiers of Science: British Dyestuffs and Synthetic Organic Chemicals 1914–39', *Minerva*, vol. 11 (1973), pp. 79–80; Reader, *Imperial Chemical Industries*, vol. 1, pp. 259–66. On the development of the chemical industry since the nineteenth century, see Lutz F. Haber, *The Chemical Industry during the Nineteenth Century* (Oxford, 1958, reprinted 1969); idem, *The Chemical Industry 1900–1930: International Growth and Technological Change* (Oxford, 1971); John J. Beer, *The Emergence of the German Dye Industry* (Urbana, Ill., 1959); H. W. Richardson, 'The Development of the British Dyestuffs Industry before 1939', *Scottish Journal of Political Economy*, vol. 9 (1962), pp. 110–29; Stephen Miall, *A History of the British Chemical Industry* (London, 1931).

British Optical Society, established in 1899, made several unsuccessful attempts to set up a central research institute for the industry in order to increase the quality of its products. Similar attempts by the Royal Society also failed.[109] The outbreak of war in 1914 showed that Britain's optical industry was fragmented, inefficient and largely dependent on German imports. In mid 1914, for example, 60 per cent of all optical glass used in Britain came from the Zeiss Works in Jena, 30 per cent from a French company and only 10 per cent from British companies. 'Great Britain has fallen so far behind in the development of her optical manufacture', wrote the British Science Guild in June 1914, 'that not only is she unable to supply her scientific and industrial requirements, but at the present moment she could not, unaided, produce sufficient quantities for the service of the Army and the Navy of the optical aids which are so important in modern warfare'.[110] Not until the war did state intervention make this field, of such great military significance, largely independent of foreign imports. This was achieved by consolidating the British industry and giving it financial support.[111]

The electrical industry had been similarly dependent on imports since the last quarter of the century. Despite the large number of basic discoveries made by British physicists — Charles Wheatstone, Michael Faraday, Lord Kelvin and James Clerk Maxwell, for example — the electrical industry had developed primarily in the USA and Germany. From the late 1880s, therefore, the catchphrase 'Made in Germany', which was calculated to make the British public and British industry alike uneasy,[112] was joined by fear of an

---

109. MacLeod and MacLeod, 'War and Economic Development', pp. 168–70. See the file 'N.P.L. and Optical Glass Manufacture laid before Council', 31 October 1912 (Royal Society Archives, Ms. 538).

110. *Ninth Annual Report of the British Science Guild* (London, 1915), p. 29. According to Board of Trade figures, the United Kingdom imported scientific and optical instruments to a total value of £710,341 in 1913. The main suppliers were: Germany, £362,891; USA, £182,293; France, £108,040; Belgium, £28,939; Switzerland, £19,872 (from *Nature*, 94, 1914–15, p. 523).

111. For details, see MacLeod and MacLeod, 'War and Economic Development'. 'The war has lent a great stimulus to the production in this country of optical glass, an industry which had previously tended more and more to become a German monopoly . . . Research on optical glass has now been undertaken by the laboratory, with the aid of a grant from the Privy Council Committee for Scientific and Industrial Research' ('The Work of the National Physical Laboratory during the Year 1915–16', *Nature*, 97, 1916, p. 507).

112. It is generally accepted that the purpose of marking goods with their country of origin, as required by the Merchandise Marks Act of 1887, was basically defensive. The aim was to prevent any confusion with products of superior quality produced by English industry. A few years later the discriminating mark had become a sign of quality. See Williams, *Made in Germany*; Kennedy, *Anglo-German Antagonism*, pp. 55–7.

American economic 'invasion'.[113] This referred essentially to the growing imports of American products and to the occasionally sensational successes of individual American enterprises on the British market since the last years of the nineteenth century. In certain industries, generously funded American competitors looked like driving domestic producers out of the market altogether by taking them over or establishing subsidiaries.[114] Around the turn of the century, therefore, many journalists were issuing apt, though often highly exaggerated, warnings about Britain's imminent defeat in the 'battle for commerce in almost every land on earth'.[115] *The Times*, for example, ran an extensive series on 'American Engineering Competition', in which Edwin A. Pratt, the paper's financial correspondent, investigated the success of America's engineering industry in export markets. He attributed it to five factors: (1) the size of factories and their integration into conglomerates of vertically and horizontally connected companies; (2) the modernity of factory equipment and the rationalisation of methods of production; (3) the quality of entrepreneurs; (4) the lack of union activity; and (5) the country's rich reserves of raw materials.

Like the vast majority of British entrepreneurs, Pratt regarded the difference in the influence of unions in the two countries as the decisive factor in the superior productivity and competitiveness of the American engineering industry,[116] a judgement which did not go unchallenged even then. But he did not overlook American entrepreneurs' receptivity to scientific and technical innovations. The question of the extent to which the reduction in Britain's economic growth rate was connected with its neglected education system and the outdated attitude of British industrialists to science and research was therefore always present during the lengthy debate on Britain's relative economic decline. It was claimed that a correlation existed between industrial growth and scientific research, especially through the development of new technologies and products. This claim was supported with reference to industries of the 'second industrial revolution' such as the electrotechnical and the organic-chemical industries, but it could not be

---

113. W. T. Stead, *The Americanisation of the World* (London, 1902); F. A. Mackenzie, *The American Invaders* (London, 1902).
114. 'The American Invaders have acquired control of almost every new industry created during the past fifteen years by the growing needs of modern life' (*Daily Mail*, 7 June 1901). According to Armytage (*Rise of the Technocrats*, p. 209), there were seventy-five American subsidiaries or companies with American interests in the United Kingdom around 1900, especially in engineering and chemicals.
115. *Daily Mail*, 6 June 1901 ('The American Invaders').
116. *The Times*, 11 June 1900.

proved in detail. British scientists, however, made their opinions on this subject very clear. In his presidential address to the British Association for the Advancement of Science, Norman Lockyer attributed what he regarded as Britain's negative economic development to the fact that trade no longer followed the flag as it had done in the past. Instead it followed the ingenuity of scientists and the industry which exploited their discoveries.[117] *Nature* stated categorically that German industry's success was due to the German system of education and training.[118] *The Daily Telegraph* wrote in 1906 that unlike the British government, the German government had in the past done everything to promote chemical research and the development of the chemical industry. And the Germans had reaped the rewards: 'They have created a colossal trade, much of which might have been ours'.[119] At the beginning of the twentieth century, therefore, Britain's most important industries (if the outstanding significance of the banking and insurance sector are disregarded) were still the key industries of the first industrial revolution: textiles, coal, iron and steel.

Even among the higher ranks of the British civil service it was becoming accepted that, as spokesmen for science repeatedly insisted, the modern economy was dependent on scientific research. In 1901 the British consul in Stuttgart submitted to the Foreign Office a report on chemical instruction and the chemical industry in Germany, in which he established a direct connection between the quality and intensity of the training that German universities provided in chemistry, and the expansion of its chemical industry. According to this report, academic research and marketing-orientated industry had achieved a close symbiosis in Germany.[120] In a cabinet memo on the state of the aniline dye industries in Germany and Britain, Francis Mowatt, long-standing permanent secretary to the Treasury, described the relative decline of the organic-chemical industry in Britain as 'humiliating'. He pointed to a similar trend in the production of medical and scientific instruments, an industry which was in danger of being taken over completely by German companies. According to Mowatt, the means of stopping this development was close at hand: 'It is the creation in this country of the scientific education which has given Germany her success in the

---

117. *Nature*, 68 (1903), p. 440.
118. *Nature*, 70 (1904), p. 83.
119. *Daily Telegraph*, 19 February 1906.
120. Frederick Rose, *Report on Chemical Instruction in Germany and the Growth and Present Condition of the German Chemical Industries*, Diplomatic and Consular Reports, Misc. Series, Cd. 430 (London, 1901), p. 5.

competition'.[121] Modern historiography of science has returned to this argument. Historians of science trace Britain's economic 'failure' since the end of the nineteenth century back to the diminishing innovativeness of British industry and to the country's technological and scientific 'backwardness', expressed at institutional level in the peripheral position of its universities as compared with Germany and the USA.[122]

Economic historians today see different and in part contradictory reasons for Britain's relative economic decline and the loss of its leading role in industry in the four or five decades prior to the First World War. The continuing discussion of this problem, in which much empirical evidence has been marshalled, has been comprehensively treated elsewhere and will not be discussed in detail here.[123] Explanations range from the view that Britain's relative economic retardation at the end of the nineteenth century was one of the prices paid for its long-held lead as an industrial power ('early start' theory),[124] to the sweeping theory of the failure of British entrepreneurs who, blinded by their successes early in the nineteenth century and rigid in their conservative attitudes, had by the end of the century largely lost their economic dynamism, investment confidence and willingness to take risks.[125] Derek Aldcroft has expressed a provocative opinion: that the slow-down in Britain's economic growth in the last quarter of the nineteenth century can largely be attributed to the 'failure of the British entrepreneur to respond to the challenge of changed conditions'.[126] D.C. Coleman and others, however, have put forward convincing arguments modifying this thesis.[127]

121. Aniline Dyes, CAB 37/66/63, 30 July 1903, PRO.
122. Cardwell, *Technology*, pp. 193–4.
123. Overviews in Charles Wilson, 'Economy and Society in late Victorian Britain', *Economic History Review*, vol. 18 (1965), pp. 183–98; Donald N. McCloskey, 'Did Victorian Britain Fail?', *Economic History Review*, vol. 23 (1970), pp. 446–59; Levine, *Industrial Retardation*; Landes, *Unbound Prometheus*, pp. 326–58.
124. Richardson, 'Retardation', pp. 115–16; Landes, *Unbound Prometheus*, pp. 334–5; Habakkuk, *British Technology*, pp. 189–220.
125. See Donald N. McCloskey, *Economic Maturity and Entrepreneurial Decline: British Iron and Steel, 1870–1913* (Cambridge, Mass., 1973); Landes, *Unbound Prometheus*, pp. 336–9; Peter H. Lindert and Keith Trace, 'Yardsticks for Victorian Entrepreneurs' in Donald N. McCloskey (ed.), *Essays on a Mature Economy: Britain after 1840* (London, 1971), pp. 239–74; Kindleberger, 'Germany's Overtaking', pp. 232–6; Temin, 'Relative Decline'; Levine, *Industrial Retardation*, pp. 68–73; Duncan L. Burn, *The Economic History of Steelmaking, 1867–1939: A Study in Competition* (Cambridge, 1940, reprinted 1961).
126. D. H. Aldcroft, 'The Entrepreneur and the British Economy, 1870–1914', *Economic History Review*, vol. 17 (1964–5), p. 113. Martin J. Wiener also emphasises the role of social factors in causing the slowdown in Britain's economic growth, as against a purely economic explanatory model (*English Culture*, esp. pp. 167–70).
127. Coleman, 'Gentlemen and Players'; Lindert and Trace, 'Yardsticks'; Wilson, 'Economy and Society'.

Even today, it seems, despite the voluminous literature on the subject, no completely satisfying explanation can be given for the specific course taken by British economic history in the second half of the nineteenth century. But the historian should view with scepticism — this at least can be concluded from the preceding discussion — any interpretation which attempts to reduce the explanation to a single cause. All such interpretations are inadequate. Monocausal answers cannot fully explain the many-sided problem of Britain's relative economic decline after about 1870; an adequate diagnosis must take into account a combination of factors, some empirically verified, others of a more speculative nature. They include the neglect of scientific research, a lack of application of science to industrial production and Britain's inadequate system of scientific and technical education — factors emphasised by economic historians today and already pointed out by the science lobby in the nineteenth and early twentieth centuries as explanations for the unmistakable 'industrial retardation' in certain sectors of the British economy.

This raises the question of what arguments scientists used in the debate about the role of science which had been going on since the early nineteenth century. This question gains significance when it is remembered that in comparison with earlier periods, the nineteenth century witnessed an unusually large number of important scientific discoveries and technical successes, but that contrary to expectations, this did not result in any lessening of the pressure on scientists to legitimise their profession. At this time scientists in Britain as elsewhere constantly had to justify their demand for public support for science. This forced them to ensure that their interests were represented in both organisational and real terms. In response to this permanent pressure, scientists brought forward arguments which, while remaining basically the same, were expressed in different ways and appeared in various combinations and with changing emphases. Because of their grouping and mutual interlocking, they cannot be clearly separated from each other. An attempt is made here to outline five fundamental arguments used to justify support for science, but the artificiality of this sort of isolated treatment must always be borne in mind.

Supporting science as an obligation of the modern state

By the 1830s at the latest, when the decline of science was being discussed in Britain, it had become apparent that the majority of scientists considered continuous support for science to be the duty of the state and no longer merely a favour to be granted on occasion. Scientists referred to Adam Smith's economic theory as well as to Jeremy Bentham to legitimise their claims. Both had defined supporting activities and organisations which promoted the general good, but which could not be financed by individuals or groups of individuals alone, as the third main task of the state, after defence against external threat and maintaining internal peace.[128] Both, but especially Bentham, recognised that important scientific discoveries had been made in the past by individual scientists without any claim on public funds. But behind this lay the conviction that scientific knowledge could be acquired more quickly if suitable public support was provided.[129] At the beginning of the 1870s *The Spectator* took up this line of argument developed in the early nineteenth century. It reminded readers that support for science was a natural extension of the state's traditional duties, such as printing money, standardising measures, maintaining lighthouses and protecting public monuments.[130]

In 1871, when *Nature* was making a name for itself as one of the scientific community's leading organs, the physicist and later president of the Royal Society, Lord Kelvin (Sir William Thomson, 1824–1907), demanded at the annual meeting of the British Association for the Advancement of Science in Edinburgh that basic research should 'be made with us an object of national concern, and not left, as hitherto, exclusively to the private enterprise of self-sacrificing amateurs, and the necessarily inconsecutive action of our present Governmental Departments and of casual committees'. Referring to his own subject, Kelvin expressed the conviction that England should always lead the field in physics.[131] In his opinion many scientific investigations from which the whole nation 'derives as great benefit as anything material can possibly produce', could only be carried out by the nation, of which the government was the executive committee. 'Investigations for which a large expenditure of money is necessary and which must be continued through long periods of years', he explained before the Devonshire Commission, 'cannot be undertaken by private individuals. Generally speaking, I believe that if the Government is well advised in respect to science, it will be for the good of the nation that the Government should make it part of its functions to promote

---

128. Parris, *Constitutional Bureaucracy*, p. 276.
129. See MacLeod, 'Science and the Treasury', p. 131.
130. 'Government and Scientific Investigation', *Spectator*, 44 (1871), p. 882.
131. Quoted in *Nature*, 108 (1921), p. 46.

experimental investigations in science.'[132] In its final report of 1875 the Devonshire Commission concluded that the development of research depended to a large extent on state aid. 'As a Nation we ought to take our share of the current Scientific Work of the World', the report continues, alluding to Kelvin's words. 'Much of this work has always been voluntarily undertaken by individuals and it is not desirable that Government should supersede such efforts; but it is bound to assume that large portion of the National Duty which individuals do not attempt to perform, or cannot satisfactorily accomplish.'[133]

The key term here, the 'national duty' which the government was to assume, was no longer in question in connection with science after the 1870s, when science faced enormously increased costs for experimental work. Only thirty years later, when during the debate on national efficiency *Nature* wrote in terms similar to those used by the Devonshire Commission of generous funding for research as an 'increasing need in modern times for every great nation',[134] it was expressing something accepted as a matter of course. Nobody could deny, wrote the Cambridge physiologist Sir Michael Foster at the same time, that scientific progress concerned the whole nation: 'The welfare of the State no less than the wellbeing of the single citizen is bound up with the advance of natural knowledge; the material prosperity of the country and, what is as important, the intellectual strength of the people are at stake in the right and speedy answering of the many questions clamouring to be answered'.[135] At that time the real problem was apparently that responsible politicians and the state had not fully recognised the task which had fallen to them. Science was therefore not generally receiving adequate support. The scientific community made this charge with almost monotonous regularity in the last third of the nineteenth century; it was still being made by the British Science Guild during the First World War.[136]

---

132. Quoted in Crowther, *Statesmen of Science*, p. 264.
133. *Eighth Report of the Royal Commission on Scientific Instruction*, C. 1298, p. 24.
134. *Nature*, 70 (1904), p. 136.
135. Michael Foster, 'The State and Scientific Research', *The Nineteenth Century and After*, vol. 55 (1904), p. 741. Foster (1836–1907) was a professor at Cambridge from 1883 to 1907, and Liberal Unionist MP for the University of London from 1900 to 1906; Gerald L. Geison, *Michael Foster and the Cambridge School of Physiology: The Scientific Enterprise in Late Victorian Society* (Princeton, NJ, 1978).
136. 'The State has neglected to encourage and facilitate scientific investigation, or to promote that cooperation between science and industry which is essential to

The utilitarian aspects of science

Lord Kelvin, pointing in the 1870s to the benefits which the nation derived from science, was using the most convincing, and at the same time the most venerable, argument for state support for science. Francis Bacon had described the objective of science as truth and social utility. In 1662 the Royal Society gained Charles II's patronage by defining as its aim the promotion of the natural sciences and 'useful arts'.[137] The Royal Society's interest, therefore, like that of the Paris Académie des Sciences, founded a few years later, was concentrated on areas which would today be classed as applied science. The Royal Institution in London, for decades one of Britain's most important research bodies, was founded at the end of the eighteenth century with the aim of making new scientific findings available to industry.[138] In the nineteenth century, too, the science lobby assured the government and the public that science would produce results which would be practical and useful for the community as a whole. Sir Humphry Davy (1778–1829), chemist and long-standing president of the Royal Society, made this point in a public lecture early in the nineteenth century when he presented Britain's political and economic position in a simple causal relationship with the encouragement of science:

> You owe to experimental philosophy some of the most important and peculiar of your advantages. It is not by foreign conquests chiefly that you are become great, but by a conquest of nature in your own country .... You have excelled all other people in the products of industry. But why? Because you have assisted industry by science. Do not regard as indifferent what is your true and greatest glory.[139]

It has already been pointed out that during the nineteenth century the British state gave financial support to specific research projects

---

national development' (Memorandum from the British Science Guild to the government about future relations between science and the state, no date, printed in *Nature*, 97, 1916, p. 463).

137. '[Q]uorum studia applicanda sunt ad rerum naturalium artiumque utilium' (Royal Society Charter of 1662, printed in *Record of the Royal Society*, p. 216).

138. The idea was to found a 'Public Institution for diffusing the knowledge and facilitating the general introduction of Useful Mechanical Inventions and Improvements' (Musson and Robinson, *Science and Technology*, p. 129). See also Morris Berman, *Social Change and Scientific Organisation: The Royal Institution, 1799–1844* (London, 1978); Thomas Martin, *The Royal Institution*, 2nd edn (London, New York, Toronto, 1948).

139. Quoted in Mendelsohn, 'The Emergence of Science', p. 33. Further examples of utilitarian arguments can be found in the same place.

which were of some practical use.[140] Early in the nineteenth century only few of the existing scientific institutions received regular state aid. They included the Royal Observatory in Greenwich, originally set up in 1675 to deal with the problem (of economic and military importance) of determining longitudes at sea, and the Royal Institution in London. In cases where research did not provide immediately obvious material benefits for the state or for industry, financial or institutional support was difficult to obtain. Agitation by the Endowment of Research Movement between 1860 and 1875 and by the scientists associated with *Nature* had little impact — both emphasised the importance of basic research as an essential prerequisite for and stimulus to economic prosperity. 'What struck me most in England', Justus von Liebig summed up in a letter to Michael Faraday in 1844, after participating in the British Association's annual meeting,

> was the perception that only those works which have a practical tendency awake attention and command respect, while the purely scientific works which possess far greater merit are almost unknown. And yet the latter are the proper and true source from which the others flow. Practice alone can never lead to the discovery of a truth or a principle. In Germany it is quite the contrary. Here in the eyes of scientific men, no value, or at least but a trifling one is placed on the practical results. The enrichment of Science is alone considered worthy of attention. I do not mean to say that this is better, for both nations the golden medium would certainly be a real good fortune.[141]

The state's interest in pure science remained weak in Britain until towards the end of the nineteenth century.

In presenting its arguments the science lobby took into account that neither politicians nor public fully comprehended the necessity for basic research. For obvious tactical reasons, therefore, when dealing with the government and the public, the science lobby emphasised the utilitarian aspects of science, which normally play a secondary part in scientists' own conceptions of their profession. The oft-cited 'ultimate justification' of modern science largely receded into the background: the expansion of knowledge for its own sake, the development of intellectual *curiositas* independently of external interests, 'theoretical curiosity'.[142] Only occasionally

---

140. See above, p. 70 in this volume.
141. Justus von Liebig to Michael Faraday, 19 Decmeber 1844, in Williams (ed.), *Selected Correspondence of Michael Faraday*, vol. 1, pp. 429–30.
142. Hermann Lübbe, 'Relevanz contra Curiositas' in *idem, Wissenschaftspolitik*, p. 12.

The Awareness of Crisis

did spokesmen for science point to the 'cultural value' of science as a justification for supporting it. In a vain attempt to explain the rather vague arguments for the concept of pure research to a public accustomed to utilitarian ways of thinking, the historian W. E. H. Lecky wrote, shortly before the turn of the century, that there were sciences and arts 'which can by no possibility be remunerative, or at least remunerative in any proportion to the labour they entail or the ability they require. A nation which does not produce and does not care for these things can have only an inferior and imperfect civilisation'.[143] A few years later the vice-president of the venerable Society of Arts, founded in 1754, expressed the opinion that for obvious reasons certain scientific work could only be done by the state:

> The work to which I refer is such as is not suitable, or to be expected from societies or individuals. It is work which is continuous and must expand in the flux of time, which is recognised by the public as useful, which is not and cannot be remunerative, which requires a staff larger than is required by the ordinary demands of a society, and cannot be dropped without serious detriment to the public.[144]

In a speech to the Chemical Society in 1907, the well-known chemist Raphael Meldola took stock of the state of scientific research in Britain. He pointed out that 'scientific research, like every other branch of human culture, is worthy of national homage, whether it leads to immediately "practical" results or not'. Any country 'which limited its appreciation of research to such branches of science as were likely to lead to industrial developments', was in his opinion 'on a low level in the scale of civilisation'.[145]

After the London Great Exhibition of 1851, Lyon Playfair had already attempted to point out that basic pioneering research and industrial development were interconnected. In a rather bold metaphor he called the 'cultivators of abstract science, the searchers after truth for truth's own sake', 'the "horses" of the chariot of industry'.[146] This comment on the internal dynamics of science did

---

143. W. E. H. Lecky, *Democracy and Liberty*, vol. 1 (London, 1896), p. 275. Lecky also gives the arguments used by opponents of state aid for science: 'There is weakening of private enterprise and philanthropy; a lowered sense of individual responsibility; diminished love of freedom; the creation of an increasing army of officials . . .; the formation of a state of society in which vast multitudes depend for their subsistence on the bounty of the State' (ibid., p. 276).
144. William Abney, 'Science and the State', *Nature*, 71 (1904–5), p. 90.
145. Raphael Meldola, 'The Position and Prospect of Chemical Research in Great Britain', *Nature* 76 (1907), p. 232.
146. Quoted in Cardwell, *Organisation of Science*, p. 81.

not, however, make a great impression. In 1910 *Nature* still felt obliged to point out that applied science, which produced economically useful results, could only be successful if it had a broad basis in pure research. Real progress in science, it wrote, comes 'from the pursuit of knowledge for its own sake'.[147]

Industry's dependence on science

Scientific achievement and a high level of technology are vital in deciding the fate of a nation — this was a standard part of the repertoire of after-dinner speakers and science journalists around 1900 in Britain and other industrial countries. Any nation that wanted to keep up with the imperialist great powers would have to train ever more and ever better qualified scientists, technicians and engineers.[148] At the height of the Boer War the mathematician Karl Pearson, who had supported the reform of London University in the 1890s and held the first chair of eugenics there from 1911, gave a speech which attracted a great deal of attention. He spoke of 'recent events in our commercial as well as our military experience', which gave cause for concern as to whether 'our supply of trained brains is sufficient, or, at any rate, whether it is available in the right place at the right moment'.[149] Economic achievement became the yardstick; the ways in which scientific research could stimulate industry had captured the interest of critical contemporaries. It was established that, in view of the far-reaching and long-term significance of the aniline dye industry,

> there can be no question that the growth in Germany of a highly scientific industry of large and far-reaching proportions has had an enormous effect in encouraging and stimulating scientific culture and scientific research in all branches of knowledge. It has reacted with beneficial effect upon the universities, and has tended to promote scientific thought throughout the land. By its demonstration of the practical importance of purely theoretical conceptions it has had a far-reaching effect on the intellectual life of the nation.[150]

To illustrate the benefits that society could derive from science,

---

147. *Nature*, 85 (1910–11), p. 29.
148. See, for example, the article quoted in ibid., pp. 29–32 ('Modern Scientific Research'). The author was the chemist Sir William A. Tilden.
149. Speech to the Literary and Philosophical Society in Newcastle, 19 November 1900, printed as *National Life from the Standpoint of Science* (London, 1901). It achieved a wide circulation within a short time; a second edition appeared in 1905.
150. A. G. Green, 'The Coal-Tar Colour Industry in Germany and England', *Nature*, 65 (1901–2), pp. 139–40.

scientists liked to point to public health and industry, areas in which, they thought, research and scientific findings had a crucial function. Especially since the Paris International Exhibition (1867), when the British public had first become aware of the gulf between science and industry and of its potential economic consequences, scientists had emphasised the significance of technical and scientific training in the secondary and tertiary education sectors for the social and economic development of the whole country. Industry based on the application of technology needed scientifically and technically trained experts. The chemist Edward Frankland, who had written his doctorate for Robert Wilhelm Bunsen in Marburg, told the Devonshire Commission that an ability to innovate, which was dependent on scientific knowledge, was an essential prerequisite for industrial success against foreign competition. 'In my opinion', he explained, 'there could not be any doubt but that the nation which neglected science must suffer in the end, because although it could buy scientific inventions from the other country, yet it still would always be behind, as it were, in the market . . . . It is also much more difficult to establish new manufactures upon new inventions in a country which neglects science because you cannot have either workpeople or managers competent to conduct those processes which depend on scientific principles.'[151]

The arguments put forward by Frankland were repeated in various forms during subsequent years. In 1904 the president of the Royal Society criticised the government's low allocations for research and scientific institutions, upon which the 'commercial and industrial prosperity' of society largely depended in view of the 'present state of international competition'. He noted a 'scientific deadness of the nation' and attributed it primarily to the humanistic ideal which dominated education in British schools and universities.[152] In a discussion which took place in the letters column of *The Times* in the spring of 1906, the London experimental physicist Silvanus P. Thompson (1851–1916), a friend of Norman Lockyer, warned of the serious consequences which neglecting research would have. The example of the aniline dye industry might soon be followed by other industries. The aniline dye industry had started in England in

---

151. Quoted in Crowther, *Statesmen of Science*, pp. 263–4.
152. *Nature*, 71 (1904–5), p. 108. A generally positive view of the English system of technical and scientific education appeared at the same time in Arthur Shadwell, *Industrial Efficiency: A Comparative Study of Industrial Life in England, Germany and America*, vol. 2 (London, 1906), p. 428. On Shadwell, see Cardwell, *Organisation of Science*, p. 194. On the anti-industrial and anti-modernist attitude of a large section of the English upper and middle classes, expressed particularly in the idealisation of country life, see Wiener's stimulating study, *English Culture*.

the mid nineteenth century, after William Henry Perkin (1838–1907) had made his pioneering discoveries, but it was developed on a large industrial scale in Germany. Other industries, like the electrical and steel industries, were no less dependent on the 'adequate cultivation of scientific research'. Thompson saw the possibility of other industries following this path foreshadowed in the fact that 'pioneering, as it is understood in an electrical factory in the United States or in Germany, is now almost non-existent in England'. In ten years' time, at the latest, this research deficit would lead to the collapse of the British electrical industry. According to Thompson, therefore, it was absolutely imperative that 'the commercial and educational leaders of the nation shall open their eyes to the absolutely vital nature of scientific research in its bearing on industrial prosperity'.[153]

An anonymous correspondent echoed Thompson's warning, a few days later, writing that only when the British people learned to value and encourage scientific work and scientific originality, would Britain be able to maintain its position *vis-à-vis* the better-trained specialists in Germany and the USA.[154] The author of a book on German industry saw the close connection which existed there between science and 'practical affairs' as the crucial reason for Germany's economic prosperity.[155] The problem was perceived in sections of industry itself. A submission made in 1904 by a group of industrialists including Sir John T. Brunner and Sir Christopher Furness, both Liberal MPs, to the leader of the Liberal Party and later prime minister, Sir Henry Campbell-Bannerman, made the following points:

> We hold that Government should initiate, and invest national funds in carrying out, measures directed towards the encouragement of scientific research, such as is not likely to be undertaken by private enterprise, and towards the development of higher education . . . . This is a matter in which we might well have taken the Germans as our pattern, but we have neglected the whole question in the most deplorable fashion.[156]

At the British Science Guild's annual meeting in 1909, Haldane

---

153. *The Times*, 3 March 1906. A few years earlier, *Nature* had written: 'Already we have lost supremacy in several branches of industry, and we shall probably be surpassed in others by America and Germany unless our commercial men learn to realise that science is the source of energy of all sustained industrial movements' (*Nature*, 65, 1901–2, p. 44).
154. *The Times*, 10 March 1906.
155. E. D. Howard, *The Cause and Extent of the Recent Industrial Progress of Germany* (London, 1907), p. 145.
156. Memorandum of 6 May 1904, Herbert Gladstone Papers, Add. MSS 45 988, British Library.

obviously considered it an established fact that countries which supported their universities and technical colleges most effectively and offered their youth a broad education were in the best position to dominate world markets. 'There is no investment', he concluded, 'that will produce such a return, not to the investor, but to the generations to come, as the endowment of higher education'.[157] A few years earlier, Haldane — described benevolently by *Nature* as 'an enthusiast for higher education'[158] — had explained that only if the state gave science more support than hitherto by developing the education system in terms of both quantity and quality, did Britain stand any chance of holding its own in the international competition, 'which is more and more coming to depend on the application of science to industry'. The decline in exports of various British industrial products indicated the necessity 'for more mind in the process of manufacture, that is, for the improvement of higher education in this country'.[159] Apparently, however, neither Haldane's public warnings about neglecting science nor his work as one of the most prominent science lobbyists of the time had any effect. A decade later they had lost none of their urgency. *Nature* complained that science and industry were still separated in Britain.[160] At the Royal Society's annual meeting in 1915 it was claimed that if the value of science continued to be ignored in Britain, the foreseeable consequence would be 'total defeat in the industrial war which must of necessity follow upon the conflict of arms now raging'.[161]

These quotations show clearly that the dependence of modern industry on research was perceived in Britain, and that the economic boom in countries such as Germany and the USA was, as a matter of course, seen as related to the massive investment these countries made in science. From the 1870s on, indeed, indicators were developed to quantify and thus compare the scientific capacity

---

157. *Nature*, 79 (1908–9), p. 379.
158. *Nature*, 72 (1905), p. 184. On Haldane's work in education policy, see Eric Ashby and Mary Anderson, *Portrait of Haldane at Work on Education* (London, 1974).
159. *Nature*, 72 (1905), p. 185. On the interdependence between the education system and economic development, see B. F. Kiker (ed.), *Investment in Human Capital* (Columbia, S.C., 1971); E. A. G. Robinson and J. E. Vaizey (eds.), *The Economics of Education* (London and New York, 1966); Robert R. Locke, 'Industrialisierung und Erziehungssystem in Frankreich und Deutschland vor dem 1. Weltkrieg', *Historische Zeitschrift*, vol. 25 (1977), pp. 265–96.
160. *Nature*, 85 (1910–11), p. 30. The argument that British industry did little research has recently been questioned. Michael Sanderson points out that several companies had laboratories, but considered the research done there a company secret (*Universities*, passim; idem, 'Research and the Firm in British Industry, 1919–1939', *Science Studies*, vol. 2, 1972, pp. 107–52).
161. *Nature*, 96 (1915–16), pp. 374–75.

of different countries: the number of students studying science and the number of scientists graduating per year, the number of scientists employed by industry, the nationality of the authors of scientific publications, the frequency with which scientific works were translated and, after the turn of the century, the geographical distribution of Nobel Prize winners. It is not surprising, therefore, that a related argument was frequently introduced into discussions of the connection between science and industrial prosperity: a country's scientific significance, even superiority, was equated, often in a highly exaggerated form, with its political significance or superiority. According to the historian of science Brigitte Schroeder-Gudehus, scientists have used this argument since the time of Francis Bacon to justify their financial demands.[162]

Nature's political agitation on this subject contained an unmistakably national dimension as early as the 1870s, and this trend was also obvious in comments made by Lord Kelvin and the Devonshire Commission, quoted above.[163] A letter published in *The Times* in 1906 pointed out that a nation which neglected to give science an appropriate place in society would be reduced to economic and political insignificance.[164] In 1907 the astronomer Sir David Gill, in his capacity as president of the British Association for the Advancement of Science, referred to the close interrelationship between a nation's general progress and its scientific capacity. He insisted that science should play a greater part 'in the education of our youth' and that a 'larger measure of the public funds' should be devoted 'in aid of scientific research'.[165] According to *The Times*, muddling through had been acceptable as long as comparable countries had done the same in the competition against other nations. But, it continued, this traditional practice 'will not serve against rivals who practice intelligent organization of intellect'.[166] Celebrations of the Royal Society of London's 250th anniversary gave *The Times* occasion to point out that science was no longer a toy: 'It [science] has become, or is fast becoming, the dominant factor in human affairs; it will determine who shall hold the supremacy among nations'.[167]

Germany seemed to provide the best illustration of this theory in

---

162. Brigitte Schroeder-Gudehus, 'Science, Technology and Foreign Policy' in Ina Spiegel-Rösing and Derek J. de Solla Price (eds.), *Science, Technology and Society: A Cross-Disciplinary Perspective* (London and Beverly Hills, 1977), p. 473.
163. See above, p. 117–18 in this volume.
164. *The Times*, 10 March 1906.
165. *Nature*, 76 (1907), p. 319.
166. *The Times*, 17 July 1912 ('The Royal Society Celebration').
167. *The Times*, 16 July 1912 ('The Royal Society: A Retrospect of 250 Years').

the decades prior to the First World War. Contemporary newspapers and journals repeatedly pointed out — admiringly or with dismay, according to the journalist's political intent — that the increase in Germany's political, and ultimately military, strength was based on its economic power, and that this in turn was based on the quality of German science. Germany's *Weltpolitik* and its part in the war after 1914 would have been unthinkable without the cooperation of scientists in both military and civilian sectors. This experience was widely discussed in England and France before and during the war.[168] Some of the provisions of the Versailles Treaty were the results of this line of thinking. For the first time a peace treaty was used to try to limit a nation's scientific and technical capacity.[169] These provisions were intended to place a check on certain areas of scientific research and on certain sectors of industry in Germany in order to curtail its political and military power. They were also designed to break the hegemony of science which Germany had undoubtedly enjoyed before the First World War.

Science in the 'struggle for existence' among nations

The exemplary nature of Germany's science organisation and the problem of the political and military pre-eminence achieved by certain countries were topics which increasingly came to dominate the debate on science and its promotion in Britain at the end of the nineteenth century. Since the 1880s the often-cited idea of a 'struggle for existence' had underlain, on a political and economic level, the image of free competition or a sporting contest between individuals and nations. According to this idea, only the fittest would survive. Fitness was demonstrated not only by surviving but also by dominating others — individuals, social groups, nations and races. Lord Salisbury reduced this dangerous mixture of social Darwinism and imperialism to a handy catchphrase for everyday political use when, in a policy speech given to the Primrose League in the Albert Hall in 1898, he spoke of the living and dying peoples into which the nations of the world could be divided. With the help

---

168. See Schroeder-Gudehus, 'Science, Technology and Foreign Policy', p. 474. In general, see also Lawrence Badash, 'British and American Views of the German Menace in World War I', *Notes and Records of the Royal Society of London*, vol. 34 (1979), pp. 91–121.

169. See here in particular the military restrictions in §§168, 171, 172, 177, 190, 191, and the provisions relating to German patents and the use of factory and trade marks in §§297, 298, 306, 307, 310. In the USA, German patents were seized in November 1918. In return for a small payment to the German patent owners, American industrialists were able to use all German patents (Haber, *The Chemical Industry 1900–1930*, pp. 219–20).

of this simplistic distinction, he justified acts of foreign and domestic policy.[170] A few years later, in an atmosphere of demagogy and rising hysteria about the crisis in Britain's world position, David Lloyd George claimed that never had a people faced greater problems than the British people at the present time. 'We are living in an age of keen competition. There is a great struggle for life, not merely amongst individuals, but amongst nations, for commerce, trade, supremacy.'[171] The British science lobby felt that it was its particular duty to preserve Britain from the danger of falling behind in this political and economic competition, perhaps even from national decline, and to offer a way out of the social and economic crisis which threatened to put Britain among the 'dying peoples'. According to the theory of selection, which provided an important model for social and political action around the turn of the century, it was necessary to mobilise all Britain's forces under the banner of the popular catchphrase 'national efficiency' — and in the opinion of the spokesmen for science these forces consisted, in the first instance, of the country's hitherto imperfectly exploited scientific resources.

The social-Darwinist attitudes of contemporary politicians are also evident in the arguments put forward by the scientists and journalists who, from the last quarter of the nineteenth century until well into the First World War, advocated more support for science. On the fiftieth anniversary of Queen Victoria's accession to the throne, T. H. Huxley, one of the most influential popularisers of social Darwinism in Britain, wrote a letter to *The Times* in which he claimed that 'we are entering, indeed we have already entered, upon the most serious struggle for existence to which this country was ever committed. The latter years of the century promise to see us in an industrial war of far more serious import than the military wars of its opening years.'[172] *Nature* celebrated the new century with an editorial arguing that 'the enormous and unprecedented progress in science during the last century has brought about a perfectly new state of things, in which the "struggle for existence" which Darwin studied in relation to organic forms is now seen, for the first time, to apply to organised communities, not when at war with each other but when engaged in peaceful commercial strife'.[173] In 1910 the British Science Guild held an essay competition on the topic: 'The best way of carrying on the struggle for existence and

---

170. The speech is printed in *The Times*, 5 May 1898.
171. Speech in Newcastle, 4 April 1903 (David Lloyd George Papers, A 11/1/26).
172. *The Times*, 21 March 1887.
173. *Nature*, 63 (1900–1), pp. 22–3. The article was written by Norman Lockyer.

securing the survival of the fittest in national affairs.'[174]

The famous speech given by Norman Lockyer as president of the British Association's annual meeting in Southport should be located between these cornerstones of social Darwinism among British scientists in the age of imperialism. Lockyer gave his speech in 1903, shortly before the end of the Boer War and in response to a widespread readiness for reform. His speech received an unusual amount of publicity all over Britain largely because he convincingly argued the connection between the struggle for existence, on the one hand, and politics and encouraging science, on the other.[175] Its title, 'The Influence of Brain-power on History', was a deliberate allusion to an important book, *The Influence of Sea Power upon History*, written by the American naval theoretician Alfred Thayer Mahan and published in 1890. Lockyer's reference to sea power is more than an expression of the general enthusiasm of the time for naval affairs. It is also an expression of the idea that military power, and industrial and economic power were equally important at a time when industrial nations were competing and expanding overseas — both types of power were indispensable if claims to imperial power were to be maintained. Political power could no longer be based on military power alone. Lockyer in fact pushed the equation of 'sea-power' and 'brain-power' so far as to demand equal expenditure on both areas, thus arriving at astronomically high figures for education and science.[176]

Lockyer's reasoning in this speech is typical of numerous contemporary attitudes. In it he deplored the neglect of science in Britain, both idealistic and financial, and the economic consequences of this neglect. He presented encouraging science as an act of national assertion, and suggested that it had much in common with armament:

---

174. See above, pp. 93–4 in this volume. Generally on this topic see also. H. W. Koch, 'Social Darwinism as a Factor in the "New Imperialism" in idem (ed.), *The Origins of the First World War. Great Power Rivalry and German War Aims* (London and Basingstoke, 1972), pp. 329–54; Semmel, *Imperialism*, pp. 29–52; Hans-Ulrich Wehler, 'Sozialdarwinismus im expandierenden Industriestaat' in idem, *Krisenherde des Kaiserreichs 1871–1918*, 2nd edn (Göttingen, 1979), pp. 281–9.
175. Lockyer and Lockyer, *Life and Work*, p. 186; Meadows, *Science and Controversy*, p. 266. When a deputation from the universities and larger scientific and learned societies, led by Lockyer, succeeded in getting state aid for British universities doubled in 1904 (see above, p. 34 in this volume), the prime minister, Balfour, suggested that this decision was made partly as a result of Lockyer's speech (see the obituary notice for Lockyer in *Nature*, 105 [1920], p. 784).
176. Lockyer's speech is printed in *Nature*, 68 (1903), pp. 439–47, and in Lockyer, *Education*, pp. 172–215. See also Meadows, *Science and Controversy*, pp. 265–6 and 269–70.

We have lacked the strengthening of the national life produced by fostering the scientific spirit among all classes, and along all lines of the nation's activity; many of the responsible authorities know little and care less about science; we have not learned that it is the duty of a State to organise its forces as carefully for peace as for war; that universities and other teaching centres are as important as battleships or big battalions; are, in fact, essential parts of a modern State's machinery, and as such to be equally aided and as efficiently organised to secure its future well being.

Lockyer thought that compared with these tasks and the necessity of making technical and scientific discoveries available to industry, the debate on free trade and tariff reform instigated largely by Joseph Chamberlain, was of only secondary significance for the future of the British economy: 'A knowledge and utilisation of the forces of Nature are very much further reaching in their effects on the progress and decline of nations than is generally imagined'. In future, statesmen and politicians would have 'to pay more regard to education and science, as empire-builders and empire-guarders, than they have paid in the past'.[177]

What was necessary for Britain therefore was 'a complete organisation of the resources of the nation, so as to enable it best to face all the new problems which the progress of science, combined with the ebb and flow of population and other factors in international competition, are ever bringing before us'.[178] It was all the more necessary to organise all resources, as 'every scientific advance is now, and will in the future be more and more, applied to war'. Lockyer continued: 'It is no longer a question of an armed force with scientific corps, it is a question of an armed force scientific from top to bottom'.[179] Barely a year later Lockyer wrote an article entitled 'The National Need of the State Endowment of Universities', in which he once again presented the struggle for existence among imperialist powers in unmistakably military terms: 'We are in the midst of a struggle in which science and brains take the place of swords and sinews; the school, the university, the laboratory and the workshop are the battlefields of this new struggle'.[180]

---

177. *Nature*, 68 (1903), p. 439.
178. Ibid., p. 440.
179. Ibid., p. 446.
180. Printed in Lockyer, *Education*, p. 216.

The Awareness of Crisis

Germany as a model of science promotion

Lockyer's speech to the British Association, which provided the ideological foundation for the establishment of the British Science Guild shortly thereafter, is more than merely another example of British scientists' traditional complaints about the lack of support for science which, by the late nineteenth century, had allowed Britain to become, in terms used by sociologists of science, a 'peripheral' country in relation to the 'centres of science' such as Germany and the USA.[181] It is also an expression of the increased self-confidence felt by scientists since the end of the nineteenth century, and especially during Britain's social crisis before the First World War. Comments on 'the neglect of science by the British nation in the past',[182] 'this national weakness', namely 'the general want of appreciation of research here',[183] 'the limited appreciation of technical education by the English manufacturing world as a whole',[184] and about the lack of a 'scientific spirit among our administrators and teachers',[185] were made more frequently than ever before at this time.

The scientific community agreed unanimously on these issues, which were raised more emphatically than previously, in discussions of the relationship between science and society in Britain. The president of the Chemical Society regarded the problem of how to remedy the unpleasant situation of 'general public ignorance of and apathy towards research' as 'a knotty question', which, it seemed to him, could not be answered at the present time. But he did not exclude the possibility that sooner or later, 'persistent attack from within' and 'the presence of competition from without' would make the country aware of its own disastrous situation, and that reforms in science organisation would be introduced.[186]

Despite the standard complaints and the fact, referred to above, that government authorities took little part in the debate about supporting science in Britain, in retrospect, the question arises as to

---

181. These terms are used by Joseph Ben-David, *Centers of Learning*, pp. 5–7; idem, 'Scientific Growth: A Sociological View', *Minerva*, vol. 2 (1964), pp. 455–76; idem, 'The Rise and Decline of France as a Scientific Centre', *Minerva*, vol. 8 (1970), pp. 160–79. See also Gizycki, 'Centre and Periphery'.
182. Richard B. Haldane at the annual meeting of the Chemical Society in 1905 (*Nature*, 71, 1904–5, p. 589).
183. President of the Chemical Society, Professor Raphael Meldola, at the Society's annual meeting in 1907 (*Nature*, 76, 1907, p. 231).
184. President of the Association of Teachers in Technical Institutions on 17 June 1910 in Birmingham (*Nature*, 83, 1910, p. 508).
185. C. A. Buckmaster in an essay, 'State Aid for Science', *Nature*, 94 (1914–15), p. 547.
186. Raphael Meldola on 22 March 1907 (*Nature*, 76, 1907, pp. 231–2).

whether the government did in fact ignore the demands put forward so persistently by scientists, or whether they were recognised by the government and were considered when political decisions were made and institutional reforms were implemented. Is there any evidence that a gradual change took place in the government's attitudes, especially after the turn of the century? For the moment, we will defer a definitive answer to the question of whether Lockyer's polemical speech to the British Association presented an accurate picture of the relationship between state and science in Britain at the beginning of the twentieth century. There are indications that the state's attitude towards science was already beginning to change. This change was reflected in increased government interest in the science organisation of other countries. Since the end of the nineteenth century, royal commissions and individual scientists, but also members of the Foreign Office such as the British consul in Stuttgart, Frederick Rose, for example, had been studying scientific institutions set up both by the state and the new research-intensive growth industries (especially the chemical and electrical industries) in the USA and on the Continent. They had also attempted to publicise in Britain the models of organisation used abroad, and how they worked. Germany in particular, Britain's great political rival, became a yardstick for British planning. The British image of conditions under which scientific research took place in Germany was undoubtedly idealised to a very large degree, with the political purpose of making the negative aspects of the situation in Britain stand out even more sharply by contrast.[187]

Britain's supposed backwardness in scientific research and Germany's allegedly greater modernisation cannot easily be demonstrated empirically, especially as the criteria used by contemporaries in arriving at their — mostly impressionistic — judgements cannot always clearly be determined. Today, therefore, it remains debatable whether, and to what extent, Britain did in fact fall behind comparable countries during the nineteenth century. Nevertheless, British

187. A few years later Adolf Harnack pursued the same tactics in a memorandum of 1909, which led to the foundation of the Kaiser-Wilhelm Society. In it, he refers to the amounts spent by other countries on scientific research and speaks of science in Germany being in a 'state of emergency'. According to Harnack, Germany 'had fallen behind other countries in important aspects of natural science, and its competitiveness was under great threat' (translated from Adolf Harnack, 'Begründung von Forschungsinstituten' in idem, *Aus Wissenschaft und Leben*, vol. 1, Giessen, 1911, p. 43). On this see Lothar Burchardt, 'Deutsche Wissenschaftspolitik an der Jahrhundertwende. Versuch einer Zwischenbilanz', *Geschichte in Wissenschaft und Unterricht*, vol. 26 (1975), pp. 271–89; idem, *Wissenschaftspolitik im wilhelminischen Deutschland. Vorgeschichte, Gründung und Aufbau der Kaiser-Wilhelm-Gesellschaft zur Förderung der Wissenschaften* (Göttingen, 1975); Schieder, 'Kultur, Wissenschaft und Wissenschaftspolitik im Deutschen Kaiserreich', pp. 9–34.

scientists in the second half of the nineteenth century certainly orientated themselves by conditions in Germany. The high regard in which German work in the humanities and the sciences was held at this time was still being expressed a few days before the beginning of the First World War, when British scholars made a joint declaration that 'we regard Germany as a nation leading the way in the Arts and Sciences, and we have all learnt and are learning from German scholars'. It continued: 'A conflict with a nation so near akin to our own, and with whom we have so much in common would be a sin against civilisation'.[188] The influence of the German scientific system on Britain, important aspects of which have been investigated by George Haines, Günter Hollenberg and Frank R. Pfetsch,[189] can be seen in the fact that since the 1830s it had become practically obligatory for British scientists, especially chemists, to spend some time studying in Germany. 'Virtually every professor of chemistry in a British university before 1914 held a German doctorate.'[190] German universities were no less attractive to young American scientists and scholars, particularly in the period from the Civil War to the turn of the century.[191]

While the social-Darwinist element in scientists' arguments jus-

188. *The Times*, 1 August 1914 ('Scholars Protest against War with Germany'). See Badash, 'British and American Views', esp. pp. 98–104.

189. Haines, *German Influence*; Günter Hollenberg, *Englisches Interesse am Kaiserreich. Die Attraktivität Preußen-Deutschlands für konservative und liberale Kreise in Großbritannien 1860–1914* (Wiesbaden, 1974) emphasises constitutional and sociopolitical aspects; Pfetsch, *Wissenschaftspolitik in Deutschland*, pp. 314–47. Also important are Searle, *National Efficiency*, and Kennedy, *Anglo-German Antagonism*, pp. 103–23. Still worth reading is Percy Ernst Schramm, 'Englands Verhältnis zur deutschen Kultur zwischen der Reichsgründung und der Jahrhundertwende' in Werner Conze (ed.), *Deutschland und Europa. Historische Studien zur Völker- und Staatenordnung des Abendlandes. Festschrift für Hans Rothfels* (Düsseldorf, 1951), pp. 135–75.

190. Cardwell, *Technology*, p. 192. See also Haines, *German Influence*, pp. 21 and 60–7; Meadows, *Science and Controversy*, p. 215. It has been claimed that between 1836 and 1850, fifty-nine Britons studied with Justus von Liebig in Giessen alone. Forty Britons studied with Wilhelm Ostwald in Leipzig between 1887 and 1906. Between 1849 and 1914, about 9,000 British students were allegedly enrolled at German universities (Gizycki, 'Centre and Periphery', p. 483). See also Hollenberg, *Englisches Interesse am Kaiserreich*, pp. 148, 151, 294–5. Hollenberg lists the English professors who spent some of their student days in Germany (pp. 297–9).

191. See Hollenberg, *Englisches Interesse am Kaiserreich*, esp. pp. 151–2 and 156–8; Haines, *German Influence*, pp. 131–2; Cardwell, *Organisation of Science*, pp. 64, 163. Carl Diehl estimates that during the period 1815–1914 the number of American students at German universities was between 9,000 and 10,000 (*Americans and German Scholarship 1770–1870*, New Haven and London, 1978, p. 1). An informative source relating to Göttingen's attractiveness to American and British students (Haldane studied there for several months in 1874) is provided by Paul G. Buchloch and Walter T. Rix (eds.), *American Colony of Göttingen: Historical and Other Data Collected between the Years 1855 and 1888* (Göttingen, 1976). The influx of American and British students into German universities did not decrease until the end of the nineteenth century when universities in the Anglo-Saxon countries improved their facilities for advanced study and research.

tifying financial support for research thus unmistakably gained ground at the end of the Victorian Age, their position also changed in as much as a comparative international perspective, going back to arguments put forward by Charles Babbage around 1830, also became more prominent again. We have already seen several references — some admiring and some admonitory — to the economic and scientific advance of both Germany and the USA after 1870. Between 1880 and 1914 Germany's method of organising and supporting science became a model for British reform attempts. There were several reasons for this. First, Germany had an efficient, socially open and secular secondary and tertiary education system (special reference was made to the fact that research and teaching were equally emphasised in German universities). Secondly, science was generously funded by the state in Germany. Thirdly, there was a fruitful exchange between science and industry in Germany, which meant that scientific knowledge was rapidly absorbed and utilised by industry. Fourthly, Germany possessed a large variety of scientific research institutions, both within the framework of the universities and *Technische Hochschulen*, and increasingly towards the end of the nineteenth century in the non-university sector.

In short, according to the British science lobby, the German scientific system deserved unreserved admiration, especially when compared with conditions in their own country. 'The German nation deserved scientifically all the admiration we could give to it', explained a speaker at a meeting organised by the British Science Guild: 'It had recognised the relation of science, not only to industries, but to methods of government and the general education of the community'.[192] A combination of scientific research and business sense 'has been applied in Germany with marvellous success', wrote *Nature*,[193] and the lesson to learn from conditions in Germany could only be 'to make industrial science a sufficiently attractive career for those who have received a superior general education'.[194] In a speech in the House of Commons in 1902, Richard Haldane pointed to the quality of the German universities and described the British education system as 'backward',[195] while Richard T. Glazebrook, director of the National Physical Laboratory, described it as 'far behind Germany in very many vital respects'.[196]

192. *Nature*, 83 (1910), p. 350.
193. *Nature*, 74 (1906), p. 319.
194. *Nature*, 69 (1903–4), p. 164.
195. Hansard, Parl. Deb., H.C., 4th Series, vol. 107, col. 708 (5 May 1902).
196. *Nature*, 83 (1910), p. 85. It hardly needs to be mentioned that German scholars and experts in the field did not share the optimistic assessment of many

The Awareness of Crisis

This represents only a selection of contemporary opinion: one could cite many similar judgements.

Naturally enough, the constant reference to Germany, which was held up as exemplary in matters of science, was frequently ridiculed by the British public. Often, too, it provoked irritation and opposition. Apparently, high government officials were often annoyed by scientists drawing comparisons with Germany. Sir Archibald Geikie, geologist and former president of the Royal Society, recalled, looking back to the time before 1914:

> The example of that country [Germany] was often cited here, and contrasted with the unsympathetic attitude and stingy support of our authorities, much to the surprise and annoyance of the permanent officials of the Treasury, who rather seemed to think that their grants to science were remarkably liberal. I remember an occasion when I had to go to the Treasury about a matter connected with the Geological Survey. The official on whom I called was one of the heads of the Department . . . . He began the interview by saying that he would be glad to hear me, but begged that the example of Germany might not be mentioned.[197]

Spokesmen for science in Britain frequently saw conditions in Germany in very general and uncritical terms, as the British response to the founding of the Kaiser-Wilhelm Society in 1911 shows. This was the most important innovation in German science organisation before the First World War, but despite its pioneering conception, it aroused little interest in Britain. There could be no question of the public receiving expert information about this new organisation for the promotion of science. The daily papers published only isolated and brief reports of the event. Predictably, the most detailed reports appeared in *Nature*, including long extracts from speeches made at the opening, and a leading article which commented in general terms on the new society's aims and organisational structure. Special emphasis was put on the close financial involvement of German industry in the society's research work, laid down in its statutes. A journal like *Nature*, devoted to promoting science, could not let this opportunity pass without sternly

---

Britons. Comparing the German and English university systems early this century, Ludwig Curtius, the classical archaeologist, for example, comes to the following conclusion: 'Das englische System leistete als rein wissenschaftliches vielleicht weniger als das deutsche, aber für die geistige Gesamterziehung der Nation leistete es ungleich mehr' (*Deutsche und antike Welt. Lebenserinnerungen*, 2nd edn, Stuttgart, 1958, p. 221).

197. Archibald Geikie, 'Retrospect and Prospect', *Nature*, 104 (1919–20), p. 196.

admonishing the British public. The leader suggests that it was instructive 'to note the difference between their method and ours' of encouraging research. The Kaiser-Wilhelm Society was to serve progress in science alone, without being too concerned about how scientific knowledge could be put to practical use, because 'our neighbours have learned the lesson that science, like virtue, brings its own reward'. It was 'wonderful how deeply the spirit of trust in science has penetrated the whole German nation . . . . This spirit, which permeates the German people, from the Emperor on his throne to the representatives of the peasants, causes admiration; would that it could inspire imitation!'

The unnamed author went on to speculate about why Germans have 'such sympathy for scientific endeavour'. The reasons for this, he concluded, were many: a long tradition of education, discipline instilled by universal compulsory military service, a widespread understanding of the use of science for industry, expressed in the social prestige accorded to scientists and the salaries they were paid, and the close interrelationship between science and industry, which put few obstacles in the path of a scientist who wanted to switch from an academic career to industry or *vice versa*. In Britain, by contrast, the author concluded that in times of good trade the industrialist believed 'that he has no need of scientific assistance; in times of bad trade he believes that he cannot afford it'.[198]

The black-and-white picture painted by the British science lobby (the report in *Nature* about the Kaiser-Wilhelm Society is another example of this), was intended primarily for the domestic market, where, indeed, it made some impact. Despite a certain amount of opposition in the British civil service, Germany was adopted as the yardstick for Britain's emerging science policy after the turn of the century. To contemporaries it seemed that the threat to vital interests posed by German policies and German industry in the years before the First World War could only be met by adopting German principles of organisation and of work, German 'scientific methods' (whatever this was understood to mean) on a large scale.[199] 'It is not the German Dreadnoughts we have to be afraid of', said the speaker, with a touch of unintentional humour, discussing German universities at the British Science Guild's annual

---

198. 'The Kaiser-Wilhelm Society for the Promotion of Science', *Nature*, 86 (1911), pp. 69–70.
199. See Norman Lockyer, speaking at the first annual meeting of the British Science Guild on 28 January 1907: 'Germany is strengthening its universities just as thoroughly as it is strengthening its Fleet, a reminder that we ought to be able to compete with other nations in the preparation and equipment for industrial progress, as well as for war' (*Nature*, 75, 1906–7, p. 327).

meeting in 1910, 'but the German schoolmaster. He is the man who is doing the damage.'[200] The conception, organisational structure and efficiency of the German scientific system made it attractive for selective emulation by the British. To what extent British planning after the turn of the century was in fact based on the German system, to what extent parts of the German system were taken and adapted to British conditions, and to what extent British science policy developed new elements and approaches as a result of being confronted with German models, is one theme of the next chapter.

200. *Nature*, 83 (1910), p. 100.

3
Science Policy Decisions in Britain: Institutional Innovations after 1900

It is commonly claimed in the modern historiography of science that 'institutional innovations' have not received the attention they merit.[1] Innovations of this sort, however, are of special interest for this study because an analysis of their origins and implementation is most likely to confirm the argument already put forward that in Britain the state's attitude towards science began to change after the turn of the century. For this reason, three outstanding initiatives between 1890 and 1914 designed to improve the facilities for research, each of a different nature, will serve as examples to describe more precisely the beginnings of a public science policy in Britain. This approach is based on the assumption that the history of institutional foundations reveals more about the relationship between science and the state than does the debate carried on by scientists, a debate generally comprising one-sided arguments about the state's omissions in the past. The following three case-studies — the founding of the National Physical Laboratory, Imperial College of Science and Technology, and the Medical Research Committee — will deal in the first instance with the course of events leading to the foundation of the respective institution, and with the part taken by the state. However, questions relating to the founders, the organisational patterns on which the institutions were modelled and their financing will also be pursued.

The National Physical Laboratory

The founding of the National Physical Laboratory (NPL) in Teddington, south-west of London, is probably the most successful

---

1. Frank R. Pfetsch and Avraham Zloczower, *Innovation und Widerstände in der Wissenschaft. Beiträge zur Geschichte der deutschen Medizin* (Düsseldorf, 1973), p. 12.

example in the history of British science in the nineteenth and early twentieth centuries of an initiative solely by individual scientists. The NPL, whose research and development work was extremely important to British industry and the British war effort during the First World War, was founded after a long period of gestation at the end of the nineteenth century. As early as 1871 the physicist Lord Kelvin had suggested founding a large state-run laboratory to do pioneering experimental work in physics, undertake routine measurements and testing of materials, and standardise units of measurement.[2] This idea had evoked no response at the time but was raised again in 1891 by the physicist Oliver J. Lodge at the annual meeting of the British Association for the Advancement of Science in Cardiff. The proposed laboratory was to be a specialist institute outside the university system, like those established or planned in various countries since the 1880s, and was to do both 'pure' and applied research, thus serving science as well as industry. It was to ensure, as the president of the Royal Society wrote a few years later, 'a more systematic application of scientific methods, both in theory and in practice, to our manufactures and industries'.[3]

This time the proposal met with some support. The British Association set up a committee, which included the leading British physicists of the time, to work through the idea of a state-run laboratory, and especially to sort out the question of finance.[4] The committee was fully aware of the difficulties it would face. Many years later Richard Glazebrook, secretary of the committee and the outstanding first director of the NPL, spoke of a feeling of resignation when faced with the task of convincing the government of the necessity for a large physics laboratory. For it was clear from the start that 'without Government aid there were no funds'.[5] *Nature* wrote, along similar lines, of the 'feeling of hopelessness with which [Lodge's] suggestion was received'.[6] For the second time it seemed that the plan to set up a National Physical Laboratory would not progress beyond the stage of vague discussions.

The final impetus towards the creation of a state laboratory was provided by the British Association's annual meeting four years

---

2. Cardwell, *Organisation of Science*, p. 177. Kelvin, for his part, was building on suggestions made by Alexander Strange and Norman Lockyer.
3. President of the Royal Society in his report for the year 1902 (*Nature*, 67, 1902–3, p. 108).
4. Russell Moseley, 'The Origins and Early Years of the National Physical Laboratory: A Chapter in the Pre-history of British Science Policy', *Minerva*, vol. 16 (1978), pp. 224–5.
5. Quoted in *Nature*, 64 (1901), p. 290. Richard T. Glazebrook, *Early Days at the National Physical Laboratory* (Teddington, 1933), p. 3.
6. *Nature*, 58 (1898), p. 565.

later in Ipswich. It was a measure of the growing awareness of technology's dependence on basic research. The president, Sir Douglas S. Galton, who had been the Association's secretary for many years, took this opportunity to repeat the call for the state to set up a British counterpart to the Physikalisch-Technische Reichsanstalt in Charlottenburg near Berlin, which he had visited in the spring of 1895.[7] Galton justified this demand with well-worn arguments, pointing out that the British government had to take a more active part in the science sector and not leave everything to private initiative as it had done in the past. According to Galton, 'our neighbours and rivals' depended 'largely upon the guidance of the State for the promotion of both science teaching and of research',[8] and in his opinion, this contained an urgent lesson for Britain. At about the same time as Galton, who as a civil servant dealt with science organisation, a group of British scientists, members of the Royal Commission on Technical Instruction,[9] visited Germany and pointed to the necessity of setting up in Britain an institute comparable to the Physikalisch-Technische Reichsanstalt.[10] This institute, not attached to a university but a sort of 'auxiliary' institute, to use Wilhelm von Humboldt's term, was financed by the state. No precedent existed for it anywhere in the world at that time. It was founded after protracted preparations in 1887 on the initiative of Werner von Siemens, as a research institute intended to mediate between science and technology.[11] The practical needs of the state, science and industry had made this innovation possible.

Galton's suggestion, which was taken up by the British Association and, a little later, by the Royal Society, initiated the process leading to the foundation of the National Physical Laboratory. It can be described only in broad outline here.[12] Three factors in the

---

7. *Report of the Committee Appointed by the Treasury to Consider the Desirability of Establishing a National Physical Laboratory*, C. 8977 (London, 1898), p. 1.
8. *Report of the Sixty-Fifth Meeting of the British Association for the Advancement of Science* (London, 1896), p. 34.
9. See above, p. 103 in this volume.
10. Philip Magnus et al., *Report on a Visit to Germany, with a View of Ascertaining the Recent Progress of Technical Education in the Country, Being a Letter to his Grace the Duke of Devonshire, Lord President of the Council*, C. 8301 (London, 1896), pp. 8–9. See also Cardwell, *Organisation of Science*, pp. 177–8.
11. Pfetsch, *Wissenschaftspolitik in Deutschland*, pp. 103–28; Walter Ruske, 'Außeruniversitäre technisch-naturwissenschaftliche Forschungsanstalten in Berlin bis 1945' in Reinhard Rürup (ed.), *Wissenschaft und Gesellschaft. Beiträge zur Geschichte der Technischen Universität Berlin 1879–1979*, vol. 1 (Berlin, Heidelberg and New York, 1979), pp. 231–63; Ludolph Brauer et al. (eds.), *Forschungsinstitute. Ihre Geschichte, Organisation und Ziele*, vol. 1 (Hamburg, 1930), pp. 175–7.
12. For details, Moseley, 'National Physical Laboratory', and Edward Pyatt, *The National Physical Laboratory: A History* (Bristol, 1983). See also Howarth, *The British Association*, pp. 168–9; *Record of the Royal Society*, pp. 200–13; Peter Alter, 'Staat und

institute's genesis are of particular interest. Firstly, the British Association's annual meetings provided the setting within which the idea of a non-university physics research institute was promoted, not only in 1891 and 1895 but as early as 1871. These meetings were the largest regular gatherings of scientists in Britain at the time, and could therefore ensure that the initiatives of individuals received wide publicity, sometimes reaching beyond scientific circles. Secondly, plans for the NPL did not begin to take on definite shape until after the mid 1890s, when the Physikalisch-Technische Reichsanstalt in Charlottenburg — 'probably the most complete institute in Europe for physical research'[13] — was introduced into the discussion more persistently as a point of reference by the British science lobby. Planning for the NPL was therefore closely modelled on the Reichsanstalt, which Galton in particular repeatedly pointed to as a suitable precedent to follow. For this reason, it was studied in detail during the NPL's planning phase.[14] According to the president of the Royal Society at the time, the projected British institute's aims and the areas in which it was to work could be described in words 'which are little more than a paraphrase of those used in official documents with respect to the Reichsanstalt'.[15] It seemed that the future NPL and the Reichsanstalt would have identical functions and aims. After the NPL began work in 1900, contact between the two institutes was maintained. The Reichsanstalt's reports after 1900 contain frequent references to visitors from the NPL, as well as from France and the USA, where similar institutes were established early this century: the National Bureau of Standards in 1901 and the Laboratoire d'Essais in 1902. In the first years of this century the British and German institutes had broadly similar functions and worked in more or less the same fields. It will become apparent, however, that this similarity did not extend to their relationship with the state.

Wissenschaft in Großbritannien vor 1914' in Helmut Berding et al. (eds.), *Vom Staat des Ancien Régime zum modernen Parteienstaat. Festschrift für Theodor Schieder zu seinem 70. Geburtstag* (Munich and Vienna, 1978), pp. 369–83.

13. Magnus, *Report on a Visit to Germany*, p. 8.

14. Moseley, 'National Physical Laboratory', pp. 225–6. *Report of the Committee Appointed by the Treasury*, C. 8976, p. 1. See also the annual report of the Physikalisch-Technische Reichsanstalt (PTR): 'On 1 and 2 April 1898, Prof. Rücker, Sir Andrew Noble, Mr Alexander Siemens . . ., members of the National Physical Laboratory Committee, visited the Reichsanstalt on behalf of the British government in order to gain information about its organisation and work' (translated from 'Die Tätigkeit der Physikalisch-Technischen Reichsanstalt in der Zeit vom 1. Februar 1898 bis 31. Januar 1899', *Zeitschrift für Instrumentenkunde*, vol. 19, 1899, p. 206). Richard T. Glazebrook, the NPL's first Director, visited the PTR in 1899, and again in February of 1900.

15. *The Times*, 1 March 1897.

The third factor which emerges in the history of the National Physical Laboratory is the rapidity with which plans for the institute were realised after the British Association's annual meeting in 1895. The fact that the NPL could draw upon the support of the Observatory at Kew, where some of the work that it was to take over was already being done, contributed to this. Galton had pragmatically recommended that the Observatory, financed and run by the Royal Society since 1872, be extended on the model of the Reichsanstalt. This would obviate the high costs necessarily incurred in founding a completely new institution. Another contributing factor in the rapid establishment of the NPL was that the committee set up by the British Association in 1891 to study the project could be reconvened. One year later, at the British Association's annual meeting in September 1896 in Liverpool, this committee made a detailed submission about the organisation, administration and functions of the projected laboratory. Referring to the provisions made for the Reichsanstalt, it made an appeal for state aid on the scale required. 'If England is to keep pace with other countries in scientific progress', ran the time-worn argument, 'it is essential that such an institution be provided; and this can scarcely be maintained continuously on an adequate scale, except as a national laboratory supported mainly by Government.'[16] What was envisaged at this time was a non-recurring grant of between £20,000 and £25,000 for buildings, plus a sum of £5,000 for basic apparatus.

Backed by the broad consensus of the scientific community, especially the physicists,[17] Galton approached the prime minister, Lord Salisbury, at the end of 1896. Lord Salisbury's own scientific interests and the support for public funding of science that he expressed before the Devonshire Commission made him seem a more suitable negotiating partner than any of his predecessors. It was agreed that the prime minister would receive a deputation led by the president of the Royal Society. On 16 February 1897 the deputation submitted plans for the National Physical Laboratory to the prime minister, taking the opportunity to point out the advantages which German industry derived from the work of the Physikalisch-Technische Reichsanstalt. Salisbury's reaction, how-

16. On the Establishment of a National Physical Laboratory, p. 2 (Royal Society Archives, M.C. 16, 335, Annex No. 1, n.d.).
17. 'A Memorial to the Government . . . is now in course of signature by all the leading Physicists, Chemists, Geologists and Engineers in Great Britain, and it is proposed that when ready the Government should be asked to receive the Deputation from the Royal Society and kindred societies and the British Association early in November' (Douglas Galton to the permanent secretary of the Treasury on 16 October 1896, Royal Society Archives, M.C. 16, 330).

ever, was less positive than expected, probably because of the long-term and incalculable financial commitment the state was being asked to make. The prime minister suggested limiting the proposed institution's functions to that of a testing and standardising laboratory, instead of approving the physical research institute which scientists had envisaged.[18]

In this critical situation, the fact that the government did not have a uniform attitude towards the question of the NPL aided the progress of the plans. Some members of the cabinet, Arthur Balfour at their head, were much more positively disposed than the prime minister towards the idea of a physical research institute.[19] The problem, which in the context of the government's total policy was of only marginal significance, was solved in the summer of 1897 by a well-established strategy. In view of the different opinions prevailing in cabinet, the Treasury appointed a committee of enquiry. Its brief was defined precisely:

> To consider and report upon the desirability of establishing a National Physical Laboratory for the testing and verification of instruments for physical investigation; for the construction and preservation of standards of measurement; and for the systematic determination of physical constants and numerical data useful for scientific and industrial purposes — and to report whether the work of such an institution, if established, could be associated with any testing or standardising work already performed wholly or partly at the public cost.[20]

The committee was composed of five Fellows of the Royal Society, the industrialist Alexander Siemens (a nephew of Werner von Siemens) and a representative each from the Board of Trade and the Treasury. At the suggestion of the chancellor of the exchequer, Sir Michael Hicks Beach, the physicist Lord Rayleigh was appointed the committee's chairman. He was a former director of the Cavendish Laboratory in Cambridge and, as Balfour's brother-in-law, had good connections in the highest government circles.[21] The composition of the committee left little doubt as to the

---

18. *The Times*, 17 February 1897.
19. See Young, *Arthur James Balfour*, p. 19.
20. *Report of the Committee Appointed by the Treasury*, C. 8976, p. III.
21. John William Strutt, Baron Rayleigh (1842–1919), Cambridge professor and director of the Cavendish Laboratory 1879–84, professor at the Royal Institution 1887–1905, president of the Royal Society 1905–8, awarded Nobel Prize 1904, chancellor of Cambridge University 1908–19. See Robert J. Strutt, *The Life of John William Strutt, Third Baron Rayleigh* (London, 1924, 2nd edn, Madison, Wisc., 1968), and the obituary in *Proceedings of the Royal Society*, vol. 98 A (London, 1921), pp. I–L.

results of its deliberations. But differences of opinion existed about the extent to which the NPL should pursue basic experimental research. 'Pioneering experimental work, such as Faraday's, for instance, which might lead to great discoveries', in the opinion of Lord Kelvin and other university physicists on the committee should be done by existing university laboratories and individual scientists.[22] In their view, Britain was suffering from a deficit not of 'experimental work' but of 'basic industrial work'. As this idea was ultimately accepted by the circle of experts, the committee of enquiry recommended in its report, published in 1898, that 'the proposed institution should be established at the national expense on lines similar to, though not yet at present on the scale of, the Physikalisch-Technische Reichsanstalt . . . . The possibility of future extension should, however, be kept in view from the first'.[23] The government accepted these recommendations.

Thus the course was set for the establishment of the NPL. The committee of enquiry also developed important principles for the organisational and administrative structure of the new institute. In the following decades the basic ideas behind these principles acquired the status of guidelines for British science policy. Although the laboratory was from the start financed by public money, it was not put under the direct administrative and financial control of a ministry. At the insistence of scientists, it was controlled by an autonomous body in which the state, science and industry (in the form of scientific and technical societies) were represented. What was created, therefore, was not a 'Reichsanstalt', but an institute under private law. When he gave evidence to the committee of enquiry, Oliver Lodge had already been against 'immediate Government supervision — in a Government office, for example'. 'I do not take the word "national" as implying that. I take it as implying permanent and organised existence with national pecuniary aid, but under the immediate government of, say, the Royal Society or other suitable body.' Another expert had expressed the scientists' position even more directly to the committee: he wanted the Laboratory 'to be supported by the State but to be administered by a committee of experts'.[24]

The committee of enquiry's report, accepted in full by the government, therefore recommended that the Royal Society, already involved at many levels in the founding of the NPL, both as a

22. *Report of the Committee Appointed by the Treasury*, C. 8977, p. 47.
23. Ibid., C. 8976, p. 5.
24. Ibid., C. 8977, pp. 35, 38.

Science Policy Decisions after 1900

scientific corporation and through individual members, be invited 'to control the proposed institution, and to nominate a Governing Body, on which commercial interests should be represented'.[25] The organisation statutes, based on these recommendations and drafted early in 1899 by the Royal Society in cooperation with the Treasury and the Board of Trade,[26] placed the NPL under the control of a General Board comprising five representatives of the Royal Society, twenty-four scientists nominated by the Royal Society and twelve representatives of various professional associations. This relatively large body elected from among its members an Executive Committee of twelve, which had direct responsibility for supervising the NPL's work, and appointed a director proposed by the Royal Society. The president of the Royal Society and the permanent secretary of the Board of Trade were *ex officio* members of the General Board and the Executive Committee. In the initial stages of setting up the laboratory, the Executive Committee played an important part; subsequently, the director's influence increased. At all events, this arrangement largely prevented direct state intervention in the administration and the work of the NPL. The statutes, however, could only be changed in agreement with the Treasury.

In 1900 the Laboratory began its work, which was described as being 'to break down the barrier between theory and practice, to effect a union between science and commerce'.[27] In 1902, when it was officially opened, the NPL had a staff of twenty-six. By contemporary standards, it had been conceived on a grand scale. The question of what moved the British government to display such relative generosity towards a scientific institution at this time, can probably best be answered by pointing to the Physikalisch-Technische Reichsanstalt, which was repeatedly cited as a model. In the short period of its existence, this institute had proved its usefulness to German industry. Imitation therefore suggested itself, and scientists gained the government's agreement more easily because setting up the proposed institute did not involve breaking new organisational ground. In addition, the Laboratory's practical orientation meant that it fitted without difficulty into the utilitarian tradition of science support in Britain.

The Physikalisch-Technische Reichsanstalt was established in

25. Ibid., C. 8976, p. 6.
26. The NPL's statutes are printed in *Nature*, 60 (1899), pp. 25–7, and in *National Physical Laboratory: Minutes of the Executive Committee*, vol. 1: *1899–1904*, pp. 1–5. See also the preparatory correspondence in Royal Society Archives, MS 538.
27. The prince of Wales, later to become King George V, at the NPL's opening (*The Times*, 20 March 1902; *Nature*, 65, 1901–2, p. 487).

145

1887, and at the time of the NPL's foundation had a staff of ninety-six, of which thirty-five were scientific staff. The NPL's arrangement seemed extremely modest by comparison, in terms of both staff and finances. The government had coupled its request that the Royal Society take responsibility for the Laboratory with a promise to provide financial support for five years at the rate of £4,000 per annum. But it was made clear that, provided the institute was a 'reasonable success',[28] long-term financial support would be forthcoming. In addition, the Treasury had allocated a non-recurring grant of £12,000 for remodelling the Observatory at Kew and erecting the new buildings required by the NPL. The Royal Society's reply to the Treasury made it clear that these sums did not meet scientists' expectations by a long way, and suggested that the money allocated was insufficient to do more than extend the Observatory at Kew:

> The President and the Council [of the Royal Society], while fully recognising the interest thus displayed by H. M. Government in the project for the establishment of a National Physical Laboratory, cannot conceal from themselves the fact that the sums named would not be sufficient to found a new institution fulfilling all the requirements indicated in the Report of Lord Rayleigh's Committee . . . As, however, it is recommended that the institution should be an extension of the Kew Observatory, the President and Council believe that the proposed endowment will greatly increase the utility and range of the work there done, and they are willing to accept in principle the suggestion of H. M. Government that the Royal Society should undertake the management of the enlarged Kew Observatory.[29]

The Reichsanstalt, it was said, had been set up at a cost of £200,000, and its annual budget was almost £15,000.[30] It was argued that the National Bureau of Standards in the USA[31] had an annual budget of £19,000 and had been founded at an initial cost of £70,000. While scientists continued for years to believe that the NPL was inadequately funded, the government felt that it had introduced a qualitative change into the relationship between science and the state. This was expressed most clearly at the opening of the NPL, when the Prince of Wales said that the state's financial involvement

---

28. Letter from the Treasury to the Royal Society, 9 June 1899 (Royal Society Archives, MS 538).
29. Letter from the Royal Society to the Treasury, 28 November 1898 (ibid.).
30. See the appendix to Douglas Galton's letter to the Royal Society, 21 October 1896 (ibid., M.C. 16, 335).
31. *The National Physical Laboratory: Report for the Year 1903* (London, 1904), p. 9.

in the project made it 'almost the first instance of the State taking part in scientific research'.[32]

In practice, the NPL was financed from a mixture of public and private sources. From the start, therefore, it had a higher budget than the £4,000 granted by the government. It received fees for work done for industry and government agencies. In addition, it received proceeds from the Gassiot Fund which had previously gone to the Kew Observatory. Thus the NPL's total budget for the financial year 1903/4 was made up of the following items:[33]

	(£s)
State subsidy	4,000
Gassiot Fund	426
Fees for meteorological surveys	400
Proceeds of endowments	1,160
Fees etc.	4,214
Total	10,200

'Proceeds of endowments' often included income from investigations carried out for scientific societies and other scientific bodies, and so this figure is misleading. The NPL did in fact receive money from endowments before the First World War,[34] but never as much as some other scientific institutions at this time. Despite all the efforts of the NPL's management, industry rarely made large donations before the First World War. One exception was a donation made by the Liverpool shipbuilder and shipyard owner, Sir Alfred F. Yarrow, who gave the NPL £20,000 in 1908 to set up expensive experimental plant for testing ships' hulls. Mention has already been made of Sir Julius C. Wernher's gift of £10,000 for the buildings of the metallurgical department which was to be set up in the NPL, but this cannot be ascribed to the patronage of British industry, as Wernher was neither British by origin, nor an industrialist.[35] Patterns of private patronage in Britain in the nineteenth and early twentieth centuries show that British industry had a very strange attitude towards an institution which, after all, was intended primarily to serve its interests. Whether the state's part in founding and financing the NPL contributed to patrons' reserva-

32. *The Times*, 20 March 1902; *Nature*, 65 (1901–2), p. 487.
33. NPL Executive Committee memorandum, 5 April 1904 (Royal Society Archives, MS 538).
34. See, for example, the list in *The National Physical Laboratory: Report for the Year 1902* (London, 1903), p. 9, and Table 1 for the years 1900–14 in Moseley, 'National Physical Laboratory', p. 239.
35. *Record of the Royal Society*, p. 203. The metallurgical department's costs totalled £30,000. After Wernher's bequest, the Treasury finally paid £15,000 of this.

tions, or whether these were caused by the Laboratory's semi-commercial character as a fee-charging testing and standardising institution, are questions for which available sources provide no answers.

Even before the first five-year term of government financing expired in September 1904, the Royal Society had pressed for a substantial increase in the annual allocation for the NPL. Scientists added their voices to this appeal in order to lend it more weight. Negotiations beginning at this stage were eased by the fact that many personal contacts had been established between civil servants and scientists as a result of the Treasury's and the Board of Trade's involvement in the founding and administration of the NPL. The Treasury initially reacted evasively to the 'constructive proposals' made at its request by the NPL's Executive Committee.[36] A meeting was thereupon arranged between the prime minister, the president of the Royal Society, the chancellor of the exchequer and the president of the Board of Trade. It took place in August 1904, and while nothing concrete was achieved, it became apparent that the government's attitude was softening. In the summer of 1905 a group of prominent MPs, comprising Richard B. Haldane, Sir John T. Brunner, Sir Joseph Lawrence and Joseph Chamberlain, organised an unprecedented campaign which resulted in a memorandum, signed by 150 MPs, being presented to the then chancellor of the exchequer, Herbert Asquith, in December 1905. Referring to the financial provision made for similar institutes in other countries, it called for 'further and immediate help' for the NPL. Otherwise, the Laboratory would no longer be able to fulfil its function, which was so important to trade and industry: 'Opportunities are lost by delay'. The memorandum suggested increasing the annual subsidy to £10,000, which was the sum considered necessary by the NPL's Executive Committee.[37]

The result of all these efforts on the part of the NPL and the science lobby was that more money was made available for build-

36. See the letter from the Treasury to the Royal Society, 16 November 1903 (*National Physical Laboratory: Minutes of the Executive Committee*, vol. 1: *1899–1904*, p. 179). The Executive Committee's suggestions are listed in 'The National Physical Laboratory: Memorandum on the Future Organisation and Expenditure', 19 February 1904 (Royal Society Archives, MS 538). The Treasury's answer of 21 March 1904 (ibid.), explicitly warns the Executive Committee that 'the expenditure on the laboratory should be rigidly confined within the limits of the income now accruing, and it must be clearly understood that no claim to an increased grant can be found on a deficit already incurred'.

37. 'Memorial Presented by Members of the House of Commons to the Chancellor of the Exchequer, December 1905' in *The National Physical Laboratory: Report for the Year 1905* (London, 1906), pp. 53–5 (quotation on p. 55). See also *Nature*, 73 (1905–6), pp. 512–13.

ings and apparatus. The annual state subsidy, which contributed to day-to-day administrative costs and salaries, was progressively increased to reach £7,000 in 1908. Further expansion resulted in the number of staff rising to 150 in 1912,[38] but financial problems remained. As the Royal Society had to make up the NPL's budgetary deficit, it asked to be released from 'this serious liability';[39] that is, from the administrative, financial and scientific responsibility for the Laboratory. 'While the rapid growth of the Laboratory had been a matter of great satisfaction', the Royal Society's treasurer explained to the General Board of the NPL, '. . . and had clearly demonstrated the national importance of the work of the Laboratory, the considerable extension of its operations had rendered the financial position of the Laboratory in relation to the Royal Society a matter of no little difficulty.'[40] The Royal Society began negotiating with the government about this problem in mid-1913. It was finally solved in April 1918, when the administration of the NPL was taken over by the newly created Department of Scientific and Industrial Research (DSIR). Scientific supervision of the NPL, however, remained the responsibility of the Royal Society.[41]

Imperial College of Science and Technology

Apart from the new universities and university colleges founded in the industrial areas of England, the National Physical Laboratory and Imperial College of Science and Technology in London were the most important innovations in British science organisation before the First World War. Plans for a technical university in the capital can be traced back as far as the 1850s. The Eidgenössische Technische Hochschule founded in Zurich in 1855 was frequently held up by scientists and education reformers as a model for any proposed British institution. Deliberations on how to use the profit made by the 1851 Great International Exhibition in London pro-

---

38. By the end of 1919 the NPL had a staff of 600 (*Nature*, 104, 1919–20, p. 196).
39. 'The Anniversary Meeting of the Royal Society', *Nature*, 92 (1913–14), p. 404. See also Moseley, 'National Physical Laboratory', p. 248.
40. Minutes of proceedings at the Meeting of the General Board, 25 April 1913 (DSIR, 10/2, PRO).
41. For details, see Erich Hutchinson, 'Scientists and Civil Servants: The Struggle over the National Physical Laboratory in 1918', *Minerva*, vol. 7 (1969), pp. 373–98. On the subsequent history of the NPL see Russell Moseley, 'Government Science and the Royal Society: The Control of the National Physical Laboratory in the Inter-War Years', *Notes and Records of the Royal Society of London*, vol. 35 (1980–1), pp. 167–93.

vided the starting point for a number of projects aiming to set up an 'Industrial University' or a 'University of Mines and Manufactures'. A few years later several associations were formed, largely on the instigation of Lyon Playfair, to agitate for the creation of a technical college. One, for example, was set up in 1870 under the chairmanship of the lord mayor of the City of London.[42] In 1871 the Devonshire Commission recommended that several existing small institutions in South Kensington be amalgamated into one large scientific and technical teaching and research establishment.

None of these early suggestions for a 'great central technological institution' in London, all making stereotyped references to foreign models, progressed beyond the stage of general discussion until the turn of the century. A correspondent in *Nature* expressed surprise that 'this country, foremost as it has always been in matters of engineering enterprise, should be so behindhand in the systematic education of its engineers, there being no establishment in England devoted to that object which is recognised by the profession'.[43] While engineers had long since received a university training in Germany, Switzerland, the Netherlands and in the USA (Massachusetts Institute of Technology, 1865), a comparable education was not available in Britain even at the end of the nineteenth century. Since the eighteenth century Britain had become the leading industrial nation of the world without having need of a developed system of technical education at university level with set courses of study, examinations and titles. The men whose inventions, technical innovations and know-how had provided the basis for the reputation of British production processes and industrial products were, as a rule, autodidacts, trained mechanics or skilled artisans, in short, men with practical workshop experience. In the early phase of industrialisation and beyond, technical knowledge was gained by direct experience of the industrial economy, by a long apprenticeship in the workshop; it was not the result of a systematic scientific training in state research institutes or universities. This practice resulted in the apprenticeship system being valued highly, and also produced a negative attitude towards a primarily theoretical education for engineers. It remained dominant in Britain until well past the middle of the century, despite the fact that new technologies had radically changed the character of the industrial economy.

As late as 1910, for example, the Board of Education in London

42. Cardwell, *Organisation of Science*, pp. 118–19.
43. *Nature*, 16 (1877), p. 44.

noted with dismay 'the small demand in this country for the services of young men well trained in the theoretical side of industrial operations and in the services [= sciences?] underlying them. There still exists amongst the generality of employers a strong preference for the man trained from an early age in the works, and a prejudice against the so-called "college-trained" man'.[44] British education reformers, too, emphasised increasingly that industrialists' adherence to historical precedent was becoming a grave handicap for Britain as an industrial nation in competition with other countries in the second half of the nineteenth century. 'Our engineers have no real scientific instruction', the Taunton Commission reported as early as 1868, 'and we let them learn their business at our expense by the rule of thumb'.[45] To the science lobby, it was obvious that the entrepreneurs bore some of the blame for this situation: 'German and American manufacturers believe in technical education, while many of their competitors in this country are still blind to its advantages'.[46]

The new university colleges in the industrial centres of the Midlands and the north of England, which established scientific and technical training courses in close cooperation with industry, to some extent relieved this unsatisfactory situation, which was seen as a growing problem towards the end of the nineteenth century, when new research-dependent industries developed. Engineering as well as chemistry and physics were taught at almost all of the newly founded colleges. Owens College in Manchester in particular achieved a leading position in these disciplines.[47] After the University of Glasgow established the first chair of engineering at a British university in 1840, engineering departments were set up at University College London and King's College London. They attracted relatively large numbers of students, especially during the 1890s, but they never reached the levels achieved by the German *Technische Hochschulen*.[48] The teaching done in university engineering departments, however, frequently seems to have suffered from a lack of practical orientation. Sidney Webb, writing in *The Times* on 'The Organisation of University Education in the Metropolis', argued that 'the most serious deficiency in the London

44. *Report of the Board of Education for the Year 1908–1909*, Cd. 5130 (London, 1910), p. 90.
45. *Report of the Schools Inquiry Commission*, vol. 6 (London, 1868), p. 629.
46. *Nature*, 68 (1903), p. 274. On this see Wiener, *English Culture*, pp. 127–54.
47. See Sanderson, *Universities*, p. 90; Anthony Sampson, *The New Anatomy of Britain* (London, 1971), pp. 168–9.
48. Sanderson, 'The University of London', pp. 246–7; Ashby, 'On Universities', p. 469; W. O. Skeat, *King's College London Engineering Society 1847–1957* (London, 1957).

faculty of science is not the inadequacy of the instruction for the science degree, but the lack of anything like adequate provision for chemical, physical, and biological technology, or the application of science to industrial processes . . . . Of public provision for instruction in scientific technology there is practically none'.[49]

In Britain, university science and engineering departments and faculties had their place within the larger framework of established academic subjects and thus avoided the division between universities and *Technische Hochschulen* which existed in the German system. In addition to these departments, there were several older institutions in Britain that either developed along the lines of the continental European *Polytechnische Hochschulen*, or specialised in a limited number of subjects. An example of the first type is the College of Science and Technology which was founded in Glasgow in 1796. In 1913 it was affiliated with the university, and in 1964 became part of the University of Strathclyde. Examples of the second type are the Royal School of Mines and the Royal College of Science, both in London.

In 1901 Sidney Webb, a co-founder of the Fabian Society and long-serving chairman of London County Council's influential Technical Education Board, had deplored the lack of a large institution for applied technical disciplines in London. This was a signal that earlier plans, all more or less aiming to set up a technical college in the capital, were being revived. Webb wrote that all those 'who are interested in the great mining enterprises of South and West Africa, America, and Australasia' must consider 'whether the time has not come for the establishment of a distinct school of metallurgy and mining . . . . In applied chemistry, too . . . practically nothing in the nature of a school of chemical technology exists in the metropolis'.[50] A second article a few days later named a third area in which the capital lacked facilities for teaching and research: 'The position of London as the capital of a vast Empire, the centre of organisation for important engineering enterprises all over the world . . . seems to call for a considerable extension of the scope and variety of instruction in civil engineering'.[51] Thus the range of subjects to be covered by a future technical college in London was laid down. By and large, it was the same as that taught in a continental European *Technische Hochschule*. But at this time, nothing had yet been said about how these plans were to be realised.

Exactly one year later Webb took up again the subject he had

49. *The Times*, 4 June 1901.
50. Ibid.
51. *The Times*, 8 June 1901.

already aired in *The Times*. The university situation in London had changed fundamentally in 1900, when the University of London had become a teaching institution, and it now had more in common with academic life in other European capitals. In a long essay in the journal *The Nineteenth Century and After*, Webb pointed out that despite all the reforms which had been introduced, London still lacked an 'institute of scientific technology adapted to postgraduate work and the experimental application of science to industrial processes'.[52] Webb, who since 1900 had also been a member of the Senate of the University of London, developed a detailed plan for the organisation, financing and functions of the proposed college. At the same time, he coined the phrase which dominated the discussion of reform for several years: 'What London University wants . . . is, to put it briefly, a British "Charlottenburg" — an extensive and fully equipped institute of technology'.[53] Founded in 1879, the Königlich Technische Hochschule zu Berlin in Charlottenburg[54] had almost come to symbolise the scientific supremacy of Germany, Britain's political rival in Europe, and from now on it inspired the planning for a similar institution in London.

But Webb's articles were not the decisive factor which led to concrete preparations being made, however strong their public impact might have been. The role of catalyst was taken by Lord Rosebery, one of the most widely known British politicians of the time, a former prime minister and foreign secretary, and first chancellor of the reorganised University of London. The project progressed beyond the stage of informal discussions as a result of his efforts. At the end of June 1903, exactly a year after Webb's last comments on the university situation in London, Rosebery sent an open letter to London County Council, the authority responsible for London's technical education and whose Technical Education Board was chaired by Webb. Referring to the *Technische Hochschule* at Charlottenburg Rosebery suggested in this letter that it was 'little short of a scandal that our own able and ambitious young men, eager to equip themselves with the most perfect technical training, should be compelled to resort to the Universities of Germany or the United States'. It was not right, he wrote, that talented students from Canada or Australia, South Africa or India 'should be unable to find within the Empire the educational opportunities that they

52. Sidney Webb, 'London University: A Policy and a Forecast', *The Nineteenth Century and After*, vol. 51 (1902), p. 927.
53. Ibid., p. 928.
54. Reinhard Rürup (ed.), *Wissenschaft und Gesellschaft. Beiträge zur Geschichte der Technischen Universität Berlin 1879–1979*, 2 vols. (Berlin, Heidelberg and New York, 1979).

need'. The time had come 'for making London, at any rate so far as advanced work in scientific technology is concerned, the educational centre of the Empire'.[55]

Rosebery's answers to important questions concerning the financing, accommodation and academic status of the planned institution suggest that his public appeal had been carefully planned. He disclosed, for example, that Wernher, Beit & Co. had offered 'to place a large sum of money in the hands of trustees'. This endowment, setting an example which would be followed by other 'public-spirited London citizens', would allow a college to be built and equipped. The offer was made by two of Britain's most important patrons in the early twentieth century, whose patronage has been discussed above.[56] The company through which the endowment was offered was a financial holding in which Sir Julius C. Wernher and Alfred Beit had invested their extensive business interests. Their offer of an endowment for a technical college in London was, however, conditional upon the London County Council contributing £20,000 per annum to its budget. Rosebery's letter was designed to secure an undertaking to provide this money.

In organisational terms, the proposed college was to be a 'school' of the University of London, making it a fully recognised member of the university. As a school under the administrative roof of the university, it would be guaranteed a large degree of autonomy. Rosebery added that attaching it to the university would also contribute towards developing 'the University in such a fashion as to make it worthy to be the University of the metropolis of the Empire'.[57] The question of where the limits of the college's autonomy should be drawn, or whether complete independence would not after all be preferable, sparked off a lengthy controversy in the further development of Imperial College. It will be discussed in detail here, because it throws light on several general aspects of technical education in Britain in the nineteenth and twentieth centuries.

The positive response to Rosebery's appeal to London County Council materialised as expected. 'The magnificent proposals which Lord Rosebery laid before the County Council', wrote *Nature*, 'have roused feelings of keen interest and high hopes in many who, for years past, have been crying, as it seemed, in the wilderness ... to improve our higher technical educational

---

55. Lord Rosebery to Lord Monkswell, 27 June 1903 (Imperial College Archives, ABC/4/1). The letter was published by *The Times* on 29 June 1903.
56. See above, p. 48–9 in this volume.
57. Lord Rosebery to Lord Monkswell, 27 June 1903.

methods.'⁵⁸ *The Times* published Rosebery's letter and expressed the hope that the project would be realised quickly, praising 'the great firm of Wernher, Beit, and Co., true to the traditions of Imperialism and of educational progress, which were dear to Mr Rhodes', for the endowment it was prepared to make.⁵⁹ The link between imperialism and the functions of the proposed college, openly expressed by Rosebery and *The Times*, gives some indication of those who supported the idea of an 'Imperial College' for engineering, and of their motives in doing so. The following discussion will concentrate primarily on three questions. First, how did Rosebery's letter to the London County Council, containing precise and detailed suggestions about the proposed institutions's financing, accommodation and academic status, come to be written? Secondly, who initiated and supported the project for which Rosebery was the public spokesman? And thirdly, what was the government's role in the founding of Imperial College?

A lack of sources makes it difficult to reconstruct the steps taken and stages passed through in the two years between the publication of Sidney Webb's articles in *The Times* and Rosebery's appeal. But it seems certain that the suggestion to reconsider the idea of establishing a technical college in London was made by Webb shortly after the turn of the century, although Haldane was usually given the credit for it while Webb remained silent about the part he had played in initiating the project.⁶⁰ Webb's articles in *The Times* and *The Nineteenth Century and After*, and his chairmanship of London County Council's Technical Education Board all point to him as the originator of the plan for an Imperial College. Drafts of letters and memoranda which after 1901 settled important issues to do with the future college were generally written by Webb, while Haldane appeared in public as the driving force behind the plans for Imperial College. Rosebery's letter to the London County Council, too, was drafted by Webb at Haldane's request. 'Please say to S[idney]', wrote Haldane to Beatrice Webb early in 1903, 'that I should be glad to have his draft of the proposed letter from Rosebery to the L.C.C. — proposing the Charlottenburg

58. 'A Charlottenburg Institute for London', *Nature*, 68 (1903), p. 203.
59. *The Times*, 29 June 1903.
60. See Sidney Webb, 'The London Charlottenburg', *University Review* (October 1906), pp. 13–24; Beatrice Webb, *Our Partnership*, p. 268. Sanderson calls Imperial College a 'brain-child of Haldane' (*Universities*, p. 204). Along the same lines Crowther, *Statesmen of Science*, pp. 273–4. After Rosebery's letter was published on 27 June 1903, Haldane received a letter which stated: 'I do congratulate you . . . on the fruition of your scheme for the foundation of an Imperial Technical College' (Selborne to Haldane, 29 June 1903, R. B. Haldane Papers, MS 5906). See also Ashby and Anderson, *Portrait of Haldane*, p. 56.

Scheme.... I will then settle its form, agree it with Rosebery, and it can be sent in [February]'.[61] According to Beatrice Webb, Sidney Webb was 'the ideal draftsman'; he was 'admirable as a social engineer' and had been 'a "behind the scenes" man' all his life.[62] The close cooperation between Haldane and Webb — in Haldane's words, 'an admirable man to work with'[63] — continued until Imperial College was successfully launched in 1907.

It cannot be established today whether Webb approached Haldane with the project or whether Haldane seized upon Webb's plans without prompting. The two men, almost exact contemporaries, had long been friends and had cooperated in the reorganisation of the University of London, so that we can safely assume that they discussed the desirability of a technical college for London at an early date. Still, it is worth noting that Haldane visited the *Hochschule* in Charlottenburg, possibly even at Webb's suggestion, in April of 1901, a few weeks before Webb's two articles appeared in *The Times*.[64] On his return to England Haldane described the German institution as 'by far the most perfect University I have ever seen'.[65] After his visit Haldane identified completely with the project to set up a similar institution in Britain. In May of 1902 Haldane called it 'our big scheme' in a letter to Sidney Webb; shortly thereafter he was already speaking of 'my London University Scheme' and 'my Charlottenburg plans'.[66] The early genesis of Imperial College seems to confirm the characterisation of Haldane in a recent study, made with reference to his work in education and science policy-making: 'He would become infected with an idea, often an idea got from someone else; for he was not an imaginative innovator'.[67] But to this rather harsh judgement should be added that Haldane was an imaginative and efficient organiser and lobbyist who possessed the talent to translate an idea into a political decision.

In this light, Rosebery's role in the foundation of Imperial College seems less important than it initially appeared. His social

---

61. Haldane to Beatrice Webb, 30 January 1903 (Passfield Papers, Section II.4.b.4).
62. Beatrice Webb, *Our Partnership*, p. 6.
63. Haldane to his mother, 12 February 1903 (R. B. Haldane Papers, MS 5969).
64. On this see Ashby and Anderson, *Portrait of Haldane*, p. 49. Webb wrote the articles in March of 1901 (Richard B. Haldane, *An Autobiography*, London, 1929, p. 91).
65. Quoted in Ashby and Anderson, *Portrait of Haldane*, p. 45.
66. Haldane to Sidney Webb, 9 May 1902 (Passfield Papers, Section II.4.b.36). Haldane to his mother, 25 June 1903 (R. B. Haldane Papers, MS 5969). Haldane to his sister, 2 August 1903 (ibid., MS 6010).
67. Ashby and Anderson, *Portrait of Haldane*, p. 2.

and political position allowed him to serve as a 'frontman', organised and encouraged by Haldane, in order to lend the public appeal to London County Council some necessary political weight. Rosebery himself was not uninterested in the project for a 'London Charlottenburg', but his papers, especially his correspondence with Haldane, do not reveal a particularly strong commitment to its realisation. The 'London Charlottenburg' was simply one among many projects with which he was involved to some degree, mostly superficially.[68] Rosebery took on only honorary functions, as he had done in the founding of the London School of Economics and in the reorganisation of the University of London: he signed appeals or accepted the position of sponsor or — as in the case of Imperial College — the chairmanship of boards of trustees set up to administer endowed capital. The only part he played in writing the letter quoted above was to sign it. Shortly before it was published, Haldane wrote to Rosebery: 'The private circular containing the University of London Scheme is now in type. It has been drawn by Webb, Principal Rücker [University of London], Myself and revised by Arthur Balfour [the prime minister]. It states that you have consented to be Chairman of the Trustees . . . . I don't think, unless you wish it, that I need trouble you with the draft, for I have already told you all that is in it'.[69] Rosebery at that time possessed much greater political influence than Haldane, who was only an MP, and Rosebery's cooperation meant that a connection between science and politics could be established more easily than would otherwise have been the case. His lack of personal involvement is not, therefore, any reason to criticise him. On the contrary: the fact that he liberally allowed his name to be associated with certain initiatives in science organisation meant that their chances of realisation were increased, and in addition, that patrons were more easily found. Rosebery had proved his worth in this respect for the London School of Economics.[70] The value of his patronage for science at the turn of the century should therefore not be underestimated.

Because the main subject of this study is the beginnings of a public science policy in Britain, it is unnecessary to reconstruct in full the process which culminated in the opening of Imperial College in 1908. But three aspects deserve particular mention:

---

68. See Searle, *National Efficiency*, p. 80: 'Rosebery was a wealthy landed aristocrat, with a traditional Eton–Christchurch background, a classical education and a passion for horse-racing: superficially, the very type of "amateur-gentleman" he spent so much time in denouncing.'
69. Haldane to Rosebery, 26 May 1903 (Rosebery Papers, MS 10 030).
70. See above, pp. 46 and 59 in this volume.

(1) As in the case of the National Physical Laboratory a few years earlier, Imperial College was founded on private initiative. No precedent existed for a technical college in this form and on this scale anywhere in Britain. Webb, Haldane and supporters of the project such as Arthur Balfour, Sir Francis Mowatt, permanent secretary to the Treasury,[71] may have held public office or seats in Parliament, but in the early phase of preparations for Imperial College, they did not act in their official capacities. The initiators of the 'Charlottenburg Scheme' were well aware from the start that a college could not be founded on the projected scale without state involvement, which was therefore sought right from the beginning. Haldane's wide political connections and legal experience allowed him to play a crucial part in these negotiations. The planners of Imperial College were aided by the fact that after 1901 attempts were being made to affiliate the Royal College of Science in South Kensington, run by the Board of Education, with the University of London, and to reform the neighbouring Royal School of Mines, many of whose graduates were employed in South African gold mines. One aim of the proposed reform was 'to make the School an Imperial one which should grant an Imperial degree in mining'.[72] These considerations, which reveal that the government, too, was uneasy about the state of higher technical education in London, provided a starting point for the advocates of a new technical college for the whole of the empire. Thus contacts were established relatively quickly with the Board of Education. Haldane's most important negotiating partner in the government was Robert Morant, dynamic permanent secretary to the Board of Education,[73] and according to Beatrice Webb, 'the one man of genius in the civil service . . . [who] has done more to improve English administration than any other man'.[74] Morant, who had been in contact with Sidney Webb since 1902, supported in an exceptionally constructive manner the proposal for a technical college in London. Immediately after Rosebery's letter was published, Morant suggested amalgamating all the existing institutes in

71. Sir Francis Mowatt (1837–1919), permanent secretary to the Treasury 1894–1903, thereafter an alderman of the London County Council and a member of the Senate of the University of London. Haldane called him 'one of the largest-minded officials I have ever come across' (*An Autobiography*, p. 144).
72. 'Original Proposal to Reorganize the Royal School of Mines, South Kensington', 18 December 1902 (Ed. 24/529/3, p. 7, PRO).
73. Robert L. Morant (1863–1920), author of the 1902 Education Act, permanent secretary to the Board of Education from 1903 to 1911, and first secretary to the Ministry of Health from 1919 to his death. Bernard M. Allen, *Sir Robert Morant: A Great Public Servant* (London, 1934).
74. Beatrice Webb, *Diaries 1912–1924*, ed. by Margaret I. Cole (London, 1952), p. 98.

South Kensington which had any connection with engineering, instead of creating a new institution. This was the basis on which the London Charlottenburg subsequently developed.[75]

Morant's suggestion set the course for further developments until 1907, which cannot be discussed in detail here. The Board of Education's records show clearly that the project of a large technical college for advanced students from the whole of the empire found more support among the government than had any previous initiative in science organisation. Imperial College was supported not only by successive presidents of the Board of Education before its foundation,[76] but also by the prime minister, Balfour, who was known to have scientific interests, and by the Colonial Office, led until September 1903 by Joseph Chamberlain. The King himself contributed £100 to the fund,[77] probably because of the project's political aspects, and later repeatedly demonstrated his interest in a 'great Imperial University worthy of London'.[78] The Treasury, at most, had a more ambivalent attitude. While it initially supported the project, Austen Chamberlain, chancellor of the exchequer from September 1903, expressed concern about the high level of expenditure on London's academic institutions compared with the new provincial universities:

> I happened to see the C[hancellor] of the Ex[chequer] last night after you left me; and I find him entirely indisposed to cede anything at S[outh] K[ensington] to Charlottenburg. So I fear I must withdraw what I said to you so far as the official Treasury is concerned. His view is that

---

75. Memorandum by Morant, July 1903 (Ed. 24/529/3, p. 1). See also Morant's letter to the Treasury, dated 25 October 1903: 'I think it would be a great mistake to attempt any reorganisation . . . without *at the same time* endeavouring seriously to bring into cooperative existence a combined Technological Education Scheme in the Metropolis on the lines suggested in the Rosebery Scheme' (Ed. 24/529/3).

76. This office was held by the Marquess of Londonderry from July 1902, Augustine Birrell from December 1905, Reginald McKenna from January 1907 and Walter Runciman from April 1908.

77. See Arthur W. Rücker to Robert L. Morant, 7 February 1904 (Ed. 24/529/2). See also Haldane, *An Autobiography*, p. 142: 'I had come into a good deal of contact with the King in the course of the negotiations for the establishment of the Imperial College of Science and Technology . . . . King Edward had accepted the view that it was the desire of his father, Prince Albert, that the valuable site of the great Exhibition should be used in part for the establishment of some such institution for higher technical education . . . . He was very helpful and was instrumental in procuring for us the grant of the requisite land from the Exhibition Commissioners.' At the end of 1904 Haldane wrote to his mother: 'The King sent for me and spoke to be about my [sic!] College of Science in South Kensington in which he is keenly interested' (quoted in Dudley Sommer, *Haldane of Cloan: His Life and Times 1856–1928*, London, 1960, p. 140).

78. Sir Francis Knollys, private secretary to the king, to the president of the Board of Education, 25 February 1906 (Ed. 24/530).

London has got too much already in its University (I mean too much out of the taxpayer); and that if we do anything more for it, we shall be compelled to give the other Universities as much. He insists also that Charlottenburg — however constituted — is a local and not an Imperial concern, which means that if we help it we must do more for the other local concerns.[79]

In March of 1905, however, Chamberlain agreed to the incorporation of the Royal School of Mines and the Royal College of Science into Imperial College, but his reservations concerning finance remained.[80] Nevertheless, the stage which preparations had reached by this time was lauded in the press as a personal triumph for Haldane, although — as Eric Ashby and Mary Anderson rightly point out — 'the credit deserved to be more widely distributed: to Morant, whose energy drove the scheme forward from within the Board of Education, to Mowatt, whose goodwill helped to ensure treasury support, and to Webb, whose influence on the London County Council and the senate [of the University of London] was critically important'.[81]

(2) As in the case of the National Physical Laboratory, it has become apparent that a German institution once again provided British science lobbyists and education reformers with a model. British plans aimed to set up an institution in London 'after the lines of the world-renowned polytechnic at Charlottenburg which represents the acme of technical education'.[82] Since the beginning of the century the word 'Charlottenburg' had in Britain come to stand for the whole system of technical education in Germany, so much so that British engineers and others could, from time to time, refer critically in public to a widespread 'Charlottenburgitis' among education reformers.[83] The difference in the numbers of students enrolled in engineering subjects in Germany and Britain alone made it clear that higher technical education needed to be thoroughly reformed in Britain, and especially in the capital of the empire. In the whole of the United Kingdom there were just under 4,000 engineering students in 1901, while in Germany at the same time, Charlottenburg alone had 3,500 enrolled students (excluding

79. George Murray, permanent secretary to the Treasury, to Robert L. Morant, 31 January 1904 (Ed. 24/529/2).
80. Austen Chamberlain to Lord Londonderry, 27 March 1905 (Ed. 24/530).
81. Ashby and Anderson, *Portrait of Haldane*, p. 56.
82. *Nature*, 68 (1903), p. 203 ('A Charlottenburg Institute for London').
83. See Manfred Späth, 'Die Technische Hochschule Berlin-Charlottenburg und die internationale Diskussion des technischen Hochschulwesens 1900–1914' in Rürup (ed.), *Wissenschaft und Gesellschaft*, vol. 1, pp. 194, 197–8.

occasional students).[84] Germany's rise to industrial great-power status was attributed to its universities and *Technische Hochschulen*, rapidly expanding since the end of the nineteenth century, and the most famous of them were located in the Reich's capital. 'Not only for Prussia and Germany', was the self-confident tone of a *Festschrift* published in 1906 for the fiftieth anniversary of the Verein Deutscher Ingenieure, 'but for all civilised nations, the *Hochschule* at Charlottenburg has become an intellectual centre, a much envied model, a focal point of technical progress which sends out and receives intellectual impulses to and from all corners of the world.'[85]

But to what extent were British plans really modelled on the highly praised *Hochschule* at Charlottenburg? And to what degree was the German model of technical education adopted, imitated and developed by universities in Britain? On closer examination, the discussion of Germany's higher technical education system, which began in Britain after the turn of the century, gives the impression that in general little was known of the real model. The constitution, organisation and functions of the German *Technische Hochschulen*, and the courses they offered, were known in broad outline in Britain from reports and visits, but understanding rarely went deeper than this and it hardly provided a basis for detailed imitation. Right from the start, despite all references to 'Charlottenburg', there had been no intention simply to make Imperial College a copy of the *Hochschule* at Charlottenburg. Haldane and Webb were aware that the highly praised German model could be transferred to another social system only in a modified form, and that it would have to be adapted to prevailing conditions — in other words, that the model could not serve as a blueprint but only as a stimulus and a challenge. Haldane in fact issued several explicit warnings against any attempt to imitate the German *Technische Hochschulen* uncritically in Britain. Even *Nature* questioned their suitability as models:

> It is by no means certain that these rapidly growing and in some respects very successful [technical] universities are models to be followed. There is at least a danger of over-specialisation in the German system — a tendency to produce machine-made men possessing a narrowly specialised knowledge and capacity, but without sufficient width and individu-

---

84. See *Nature*, 68 (1903), p. 274, and Rürup (ed.), *Wissenschaft und Gesellschaft*, vol. 1, p. 572.
85. Quoted in and translated from Reinhard Rürup, 'Die Technische Universität Berlin 1879–1979: Grundzüge und Probleme ihrer Geschichte' in *idem* (ed.), *Wissenschaft und Gesellschaft*, vol. 1, p. 20.

ality. Our captains of industry have to control men, as well as to perfect processes.[86]

The problem of to what extent British developments should follow Germany's lead in technical education, or at least be guided by it, was also dealt with by the departmental committee set up by the government in response to the discussion in 1904 on reorganising technical education in London. It began work under Sir Francis Mowatt, former permanent secretary to the Treasury, and following his resignation at the beginning of 1905, it was chaired by Haldane. After surveying technical education in Germany and other Western European countries as well as in the USA, the committee developed several differentiated proposals for the proposed institution in London. Referring to the specific historical and social conditions in which the German model of technical education had been created, the committee did not recommend its wholesale adoption in Britain: 'There are certain elements in the German system which cannot, however admirable in themselves, be reproduced in this country'. The final report suggested reasons why the German model was not transferable:

> Thus we cannot require at present, as a condition of admission to our higher technical colleges, the same standard of general education as in Germany . . . . Nor can we instantaneously compel ourselves to feel the German love of education for its own sake. Neither could we, even if we desired, imitate the uniformity of the German standard, or, generally, the thoroughness and singleness with which all forms of education are regulated by and made subservient to the purposes of the State. On the other hand, there are certain elements in the German system which can, and in our opinion undoubtedly should, be transplanted here.

The elements to be taken over were, however, discussed only in very general terms:

> We have no hesitation in proclaiming this country's urgent need of a greatly increased provision for education in the Sciences applicable to industry, of a University grade, and — so far as consistent with the predominant aim — of a University type, an education concerned with principles rather than with processes, and advancing to the highest planes of specialisation and research.[87]

---

86. 'The Position of Technical Instruction in England', *Nature*, 88 (1912–13), p. 320.
87. *Final Report of the Departmental Committee on the Royal College of Science, etc.,*

The committee concluded, after two years of work, that a close examination of the German model would be useful, but that a technical college could only sensibly be established in London if, while partially drawing on the German model, it fitted into the British context and fulfilled the specific functions demanded by the situation. One result of these considerations — perhaps the most important of all — was that Imperial College of Science and Technology was not created as an institution *sui generis* and separate from the University of London, but was attached to it almost as a technical faculty, much as Rosebery had originally suggested. A separation between universities and technical colleges was thus avoided in Britain, thanks largely to its university tradition; to the present day the engineering sciences are more or less integrated into the universities.

The process by which the British side unmistakably distanced itself from the German model as preparations progressed was reflected in the search for a suitable name for the new institution in London. While in the early discussions and the correspondence of government departments it was almost always referred to only as 'Charlottenburg' or the 'new Charlottenburg', different names begin to appear for it after Rosebery's letter was published in 1903. They illustrate that finding a name for the new institution proved to be difficult. Apart from 'Charlottenburg', often coupled with 'London', 'English' or 'British', more neutral names such as 'the new Technological Institution', 'the New Institution', 'the proposed College of Technology', or 'the new Technological College' were common. The numerous versions of the royal charter drawn up after the beginning of 1906 referred to it as the 'New Institution'.[88] But such an unspecific name was generally felt to be unsatisfactory, 'as the words "New Institution" certainly cannot convey much to the mind of anyone when talking about it'.[89]

The name 'Imperial College of Technology' appears for the first time in documents after the departmental committee appointed by the government in 1904 had published its report in 1906. Drafts of the royal charter from early in 1907 in the Board of Education's records use the plural forms 'Imperial Colleges of Science and

---

Cd. 2872 (London, 1906), p. 22. On the Committee see also Ashby and Anderson, *Portrait of Haldane*, pp. 54–5. In general, see Michael Argles, *South Kensington to Robbins: An Account of English Technical and Scientific Education since 1851* (London, 1964); A. Rupert Hall, *Science for Industry: A Short History of the Imperial College of Science and Technology and its Antecedents* (London, 1982).

88. See the documents in Ed. 24/531.

89. Letter from the London County Council to Robert L. Morant, 26 July 1906 (ibid.).

Technology', 'The Imperial Colleges' or 'The Royal Schools of Science and Technology', or, less frequently, 'Imperial Technical College' or 'King Edward VII Technological College'.[90] Rumours that the new institution was to be called 'Royal Institute of Technology' led the Board of Education to produce a memorandum dealing in detail with the question of a name. It suggested including the words 'College' and 'Imperial' in the name of the new institution. 'Imperial College of Science and Technology' appears for the first time in a marginal note made in this memorandum by Morant, and it was accepted in the royal charter of 8 July 1907.[91]

Before this, however, the proposed name, which not only stated an intention but also made a claim, provoked very different reactions from the British universities. The vice-chancellor of Oxford University approved of it because it seemed to him to show that 'the Institution is not to be one of merely local or even metropolitan scope and importance, but one for the whole Empire'.[92] The new provincial universities, by contrast, were more negative in their judgements. They themselves had strong engineering departments and obviously felt that their academic status would be threatened by the new college in London: 'We fear that the adjective "Imperial" . . . may create an impression, especially in the Colonies and elsewhere outside the United Kingdom that there is some special state-recognition peculiar to this Institution, such as we hope is not contemplated'.[93] The University of London made similar objections, but the government insisted on retaining the imperial reference in the new college's name. Whether or not expectations that it would be a college for the whole empire were subsequently fulfilled is another matter. In 1911 it had just under 900 students: 24 per cent were born in London, 45 per cent in England outside London, 7 per cent in Scotland, Ireland or Wales, 11 per cent abroad and 13 per cent in the British colonies or dominions.[94]

(3) Turning the idea of a 'London Charlottenburg' into reality

90. See the drafts and letters in Ed. 24/531 and Ed. 24/540.
91. Memorandum dated 12 March 1907 (F. G. Ogilvie to Robert L. Morant, Ed. 24/540). *The Imperial College of Science and Technology. Charter of Incorporation*, Cd. 3625 (London, 1907).
92. Vice-chancellor Warren to Robert L. Morant, 18 May 1907 (Ed. 24/530).
93. Vice-chancellor of the University of Birmingham to the president of the Board of Education, 5 June 1907 (ibid.). Similarly, the Universities of Leeds, Manchester and Liverpool pleaded for 'some more appropriate and less exclusive name than that of "Imperial"' (vice-chancellor of the University of Leeds to the president of the Board of Education, 5 June 1907, ibid.).
94. *Fourth Annual Report of the Governing Body of the Imperial College of Science and Technology* (London, 1912), pp. 41–3.

required, in addition to the political will to do it, considerable sums of money. As with other initiatives in science organisation at this time, the money did not come from a single source; Imperial College was not supported solely by the state or by London's municipal administration. In its mixed financing from both public and private sources, Imperial College resembled the new university colleges in the Midlands and the north of England and differed greatly from the *Technische Hochschulen* in Germany, which were financed exclusively by the state. A Board of Education memorandum briefly summed up this typical feature of British university financing: 'The New Institution, as at first established, must, in the Nature of things, be in part local, in part National, in part Imperial, and it must owe its establishment to Imperial, local, and private funds.'[95]

Advocates of a British 'Charlottenburg' knew what sorts of amounts were necessary to get this project off the ground. Sidney Webb estimated that initial capital of half a million pounds was necessary to establish the College, which would have an annual budget of £20,000.[96] In Rosebery's letter to the London County Council in June of 1903, building costs are estimated at £300,000 (taking the Hochschule at Charlottenburg as a point of reference), and the annual budget at least at £20,000. In the context of what was being spent on British universities and science at that time, these sums must have seemed exorbitant, putting the whole project into doubt.

A large private endowment started off the actual foundation process of Imperial College, as had been the case in most of the university colleges founded in the Midlands and the north of England in the second half of the nineteenth century. The patrons in this case, lined up by Haldane, were the 'South African millionaires' Julius C. Wernher and Alfred Beit, whose philanthropy has already been discussed elsewhere.[97] An important part of Haldane's tactics was to interest potential patrons in the project at an early stage of the planning process. His papers contain frequent references to occasions on which he introduced future patrons to the planners of a technical college in London.[98] Haldane therefore described himself, with some justification, as 'a sort of Evangelist

---

95. 'The New Institution', Memorandum dated 1 March 1907 (Ed. 24/531).
96. Webb, 'London University', p. 930.
97. See above, pp. 46–50 in this volume.
98. See, for example, a letter from Haldane to his mother, 26 July 1902, in which he writes about a dinner attended by Francis Mowatt and Sidney Webb as well as by Julius C. Wernher and Alfred Beit (R. B. Haldane Papers, MS 5968).

stirring people up about this Education business'.[99] The result of his efforts, which enjoyed the support of Cecil Rhodes, was that a 'large sum of money' was made available. In his letter, Rosebery could announce that the starting capital for the 'London Charlottenburg' had been raised.

The list of patrons of Imperial College before 1914[100] shows that after the initial large endowment by Wernher and Beit, the principal benefactors came from the small circle of London financiers and bankers of South African or German origin, which we have already encountered. The amounts of money given were sometimes large: Wernher, Beit & Co.'s announced endowment came to £100,000, Alfred Beit bequeathed the College an additional £135,000, Wernher gave a further £150,000 in 1912, and the bankers Sir Ernest Cassel and Maximilian Michaelis each donated £10,000. Haldane appears in the list as donating £1,000. British industrialists, however, who were meant to be the College's principal beneficiaries, made very few large donations. £20,000 came indirectly, via the Bessemer Memorial Fund, which was set up in 1903 with donations from industry. The Goldsmiths' Company, one of the old Livery Companies of the City of London, gave a total of £137,000, but it can hardly be considered a representative of industry.

In principle, private endowments paid building costs and provided apparatus and equipment, while public money paid day-to-day administrative costs. London County Council responded to Rosebery's appeal by granting £20,000 per annum as early as July of 1903, largely thanks to the efforts of Sidney Webb, who was still a member of the Council. Added to this was the Treasury's contribution, which after considerable delay and a vigorous representation by the then president of the Board of Education, Lord Londonderry, was also fixed at £20,000 in 1905.[101] Even the thriftiest chancellor could not allow the College to shirk its imperial duties, which were constantly discussed in public. To this was added the fact that subsidising Imperial College did not involve a completely new departure in science policy but only the expansion of an existing one, as the Royal College of Science and the Royal School of Mines were already maintained by the state. These institutions had been founded in 1853 and 1851, respectively; on Robert Morant's suggestion they amalgamated with Imperial College and became departments within it. One other existing institu-

---

99. Haldane to his mother, 4 July 1903 (ibid., MS 5970).

100. See *Annual Reports of the Governing Body of the Imperial College of Science and Technology 1908–1914* (London, 1909–15).

101. Lord Londonderry to the prime minister, Arthur J. Balfour, 27 June 1907 (Ed. 24/530).

tion, Central Technical College, which had been supported by the London Livery Companies since 1884, also amalgamated with the new college. In this sense, Imperial College was not the new foundation which Webb and Haldane had envisaged at the early planning stages. Largely at the insistence of the president of the Board of Education, it was an amalgamation and extension of existing older establishments, 'a union of long-established and justly famed institutions'[102] which, like the Royal College of Science and the Royal School of Mines, had belonged to the reformed University of London since the turn of the century.

Attempts at reform, conceptual innovation and building on existing structures were not mutually exclusive in establishing a technical college in London. The state — that is, the Board of Education which was responsible for these matters — proved to be extremely flexible and cooperative in the process leading to the foundation of Imperial College. The Board of Education supported the new institution by appointing a departmental committee in 1904, by making public funds available, and by transferring to it two of the institutions it had maintained, without claiming the right to have a decisive say in its running. The Board also supported without reservation the departmental committee's recommendation, made in an interim report in 1905, that the state should take neither the full financial nor the full scientific responsibility for the new College.[103] 'It seems highly desirable', wrote the president of the Board of Education, Lord Londonderry, to the chancellor of the exchequer, Austen Chamberlain, 'that the management of a great Teaching Institution like this should not be in the hands of a Government Department, but in the hands of persons better qualified to keep abreast of the changing conditions of modern teaching in the higher ranges of Science.'[104] The state's involvement in the management of Imperial College was, however, anchored in the royal charter, in that ten of the forty members of the College's Governing Body were appointed by the government (article IV). Other members of the Governing Body were representatives of the University of London, London County Council, the Royal Society and various professional associations such as the Institution of Civil Engineers, the Iron and Steel Institute and the Society of Chemical Industry.

---

102. Vice-chancellor of Imperial College, 7 October 1908 (*Nature*, 78, 1908, p. 613).
103. *Preliminary Report of the Departmental Committee on the Royal College of Science*, Cd. 2610 (London, 1905), p. 9.
104. Lord Londonderry to Austen Chamberlain, 24 March 1905 (Ed. 24/530).

The problem of state influence or control did not play a large part in the process by which Imperial College was created. It was solved without long discussion and with the agreement of all parties. There was agreement, too, about the College's function which, according to article II of the royal charter, was 'to give the highest specialised instruction and to provide the fullest equipment for the most advanced training and research in various branches of science, especially in its application to industry'. The relationship between the new technical college and the University of London, by contrast, proved to be a much more controversial issue. It led to a protracted dispute which at times jeopardised the whole 'Charlottenburg Scheme'.[105] It began while the departmental committee was still sitting, and at its most elemental it was a conflict between industry, with its practical orientation, and science, which looked more to basic research. Rosebery's letter of 1903 had suggested that the planned technical college should cooperate with the university, while retaining its autonomy as 'a distinct "School" of the University under the management of its own committee'. Haldane and Webb supported this proposed organisational structure — both had repeatedly rejected the institutionalised dualism which existed in Germany, where basic research was the province of the universities, and applied science that of the *Technische Hochschulen*. In his autobiography, Haldane wrote that during his frequent trips to Germany he had come to view the German system as disadvantageous. He had concluded that 'it is only in the larger atmosphere of a University that technical education of the finest kind can be attained'.[106] By deciding against creating two types of university in which the two areas of science would be separated, the founders of Imperial College were also conforming to nineteenth-century British educational philosophy, which had prevented the provincial universities established since the middle of the century from excluding the humanities from their syllabuses, although attempts had been made to do just this.[107] On the contrary, the policy of British education reformers in the nineteenth century was to extend the competence of the universities beyond the traditional canon to include technical subjects. Thus it was accepted in Britain that 'the

---

105. See J. C. G. Sykes, secretary of the departmental committee, to Robert L. Morant, 28 July 1906 (Ed. 24/532). The Board of Education's voluminous files (Ed. 24/531 and Ed. 24/532) give some impression of the intensity of the debate about the status of Imperial College. On this see also Humberstone, *University Reform*, pp. 92–104, and Ashby and Anderson, *Portrait of Haldane*, pp. 101–11.
106. Haldane, *An Autobiography*, pp. 91–2. See also Ashby and Anderson, *Portrait of Haldane*, p. 55, and Crowther, *Statesmen of Science*, p. 287.
107. See above, pp. 52–3 in this volume.

manager-technologist must receive not only a vocational training: he must enjoy also the benefits of a liberal education; or at least he must rub shoulders with students who are studying the humanities'.[108]

The British principle of combining the study of technical subjects, the natural sciences and the humanities at universities was not, on the whole, challenged during the foundation of Imperial College. But differing ideas developed about the extent to which the University of London and the planned college should combine. A minority on the departmental committee set up by the Board of Education in 1904 to prepare for the foundation advocated direct and absolute control of Imperial College by the Senate of the University of London. They proposed that in terms of administrative law, the new college should be put on the same basis as University College and King's College. If this solution had been adopted, little would have remained of the financial and administrative autonomy which Rosebery had envisaged for the new institution. This proposal was supported by the representatives of the University of London on the committee. The University felt that a potentially competitive college might threaten its authority in the tertiary-education sector in London, and diminish its status, which, despite the organisational reforms of the beginning of the century, had not yet been consolidated. 'There would', argued the University, '. . . evidently be danger of friction between the University and a powerful School if each were regarded by the other as external to itself.'[109] The vice-chancellor of the University of London warned, in the autumn of 1906, that if the planned technical institution went ahead independently of the University, 'a blow to University organisation in London will [be] struck from which it will be difficult for the University to fully recover'.[110] Not quite two years later, a confidential letter to Rosebery, in his capacity as chancellor of the University of London, openly spoke of the precarious situation the University was facing at that time.

> Things are not at all well with the University at the present time . . . There can be little doubt that the University is not rising in the public estimation. If there was a chance of our obtaining the control of the new Technological College a couple of years ago, I suppose it may be accepted that our chance has quite vanished now. Financially, our

108. See Ashby, *Technology and the Academics*, p. 63.
109. *Final Report of the Departmental Committee*, p. 27.
110. Memorandum from the vice-chancellor of the University of London to the Board of Education, 26 October 1906 (Ed. 24/532).

position is anything but good, and we are not likely to obtain gifts or subscriptions so long as the various teaching institutions which are constituent colleges of the University place their own interests in front of the interests of the University.[111]

For this reason, as soon as the departmental committee's final report was published, the Senate of the University of London adopted resolutions which gave the University more direct control of the technical college.[112]

Representatives of the professional associations as spokesmen for industry on the departmental committee, by contrast, advocated the greatest possible autonomy for Imperial College. Their decisive argument against the University of London extending its hegemony over all types of academic education in London was that if the new technical college were integrated into traditional academic structures, it might lose its flexibility to adapt to the constantly changing demands of industry, for whose benefit it had been established in the first place. If the college was to maintain close contacts with industry in the United Kingdom and the empire, and be able to react to its specific interests, then

> its organisation must be free from all impeding trammels founded upon experience of the well-tried and comparatively little-changing track of an education regulated, and rightly regulated, by other aims. It must be free to adapt itself, its staff arrangements, and its methods of teaching, to the conditions of the time . . . These conditions of success appear to exclude the proposition that the control of the institution should be vested in a University.[113]

A connection with the University limited to that of a 'school' seemed most appropriate to this position as mediator between science and industry.

As there was agreement about the necessity for a technical college in London, and it was only the relationship between Imperial College and the University of London that was in dispute, the departmental committee recommended that the College should be set up quickly as a 'school' with its own governing body (on

111. Henry T. Butlin to Lord Rosebery, 3 April 1908 (Rosebery, Private Papers, MS 10 121).
112. Resolutions adopted by the Senate of the University of London on 7 March 1906 (Ed. 24/532). See also the minutes of a meeting between a deputation from the Senate and the Board of Education on 9 March 1906 (Imperial College Archives, ABC 3/2).
113. *Final Report of the Departmental Committee*, p. 26. *The Times* (7 March 1906) argued along the same lines.

which the University of London would be represented), without wishing in this way to anticipate a future decision on the relationship between the two institutions.[114] The problem was eventually solved by a royal commission into the various options available after Imperial College had opened. This royal commission, also chaired by Haldane, started work in February 1909. After extensive inquiries, it rejected the idea of establishing two universities in London and recommended that Imperial College give up its autonomy and be incorporated into the University of London.[115] But a final decision was still not arrived at; the First World War pushed the problem into the background for the time being, and it was not solved until almost fifteen years later. The war also contributed to the fact that Imperial College saw itself more and more as an independent university *vis-à-vis* the University of London. After the war, attempts were made to confer on Imperial College, which now had more than 1,300 students, the legal status of a technical university. But the protracted debate did not come to an end until 1926 when the University of London Act, harking back to the idea originally put forward by Rosebery in his letter of 1903, assigned Imperial College to the University of London as an autonomous 'school'. This Act in effect confirmed the status provisionally conferred on the College in the royal charter of 1907.

The 1926 University of London Act integrated technical subjects more firmly into the wider framework of the University, pushing further a trend which had emerged in Britain since the mid nineteenth century. A similar development did not take place in Germany until the 1960s. Berlin played a pioneering part in the process, already begun, by which *Technische Hochschulen* expanded by the accretion of non-technical subjects. The transformation of the *Technische Hochschule* in Berlin — the old 'Charlottenburg' admired by British education planners early this century — into the Technische Universität Berlin in April 1946 was a deliberate new start, politically as well as scientifically, in an almost totally destroyed city. Apart from education policy decisions for a 'new' Germany, it was essentially practical considerations which led to

114. Article VIII of the royal charter of 1907 stated: 'Subject to compliance with the Statutes of the University of London and pending the settlement of the question of the incorporation of the Imperial College with that University the Imperial College shall be established in the first instance as a School of the University.' The newly appointed president of the Board of Education, Reginald McKenna, had expressly approved this procedure in agreement with Haldane, who was pressing for Imperial College to be opened as soon as possible (McKenna to Francis Mowatt, 19 February 1907, Ed. 24/ 531. The letter is printed in *The Times*, 21 February 1907).

115. *Royal Commission on University Education in London: Final Report*, Cd. 6717 (London, 1913), pp. 32–3; also Ashby and Anderson, *Portrait of Haldane*, pp. 106–11.

this development.[116] The Technische Universität Berlin was established with considerable help from the British military government in Germany — thus, in a way, bringing full circle the mutual influence and interaction between the British and German systems of technical education in this century.

The Medical Research Committee

The last innovation in British science organisation to be discussed here, the establishment of the Medical Research Committee, was not modelled directly on a foreign institution, as were the National Physical Laboratory and Imperial College of Science and Technology. The Committee represented a trail-blazing innovation not only in medicine but also in the organisation of science as a whole. No comparable institution existed at that time in either the USA or Europe. On the contrary, the system of specialised research committees or councils for specific sciences functioning independently of directives from a third party, which developed in Britain shortly before the First World War, was subsequently imitated in other countries. For the first time since the eighteenth century, a model for the organisation of science which had an impact abroad was created in England. The Medical Research Committee, based in London, differed from the other two foundations which have been discussed here in so far as the state influenced its genesis much more strongly; in some respects it can be said to have been its originator.

The Medical Research Committee was created in August 1913 to coordinate the medical and biological research being done by various bodies.[117] Since the early nineteenth century, research of this kind was more frequently undertaken by the state as a conse-

---

116. 'Translated into another language, the term "*Technische Hochschule*" could only have the wrong connotations' (translated from Peter Brandt, 'Wiederaufbau und Reform, Die Technische Universität Berlin 1945–1950' in Rürup, ed., *Wissenschaft und Gesellschaft*, vol. 1, p. 500).
117. A. Landsborough Thomson, *Half a Century of Medical Research*, vol. 1, *Origins and Policy of the Medical Research Council (UK)* (London, 1973); idem, 'Origin and Development of the Medical Research Council', *British Medical Journal* (1963), Part 2, pp. 1,290–2. In addition to these works by the long-serving deputy director of the MRC, the older literature on the subject includes the following important titles: John Charles, *Research and Public Health* (London, 1961), esp. pp. 94–104; Henry H. Dale, 'Fifty Years of Medical Research', *British Medical Journal* (1963), Part 2, pp. 1,279–81 and 1,284–94. See also Rose and Rose, *Science and Society*, pp. 46–51. The archives of the Medical Research Council in London contain practically no material relating to the process by which it was founded. Some important information is provided by Harold P. Himsworth, 'The Support of Medical Research', typescript, 1958 (MRC, 1472 II).

quence of the expansion of its functions, although in general the principle still remained valid that promoting medical and biological research was not seen as the task of the state. Almost inevitably, the state got involved in public health services and in preventive medicine, which became increasingly expensive after the Industrial Revolution and the changes it produced in living patterns. As a result of the state's involvement in this area, the boundaries between private and public efforts gradually became blurred. Thus, for example, the General Board of Health, established in 1848, collected statistics on causes of death and investigated certain infectious diseases. Similar investigations were carried out for the Local Government Board, which based the nineteenth-century Public Health Act on similar statistics. After 1910 the Board maintained a bacteriological laboratory in London. Important medical research, especially into tropical diseases, was undertaken by the War Office, the India Office and the Colonial Office. In 1905 the Colonial Office established the Tropical Diseases Research Fund to support the research it financed, and in 1909 it set up the African Entomological Research Committee. Finally, growing state interest in general public health measures was also expressed in the establishment of royal commissions which either encouraged experimental research and clinical studies, or carried them out itself, as in the case of the Royal Commission Appointed to Inquire into the Relations of Human and Animal Tuberculosis, set up in 1901. Significantly, the Royal Commission's Final Report was published in 1911, the same year in which Lloyd George's National Insurance Act was passed by Parliament, laying the basis of the British welfare state.[118]

The concept of the Medical Research Committee was worked out by a departmental committee on tuberculosis, chaired by Waldorf Astor, MP (later Lord Astor). It was set up by the Local Government Board in 1912 and published its findings in 1913.[119] Referring specifically to tuberculosis, a very common disease, the Committee recommended establishing a central organisation to promote medical research in the United Kingdom. This recommendation was based on the National Insurance Act of 1911, which contained a clause saying that the Treasury would contribute to a Medical Research Fund (especially for research into tuberculosis)

---

118. *Final Report of the Royal Commission Appointed to Inquire into the Relations of Human and Animal Tuberculosis*, Cd. 5761 (London, 1911). On this also Bently B. Gilbert, *The Evolution of National Insurance in Great Britain: The Origins of the Welfare State*, 2nd edn (London, 1973).

119. *Final Report of the Departmental Committee on Tuberculosis*, Cd. 6641 (London, 1913).

one penny each year for every person insured[120] — without, however, specifying in detail how the money was to be administered. Thus the 1911 Act provided the financial basis on which the Medical Research Committee was established. The question of who introduced the all-important clause into the Act cannot, unfortunately, be answered today. Various names have been suggested. Lloyd George himself, author of the National Insurance Act, may have been responsible.[121]

In its Final Report, the departmental committee rightly called the clause 'a most important development in the attitude of the State towards scientific research into the causes, treatment and prevention of disease'.[122] Before the Medical Research Committee set up its own research institute in London,[123] it could manage without a large administration, and its annual income, guaranteed by the National Insurance Act, was more than £50,000. Later, this source still provided over 90 per cent of its budget. That endowments initially played only a small part in financing the Medical Research Committee may have been a result of the circumstances of the time, the small degree of publicity and the lack of a public appeal for funds. Not until after the First World War did the Medical Research Council, as it had become, receive more money from endowments of various kinds. The Council received payment for investigations it carried out on behalf of government departments, the armed forces and industrial firms. Large public foundations — for years the most important one was the Rockefeller Foundation — regularly financed specific projects and programmes. In the 1920s the Medical Research Council also received private benefactions, but as before the First World War, private patrons still seemed to prefer supporting hospitals or sanatoria.

The £55,000 which the Medical Research Committee received automatically each year through the National Insurance Act represented an enormous sum in the context of prewar Britain, where almost no public funds had been made available for these purposes before. Looking back to the situation pertaining before the introduction of Lloyd George's National Insurance Act, the departmen-

120. National Insurance Act, 1911, section 16 (2) (b), 1 & 2 Geo. 5, ch. 55. See also *First Annual Report of the Medical Research Committee, 1914–15*, Cd. 8101 (London 1915), p. 3.
121. See the discussion of this problem in Thomson, *Medical Research*, vol. 1, pp. 11–13.
122. *Final Report of the Departmental Committee*, vol. 1, p. 13.
123. On the National Institute for Medical Research, founded in 1914 and influenced in conception by the Rockefeller Institute for Medical Research in New York (1901), see Thomson, *Medical Research*, vol. 1, pp. 108–32.

tal committee wrote: 'Hitherto, apart from a small annual sum expended by the Local Government Board and occasional grants for particular objects, the State has, in the main, left research to voluntary agencies. The Committee welcome the fact that by the National Insurance Act a considerable sum of money is now permanently available for the purpose of research'. But it thought that private research, especially in the London medical schools, should not be impaired in any way by state intervention. 'Research under the National Insurance Act should be organised in such a way as not to discourage either voluntary contributions or voluntary research towards the same ends. The aim should rather be to stimulate and co-operate with voluntary agencies.'[124]

The distribution of funds between its own institute, individual scientists and research groups was determined solely by the Medical Research Committee, a small body of experts with executive functions. The Committee, which had extensive powers never precisely defined, has since then played the key part in promoting medical research. Its composition and its relationship with the state are therefore of some interest. Meeting for the first time in the summer of 1913 under the chairmanship of the lawyer and lord justice of appeal John F. Moulton (1844–1921), the Committee was composed of nine members: six scientists nominated by the government, two MPs and one peer. The latter was Lord Moulton; the two MPs were Christopher Addison,[125] later to become the first minister of the Ministry of Health created in 1919, and the publisher Waldorf Astor. Both had been on the departmental committee on tuberculosis in 1912. Committee members had a three-year term of office. The fact that the Medical Research Committee was not under the Treasury or any other ministry was of the greatest importance. It was placed instead under the control of the Advisory Council for Research, a body with forty-two members and also led by Lord Moulton. The government appointed members of the Advisory Council 'after receiving suggestions for suitable names from each of the universities of the United

---

124. *Final Report of the Departmental Committee*, vol. 1, p. 13.
125. Christopher Addison (1869–1951), professor of anatomy at Sheffield and London, Liberal MP 1907–22, worked closely with Lloyd George on the National Insurance Bill, parliamentary secretary to the Board of Education 1914–15, from 1916 various portfolios, Labour MP 1929–31 and 1934–5, elevated to the peerage 1937, Leader of the House of Lords 1945–51. Addison's diaries have been published as *Four and a Half Years: A Personal Diary from June 1914 to January 1919*, 2 vols. (London, 1934); See also idem, *Politics from Within; 1911–1918*, 2 vols. (London, 1924); R. J. Minney, *Viscount Addison: Leader of the Lords* (London, 1958); Kenneth Morgan and Jane Morgan, *Portrait of a Progressive: The Political Career of Christopher, Viscount Addison* (Oxford, 1980).

Kingdom, from the Royal Colleges of Physicians and of Surgeons, from the Royal Society, and from other important public bodies interested in the question'.[126] Because there was no Ministry of Health until 1919, the Medical Research Committee was financially accountable to the National Health Insurance Joint Committee, which was led until 1915 by an under-secretary of state to the Treasury and thereafter until 1919 by a higher civil servant.

This seemingly complicated administrative structure, which in similar form had already served for the National Physical Laboratory and Imperial College of Science and Technology, was intended to guarantee the independence of the Medical Research Committee and its work, and to ensure that it remained free from government interference. Practical decisions about the areas on which research was to concentrate were taken by the Committee, while the Advisory Council, composed according to more representative principles, limited itself to being informed of the Committee's scientific programme and approving it. As it met only once a year, there was no question of it exerting effective control over the Committee. Dissatisfaction with this situation resulted in the dissolution of the Advisory Council after the war. On the recommendation of the minister of health, Christopher Addison, the Medical Research Committee, like the new Department of Scientific and Industrial Research, was placed under the Privy Council led by the lord president as an autonomous, self-governing corporation. Henceforth, the lord president was answerable to Parliament for the management of the Committee's finances, but not for its research policy. According to the critical science lobby, the reform of 1920 also satisfactorily solved a difficult problem 'in the art of government': 'The preservation of the freedom and self-government of scientific research work as to both initiative and execution, with due regard to a just responsibility to Parliament in respect of State endowment'. It had obviously satisfied a demand which *Nature* had always vociferously supported: that 'scientific men themselves should decide upon the allocation of funds for research, . . . and that they should be responsible for any schemes of organised investigation'.[127] In 1920 the Committee was renamed the Medical Research Council, but its organisational structure was retained; since then, it has coordinated and promoted medical research in Britain.

126. Text in Thomson, *Medical Research*, vol. 1, p. 23. On the organisation of the Medical Research Committee: *First Annual Report of the Medical Research Committee*, esp. p. 4.
127. 'The Promotion of Medical Research', *Nature*, 105 (1920), pp. 221, 223.

The first modern scientific institution in Britain which owed its existence to an initiative made directly by the state enjoyed generous financial support in a form not previously granted in Britain; that is, with extensive autonomy in scientific questions and far-reaching responsibility for research in one scientific discipline. The reasons for the state's accommodating attitude towards the Medical Research Committee are obvious. As a result of its direct and daily confrontation with the problems of public health, and Lloyd George's introduction of National Insurance, the government became interested relatively early in supporting medical research outside the university sector. The practical benefits of advances in medical knowledge could also be demonstrated to the public, to Parliament and the administration much more easily than those, for example, of basic scientific research. Nevertheless, the attitude of the state and its officials did not escape criticism. The third annual report of the Medical Research Committee remarks: 'In the early stages of the war, while our deficiencies due to our former failures to encourage or to apply the work of science were widely deplored, it was very common nevertheless to hear that research should be laid aside "till the war is over". Bitter need in every direction of the contest, and not least upon the medical side, has shown how vital the spirit of inquiry is to success in this prolonged competition of national efficiencies'.[128] Despite this muted criticism, however, British scientists were aware of the significance of the establishment of the Medical Research Committee in 1913. *Nature* called it the 'greatest advance in the organisation of scientific effort in the service of medical science that has yet taken place in this country'.[129]

Spokesmen for Science

'These people's politics were not my politics', the scientist narrator ponders on the politicians attending a party in C. P. Snow's *Corridors of Power*: 'They didn't know the world they were living in, much less the world that was going to come'. Snow's intriguing novel on politics and high society was published in 1964, and although dealing with the Britain of the years after 1945, it refers to a relationship between scientists and politicians which seemed not to have changed much since Victorian times: the world of politics

---

128. *Third Annual Report of the Medical Research Committee, 1916–1917*, Cd. 8825 (London, 1917), p. 8.
129. 'The Work of the Medical Research Committee', *Nature*, 105 (1920), p. 43.

and the world of science separated by a deep gulf. But was that really still true early in the twentieth century?

The National Physical Laboratory, Imperial College of Science and Technology and the Medical Research Committee represent important landmarks in the struggle of scientists after the turn of the century to reform the organisation of science in Britain. Whether they did in fact contribute to improving the British science system and thus to removing the basis of scientists' complaints about the neglect of science in Britain cannot be established conclusively here. Our description of the foundation of these institutions has been guided less by considerations of this sort than by an interest in the extent to which they reveal a change in the relationship between science and the state. The foundation history of these three institutions throws some light on this issue.

The establishment of three important scientific institutions — all still prominent in Britain's scientific system today — raises the question not only of the state's attitude towards them but also of the identity of 'spokesmen for science' in Britain before the First World War. The preceding discussion has already dealt with the essential aspects of this issue, which thus requires only a brief summary here. The term 'spokesmen for science' refers to those people who worked for the interests of science in public, in Parliament or in government during the period under investigation; they were part of the science lobby in the widest sense and contributed actively to an improvement of the financial and institutional status of science in Britain. Participation in this work varied enormously in form and intensity, ranging from advising commissions of inquiry or planning committees, to promoting science organisation initiatives in government departments, or encouraging potential patrons to support specific institutional projects. Committed spokesmen for science like Richard Haldane, an 'enthusiast for higher education',[130] combined all possible functions with great flair.

But the example of Haldane also shows that many science policy initiatives made by private individuals in the second half of the nineteenth century were only possible because patrons were found to finance them. In many cases it was only private patronage which made it possible to realise these plans at all. We have already shown that private patronage was extremely venturesome and flexible;[131] in many instances the state only became involved in financing new institutions after a large private grant had been made. This process

130. *Nature*, 72 (1905), p. 184.
131. See above, pp. 57–8 in this volume.

is clearly illustrated by Imperial College. Furthermore, the analysis of private patronage in Britain revealed that from the end of the nineteenth century South African financiers living in London played a particularly important role as patrons. It was no mean achievement of Haldane's that they extended their patronage to science and universities. Beatrice Webb, a sharp-tongued observer, recognised this situation; her description of it contains a peculiar mixture of irony, admiration and gratitude: 'What a significant fact was the appearance of the South African millionaire gold-diggers, who dominated London "society" as well as the City, in the last years of the nineteenth century. I must admit that even the Webbs accepted their gracious hospitality in return for their benefactions to the London University, the London School of Economics and the Imperial College of Science!'[132]

Our attempt to establish more precisely the identity of spokesmen for science in Britain around the turn of the century allows two initial observations to be made. Firstly, the sphere in which science and politics overlapped involved a remarkably small circle of people after the late nineteenth century. The same names appear time after time as initiators and planners of new scientific institutions, as we have seen in the foregoing case studies. Secondly, while the small circle of spokesmen for science was politically and socially heterogeneous, its members shared certain political convictions which made them regard greater support for science in Britain and a closer connection between science and the state as necessary.

In broad terms, spokesmen for science at the turn of the century generally came from three areas: from scientific institutions, the higher echelons of the civil service, or Parliament. Their ranks included the leading representatives of important scientific organisations such as the presidents of the Royal Society, the British Association for the Advancement of Science and the larger specialist societies, as well as scientists whose demands and suggestions were taken up by scientific societies and who were thereupon appointed to represent those organisations on commissions or foundation committees. The best-known members of this group are Sir Oliver Lodge, Sir Douglas Galton and Lord Rayleigh. They were also able to influence science policy decisions because they worked closely with officials in key positions in the relevant ministries. From the end of the nineteenth century, government departments were much more willing to cooperate with representatives

132. Webb, *Our Partnership*, pp. 488–9.

of science. Scientists' efforts had obviously had some effect. Officials who had exerted considerable influence on the realisation of initiatives in science organisation included, for example, permanent secretaries to the Treasury Sir Francis Mowatt (1894–1903), Sir George Murray (1903–11) and Sir Robert Chalmers (1911–13 and 1916–19), as well as the long-serving permanent secretary to the Board of Education, Sir Robert L. Morant. The latter maintained his remarkable interest in science policy when he took over the chairmanship of the National Health Insurance Commission for England in 1912. Morant participated in founding the Medical Research Committee in an official capacity, and later he described this body as 'a particularly favourite child of mine since the beginning of 1912'.[133] The anatomist Christopher Addison occupied an intermediate position. First as an MP and later as the holder of important public offices, he not only stimulated British science organisation after 1911 but also contributed in imporant ways to developing and improving it, especially during the First World War.

In Parliament — that is, primarily in the House of Commons — only a handful of MPs took the opportunity 'to impress upon the nation the essential part which science and higher education must play in the polity of the modern State if progress is to be secured',[134] and attempted to translate their convictions into political action. In many cases they represented university constituencies or worked in education and were thus close to the science lobby. Sidney Webb must be counted among them. Although he did not have a seat in the House of Commons, as a member of the London County Council and co-founder of the Fabian Society, he played an outstanding role in London's political life at that time. Of the leading politicians of the time, Lord Salisbury, Arthur Balfour, Lord Londonderry, Joseph Chamberlain and later David Lloyd George, 'the father of the Medical Research Committee',[135] all supported science in both words and action from within the framework of their official positions. Their willingness to comply with the requests of scientists and to identify with their aims, however, on the whole did not go beyond certain limits. Arthur J. Balfour, who after all had been a Fellow of the Royal Society since 1888, provides the clearest illustration of this. His interest in scientific problems, reflected in the fact that he became president of the British Association for the Advancement of Science in 1904 and

---

133. Robert L. Morant to Arthur J. Balfour, 28 October 1919 (MRC, P.F. 3).
134. *Nature*, 75 (1906–7), p. 275.
135. Thomson, *Medical Research*, vol. 1, p. 41.

of the British Academy in 1921, did not mean that his period of office as prime minister between 1902 and 1905 was a golden age for the promotion of science in Britain. It may be acknowledged, however, that Balfour's support facilitated the efforts made in this field by Haldane, who was a friend, and Lord Rayleigh, a brother-in-law.

Of the politicians, Joseph Chamberlain provided the most active institutional and financial aid for science, first as lord mayor of Birmingham when the civic university was founded, and after 1902 as colonial secretary, when he suported various research programmes and institutes. But Chamberlain does not deserve to be called a 'statesman of science'. This title is reserved for Richard Haldane alone, whose advocacy of science made him an exception among the politicians of his time.[136] 'The spectacle of a Minister of the Crown who was a whole-hearted believer in the benefits of science and who could proclaim those benefits with knowledge and experience was a rare, if not a unique, phenomenon in this country.'[137] In 1926 Haldane wrote about his work in this field:

> Science had been developed and applied in Germany as it had not with us, and it was very difficult to get my colleagues to realize this, and to avoid when I approached it being put down as a pro-German enthusiast. Anyhow, it was organisation for war and organisation for industry which were the two subjects which fascinated me during the years of Liberal Cabinet life, and I did not succeed in educating my colleagues, although I got the Army reorganised, the Navy influenced, and more Universities founded.[138]

---

136. Richard B. Haldane (1856–1928), son of a Scottish lawyer, Liberal MP 1885–1911, secretary of state for war 1905–12, 1911 Viscount Haldane of Cloan, lord chancellor 1912–15 and 1924. In the summer of 1874 Haldane spent four months in Göttingen for the purpose of study. This visit left a deep impression on his thinking. Haldane, *An Autobiography*; Frederick B. Maurice, *Haldane*, 2 vols. (London, 1937, 1939; reprinted Westport, Conn., 1970); Sommer, *Haldane of Cloan*; Stephen E. Koss, *Lord Haldane: Scapegoat for Liberalism* (New York and London, 1969); Ashby and Anderson, *Portrait of Haldane*; Hollenberg, *Englishches Interesse*, esp. pp. 243–64.
137. *Nature*, 91 (1913), p. 357. *Nature* also noted critically that Haldane noticeably reduced his efforts on behalf of science as soon as he became a minister: 'Our only regret is that while Lord Haldane was a member of the Government he did not see that decided steps were taken to remedy the defects to which he refers, and thus give us the strength needed to compete successfully in the rivalry of nations. When he was president of the British Science Guild he took an active part in asserting the claims of science and scientific education to fuller recognition by the State, and we looked naturally to the realisation of these aims when he was in office' (*Nature*, 97, 1916, p. 417).
138. Note on Letters, 1926, p. 16 (R. B. Haldane Papers, MS 5923).

Haldane never held a public office in the education sector; he rejected an offer from Ramsay MacDonald in 1924 to take over the Board of Education in the first Labour government. Until he became a member of Campbell-Bannerman's cabinet in 1905 as secretary of state for war, Haldane, as a Liberal MP, had been involved in some way in almost all science policy initiatives and planning since the end of the nineteenth century. His outstanding influence as a science policy-maker rested at the time on his position as an advocate of science, as someone who issued public warnings about the consequences of the political and economic superiority other countries were achieving on the basis of their encouragement for scientific research, as a pioneer and organiser of institutional foundations and as an adviser to ministers and high government officials. Haldane was a 'courtly lawyer with a great capacity for dealing with men and affairs, and a real understanding of the function of an expert, and skill in using him'.[139]

What motivated Haldane and several other British politicians to make the promotion of science one of their main political activities from the late nineteenth century onwards? From what the meagre sources reveal about their motives, we can conclude that the socially and politically heterogeneous group of spokesmen for British science shared beliefs generally attributed to the 'social' or 'liberal' imperialists, or the 'efficiency group'. Imperialist policies, strengthening the empire, improving national efficiency in competition with other nations, social reforms — all these slogans stood for issues which politicians and journalists of various political persuasions were committed to in the atmosphere of political crisis dominating Britain at the turn of the century. They included not only Liberals such as Lord Rosebery, described by Gladstone in 1886 as the man of the future,[140] Haldane, Asquith and Grey, but also Liberal Unionists such as Joseph Chamberlain and Lord Alfred Milner, Conservatives such as Balfour, and Socialists like the Webbs. Senior civil servants such as Mowatt and Morant, who maintained close relations with both Haldane and the Webbs, were associated with this group, whose party-political allegiance was not unequivocally clear.

While the political programme of the social or liberal imperialists — exhaustively analysed by Bernard Semmel, G. R. Searle and H. C. G. Matthew — was one bond unifying spokesmen for

139. Beatrice Webb, diary entry 15 December 1905 (Webb, *Our Partnership*, p. 325).
140. H. C. G. Matthew, *The Liberal Imperialists: The Ideas and Politics of a Post-Gladstonian Elite* (London, 1973), p. 11.

science prior to the First World War, this did not mean that in supporting science, individuals did not further other interests. Thus the Webbs, declared opponents of the liberal principle of *laissez-faire*, advocated more state support for science because it, like social reform, promised to make the British people 'ever more efficient, mentally and physically'. At the same time, it corresponded to their belief in the benefits of state intervention in all areas of social life. Haldane, who can be credited with a genuine interest in scientific and educational issues, recognised early that taking up the issue of science could further his political career. In 1905, when he was elected chairman of the British Science Guild, he explained that 'nearly ten years ago, when the political party to which [I] belonged went out of office, [I] looked about for something to do, and [I] thought [I] might as well turn [my] hand to the somewhat cobwebbed state of the higher education of this country'.[141] Education and science undoubtedly helped Haldane to establish himself as a prominent politician around the turn of the century. Beatrice Webb noted in 1902: 'He [Haldane] has improved his status as a leader of opinion, has shown that he knows and is keen about the higher branches of education. And the higher branches of education are one of the coming questions.' But she could not find a convincing explanation for Haldane's interest specifically in the issues of education and science. 'It is a paradox', she reflected on Haldane, 'that a mind that is essentially metaphysical, laying stress on the non-material side of human thought and feeling, should have been, as a matter of fact, chiefly engaged in promoting applied physical science.'[142]

For the free traders among the liberal imperialists, including Haldane, the opening of the campaign for a 'London Charlottenburg' in June 1903 was also a welcome alternative to the tariff reform campaign introduced by Joseph Chamberlain only a few weeks before. 'It really is a great scheme and it is taking hold of the public mind as an alternative policy to Chamberlain's.'[143] According to Chamberlain's opponents, protection from foreign economic competition would be afforded not by the imposition of tariffs and the creation of a closed empire market, but by a British industry strengthened by its association with science and the devel-

---

141. 'The British Science Guild', *Nature*, 73 (1905–6), p. 12. Also printed in *The Times*, 31 October 1905.
142. Webb, *Our Partnership*, p. 247.
143. Haldane to his mother, 30 June 1903 (R. B. Haldane Papers, MS 5969). See also Searle, *National Efficiency*, p. 148; Ashby and Anderson, *Portrait of Haldane*, p. 73; Semmel, *Imperialism*, p. 84.

opment of new technologies. Nearly four years later, when the preparations for founding Imperial College of Science and Technology were almost complete, the *Morning Post* made the following comment, harking back to the positions articulated in the debate of 1903: 'Under modern conditions the industrial progress of a country depends more upon the extent to which it applies scientific principles to its industrial processes than upon any fiscal system it may adopt or inherit'.[144] On a party-political level, the tariff reform debate put an abrupt end to any hopes of a 'party of national efficiency' emerging. An informal club which called itself the 'Coefficients', founded by the Webbs at the end of 1902, had seen itself as a core from which a party of this type might develop.[145] Haldane and Chamberlain had occasionally worked together in promoting science and education — a field to which both had already ascribed 'chief importance' by the summer of 1902.[146] Despite the differences in their opinions about economic policy, they continued to cooperate after 1903.

Haldane's efforts on behalf of science and education were made in public: within the framework of extra-parliamentary organisations, in the press, at meetings, on commissions of inquiry and on the foundation committees of new institutions. But they also took place 'behind the scenes' in negotiations and discussions with politicians and civil servants, where important decisions relating to science policy were made. Only rarely, however, did Haldane use the House of Commons — which historians and political scientists like to see as a 'mirror of public opinion' — as a forum for his activities, correctly assessing the value attributed by the large majority of MPs to science as a political issue. For it would be misleading to conclude from occasional remarks made by a few politicians from both major parties that British MPs had a particular interest in science. Of the 670 members of the House of Commons at that time, those who worked for the interests of science and the universities formed an insignificant and largely disregarded minority. 'In Parliament', Haldane wrote later about his experiences, 'there were unfortunately very few who cared for the subject [of higher education].'[147] In late-Victorian and Edwardian Britain, science, research and the problems of universities were not subjects which stimulated political controversy. Did

144. *Morning Post*, 22 February 1907.
145. Searle, *National Efficiency*, pp. 150–2; Semmel, *Imperialism*, pp. 72–82, 128.
146. Joseph Chamberlain to Richard B. Haldane, 11 August 1902 (R. B. Haldane Papers, MS 5905).
147. Haldane, *An Autobiography*, p. 292.

Parliament, the opinion-forming centre of Britain, thus reflect only the contemporary lack of interest in science and its financial problems?

An examination of parliamentary debates in the decades before the First World War yields meagre results: MPs and political parties rarely commented on issues to do with science. It would go too far to interpret this reticence as indicating a widespread scepticism of, or even hostility towards, science among politicians. But it clearly expresses their indifference and helplessness when dealing with this increasingly important social issue. Confronted with pressing problems in domestic and imperial policy, in the economic, the social and the defence sectors, and in Ireland, neither Liberals nor Conservatives developed any strategy which outlined their attitude towards science or contained the germ of a long-term public science policy. Because of the low priority politicians gave to the concerns of science, it was left to individual MPs to put occasional questions to the government (mostly at times when the House was poorly attended). Almost without exception, these questions referred to the inadequate financing of existing scientific institutions, the limited expenditure on scientific research and the principles by which allocated moneys were distributed.[148] To some extent Parliament practically excluded itself from exercising any influence on innovations in the organisation of science.

Not until after the turn of the century did Parliament display any greater willingness to deal with issues relating to the organisation and financing of science. There was no precedent in the history of the British Parliament for the petition, mentioned above,[149] signed by 150 MPs and sent to the prime minister in the summer of 1905 on the initiative of Haldane, Joseph Chamberlain and Sir John T. Brunner, chemical industrialist and patron, demanding that more money be made available for the National Physical Laboratory. Never before had so many MPs from both sides of the House made an effort on behalf of science in this manner. Four years later the parliamentary budgetary committee, in the presence of war secretary Haldane, held a long debate on the imminent advent of aviation, the consequences for defence policy which could arise from the military use of the spectacular new airships and aircraft, and the issue of to what extent the government should intervene to

---

148. Particularly instructive examples: Hansard, Parl. Deb., H.C., 4th series, vol. 123, col. 295–6 (8 June 1903); ibid., 5th Series, vol. 19, col. 476–81 (13 July 1910); ibid., 5th Series, vol. 22, cols. 1815–18 (10 March 1911) and cols. 2627–2630 (17 March 1911).

149. See above, p. 148 in this volume.

promote the development of these new technologies.[150] The construction of airships and aircraft was an extremely popular topic at the time and caught the general public's imagination much more than did other technological and scientific developments. It also aroused interest in Parliament for this reason, and not only because of its military applications.

After this debate, however, six years passed before MPs again dealt in any detail with scientific issues. The British chemical industry's totally inadequate production capacity — especially in the areas of synthetic dyes, medicines and explosives — revealed in the first few months of the First World War, provided the background in February and March 1915 to the most intensive debate so far among British MPs about the relationship between science and industry, and the problem of state aid for science. It is not necessary to report the details of this debate here, because essentially it consisted of restatements and variations of the arguments already put forward by scientists. Parliament was united in believing that in respect to the promotion of science, 'a great many precedents and some principles . . . will have to be thrown to the winds'.[151] In view of the war and the difficulty of access to essential goods, the government intended to make public funds available to industry and universities for research in organic chemistry. It also wanted to improve the production capacity of the aniline dye industry by implementing state measures. This provoked almost no opposition. 'When you are engaged in a great war', was the brief conclusion, 'you have got to organise your civil side quite as much as your military and naval side if you wish to be successful.'[152]

Making the traditional references to the efficiency of Germany's science organisation and Britain's dangerous dependence on the German chemical industry, the House of Commons approved the government's plan to secure the country's supply of synthetic dyes by founding British Dyes Ltd. British industry was obviously not capable of filling by its own efforts the gap in supply created at the beginning of the war by the loss of German imports. The establishment of British Dyes Ltd, half of the original capital for which was provided by the British government, thus represents one of the first large interventions in private industry made by the state in Britain in order to regulate the market. Parliament also approved the

---

150. Hansard, Parl. Deb., H. C., 5th Series, vol. 8, cols. 1564–617 (2 August 1909).
151. Hansard, Parl. Deb., H. C., 5th Series, vol. 70, col. 52 (22 February 1915).
152. W. A. S. Hewins, former director of the London School of Economics, and Conservative MP from 1912 (ibid., col. 60).

annual sum of £10,000 which the new company was to have for ten years to fund research in its own laboratories and at universities. A few MPs, among them Sir Philip Magnus (Member for the University of London) and Sir Alfred Mond (son of the chemist and patron Ludwig Mond), thought this sum was 'absolutely inadequate',[153] but their efforts to increase British Dyes Ltd's research budget were unsuccessful.

The establishment of British Dyes Ltd and the Department of Scientific and Industrial Research a little later showed that under the pressures of war, the British Parliament was more willing than before to provide public money for science and research. Neither did it place any obstacles in the way of a reform and improvement of British science organisation introduced soon after the beginning of the war. But anyone who was led by this attitude on the part of Parliament to believe that there had been a fundamental change in the relationship between science and politics in Britain was disappointed on several counts after 1918. The relative generosity towards science and the increased readiness to support institutional innovations displayed by both government and Parliament during the war were replaced by restrictive policies in the difficult economic situation after the war. In 1921 Sir Alfred Mond, who became the first chairman of the newly constituted Imperial Chemical Industries (ICI) in 1926, launched a bitter attack on the government's science policy. He described the House of Commons attitude towards scientific research as 'much the same as that which led to the loss of the dye industry to this country'. In his opinion it was obvious that 'there are still people in positions of authority who do not understand the significance of research, and prefer the experience of a practical man to the results of the most careful scientific inquiry'.[154] Mond was a Liberal MP from 1906 to 1922, and in 1921 took over the Ministry of Health for a short time.

The British Parliament attempted neither to influence the processes of science policy decision-making to any great extent nor to introduce science policy initiatives, even as late as the period spanning the First World War. This was undoubtedly related to the fact that few MPs had a scientific or even technical training. Political parties may have been influenced by the image of the scientific expert who — as G. R. Searle suspects — was seen as a 'monomaniac', whose 'enthusiasms made him incapable of judging a situation from a general, all-round point of view'.[155] For this

153. Ibid., cols. 70, 114.
154. *Nature*, 107 (1921), p. 97.
155. Searle, *National Efficiency*, p. 22. See also a comment made by Cecil Harms-

reason, he was unlikely to be selected as a candidate for Parliament. From the point of view of MPs, moreover, the scientific expert threatened the freedom of politicians to make decisions. He represented a threat to 'constitutional government' because he possessed 'knowledge and techniques that escaped the comprehension, and thus the control, of the most intelligent "layman"'.[156] Therefore, as Lord Salisbury had explained to the Devonshire Commission in 1874, the British Parliament, ever aware of its total sovereignty, would never hesitate to reject or ignore proposals made by a scientific advisory body, or recommendations made by scientific experts.[157]

At least since the turn of the century, but especially after the British Science Guild was founded in 1905, the science lobby was fully aware of the inadequacy of the representation of science at Westminster. Scientists in the House of Commons were rare. The few well-known figures included the Cambridge physiologist Sir Michael Foster, and the director of the City and Guilds Institute in London, Sir Philip Magnus. Both represented the University of London as Liberal Unionists — Foster from 1900, and Magnus from 1906. In descriptions of the professions of MPs there was no category for 'scientist' at this time.[158] In 1913 Sir David Gill, astronomer and former president of the British Association for the Advancement of Science, summed up the situation at a banquet held by the British Science Guild, saying that it was sad to think

> how very few of our leading politicians — how very few, indeed, of our members of Parliament — have any serious knowledge of science; and

---

worth, Liberal MP: 'If we are to judge by the many Debates in this House, experts are at an entire discount amongst us' (Hansard, Parl. Deb., H. C., 5th Series, vol. 8, col. 1583, 2 August 1909).

156. Searle, *National Efficiency*, p. 24. On this see also *Nature*: 'The belief that the expert – whether scientific or industrial – has to be controlled or guided by permanent officials having no special knowledge of the particular subject in hand is typical of our executive system' ('Science in National Affairs', *Nature*, 96, 1915–16, p. 195). On the issues, occasionally raised today, of to what extent scientific work evades the control of Parliament, and how parliaments can meet the threat of domination by experts and the abolition of politics through the force of circumstances, see Gerhard A. Ritter, 'Die Kontrolle staatlicher Macht in der modernen Demokratie' in idem (ed.) *Wohlfahrtsausschuβ*, pp. 69–117; and Paul G. Werskey, 'The Perennial Dilemma of Science Policy', *Nature*, 233 (1971), pp. 529–32.

157. *Royal Commission on Scientific Instruction and the Advancement of Science, Minutes of Evidence*, vol. 2, C. 958 (London, 1874), p. 345.

158. See J. A. Thomas, *The House of Commons 1906–1911: An Analysis of its Economic and Social Character* (Cardiff, 1958), pp. 16, 22–24. The House elected in December 1910 contained seven doctors and sixteen 'academic people', who 'either were or had been engaged in teaching at school or university level' (ibid., p. 23). See also Frank Foden, *Philip Magnus: Victorian Educational Pioneer* (London, 1970).

yet it is upon science ... that the whole progress of our modern civilisation depends.... Since science is so important to our existence as a nation, is it not strange that amongst our leading legislators there are so few who have any reasonable acquaintance with science?[159]

But the science lobby did not start considering what could be done effectively to remedy this unsatisfactory situation until towards the end of the First World War. A few weeks before the elections of November/December 1918, the first general elections since 1910, medical professional organisations discussed the possibility of achieving a 'more adequate representation of the medical profession in Parliament'. *Nature* expressly supported this discussion.[160] Efforts towards a better representation of science in Parliament were, however, unsuccessful in the 'khaki' elections of 1918. Referring to an article in *The Times* about the social composition of the newly elected House of Commons, *Nature* noted bitterly 'the practical absence of leading representatives of scientific knowledge and research in the new Parliament'. Among its 707 members there were, for example, only two Fellows of the Royal Society: Arthur J. Balfour and Sir Joseph Larmor, a physicist and former secretary of the Royal Society, 'neither of whom can be considered specifically to represent science', continued the unnamed leader-writer in *Nature*. He then sharply criticised Britain's parliamentary system:

> The work of Parliament is more and more coming to be a sordid scrimmage of hereditary, vested, class and sectional interests. Out of the base-metal of the various self-seeking coteries represented — agrarian, commercial, financial, professional, proletarian, and so on — by some obscure alchemy too absurd for belief, Westminster is supposed to effect a synthesis of the pure gold of wisdom, and in its odd moments from this conjuring entertainment to administer the affairs of an Empire on which the sun never sets.

Scientists, he wrote, were 'segregated, to their own and the nation's detriment, from any share in the solution of the vast and overwhelming problems which their activities in the first instance create'. *Nature* suggested the creation of an election campaign fund, to be administered by the Conjoint Board of Scientific Societies. Scientists were not to present themselves to the electorate as representatives of any particular social interest, but 'on the broad

159. *Nature*, 91 (1913), p. 358.
160. 'Science and Parliamentary Representation', *Nature*, 102 (1918–19), p. 144.

and elementary ground that their life-work has given them special knowledge and insight into the scientific discoveries which in the short space of a few generations have revolutionised the whole world, and which the Mother of Parliaments will ignore and continue to run counter to only at the nation's peril'.[161]

After 1918 science continued to be represented in the House of Commons fundamentally only by the MPs for university constituencies. The existence of special university constituencies in practice gave British academics a plural vote; under a system of equal and universal suffrage, widely implemented in Britain by the electoral reform of 1918, they were an anachronism whose continued existence had to be justified. Consequently, scientists defined them as a way 'of getting into the House of Commons men of science, men of scholarship, men of special and peculiar gifts quite alien from the ordinary working politician'.[162] In 1948 university constituencies were abolished by Clement Attlee's Labour government, thus further limiting opportunities for a modest representation of the interests of science. The fundamental under-representation of scientists in Parliament in comparison with other professions such as lawyers and teachers has not changed much since then. The pathetic demand made by the Association of Scientific Workers in the 1930s, that 'science must storm the portals of Parliament *en masse*', went unheard. Almost fifty years later the House of Commons, now numbering 635 members, still had only eight doctors and six graduate scientists,[163] including — and this too is something new in the history of this office — the prime minister.

---

161. 'Science in Parliament', *Nature*, 102 (1918–19), pp. 421–2. The article had appeared in *The Times* on 21 January 1919.
162. 'University Representation in Parliament', *Nature*, 110 (1922), p. 626.
163. 'Science in the House', *Economist*, 29 September 1979, p. 17. According to figures given by Derek J. de Solla Price, at the beginning of the 1970s less than 3 per cent of senators in the USA and MPs in Britain had a scientific or technical training (*Little Science, Big Science*, New York and London, 1963, p. 112). In general on this topic, see Vig, *Science and Technology in British Politics*.

4
Policies for Science during the First World War

'The War of Chemists and Engineers'

During the House of Commons debate on the founding of British Dyes Ltd, one speaker referred to the British public's shocked reaction to the outbreak of war in Europe. He reminded his listeners of the panic of those days, which clearly showed that in August 1914 neither British industry nor the government was prepared for a full-scale war in Europe. For the first months of the war they therefore faced grave problems. 'The fact is that at the beginning of the War we got into a panic. We thought that we should have no sugar and, let us say, no dyes in this country, and that we should want for all sorts of things.'[1] The break in hitherto lively trade relations with the Central Powers precipitated a crisis in the supply of important industrial products in Britain and exposed the almost total dependence of key British industries on deliveries from now hostile countries. This sudden revelation was a 'startling experience'[2] not only for the British public but also for politicians and MPs. Winston Churchill, first lord of the admiralty in Asquith's cabinet from 1911, attempted to counter it by making the now famous call for 'business as usual'.[3]

Churchill's seemingly cold-blooded maxim, which set the tone of the British government's propaganda during the first months of the war, effectively doused the initial public panic. But it could not disguise the obvious shortage of certain goods. Immediately after the beginning of the war, synthetic dyes were in short supply. The

---

1. Hansard, Parl. Deb., H. C., 5th Series, vol. 70, cols. 58–9 (22 February 1915), speech by the Liberal MP Thomas Lough.
2. J. A. Fleming, 'The Organisation of Scientific Research', *Nature*, 96 (1915–16), p. 694.
3. On the origin of this catchphrase, which was not coined by Churchill, see Arthur Marwick, *The Deluge: British Society and the First World War* (London, 1973), p. 39; idem, 'The Impact of the First World War on British Society', *Journal of Contemporary History*, vol. 3, no. 1 (1968), pp. 51–68.

textile industry needed them for dyeing uniforms among other things, but Britain produced no more than one-fifth of its large requirement, and part of this process was dependent on intermediate products.[4] No substantial stocks of dye existed in the summer of 1914. Suddenly, too, other chemical products could no longer be supplied: various special pharmaceutical lines, certain products such as acetone and phenol which were required in the manufacture of explosives, and photo-chemical goods. There was also a lack of optical instruments, precision instruments and ignition magnetos — all of obvious significance in time of war.

In the eyes of British scientists and spokesmen for science, this alarming situation validated the warnings and demands they had been making. The supply crisis was rooted, after all, in the inadequate development of the new science-dependent industries. *Nature* repeated what it had been saying for years: that the British government 'cares too little for the nation's need for science, which is as important for peace as for war purposes'.[5] A few months after the beginning of the war, *Nature* published an article which, harking back to the previous protracted discussion, suggested that Britain needed a scientific programme to support the war effort:

> The war now raging will at least demonstrate one thing to humanity — that in war, at least, the scientific attitude, the careful investigation of details, the preliminary preparation, and the well thought-out procedure bring success, where the absence of these leads only to disaster. After all, the necessity for research is the most evident of all propositions. But the question (which I hope will receive still more careful attention when the war is over) is, What can the State do to make the machinery of investigation the most efficient possible? The mere citing of popular misconceptions is not enough; we need to have specific programmes.[6]

December 1914 was still too early for specific scientific programmes, but two things had become clear in the first months of the war. Firstly, in contrast to its earlier much-criticised practice, the government was now taking the initiative in the science sector much more strongly than ever before. It attempted to support industry in the exceptional situation created by the war by taking stock of and coordinating the country's research capacities. This included turning more often to institutions such as the National Physical Laboratory, Imperial College of Science and Technology

---

4. Haber, *The Chemical Industry 1900–1930*, p. 188; Cardwell, *Organisation of Science*, p. 221; MacLeod and MacLeod, 'Social Relations of Science', pp. 308–9.
5. 'Science in Warfare', *Nature*, 94 (1914–15), p. 455.
6. 'Organisation of Science', *Nature*, 94 (1914–15), p. 367.

and the universities. Their research potential was thus increasingly harnessed in order to solve specific industrial problems. By 1915 the results achieved could already be summed up provisionally: 'The problem of linking university work with the scientific industries is being solved . . . with marked success, and is part of a great and growing movement to which the war has given a fresh stimulus'.[7] And early in 1916 *The Times* approvingly called the universities of Manchester, Liverpool, Leeds and Sheffield 'branches of the country's defensive forces'.[8] Secondly, in the critical first few months of the war it became clear that state initiatives were necessary because British industry at first proved unable to remedy shortages effectively by its own efforts. Traditional market mechanisms failed. The problems of British industry during the war meant that there could be no question of 'business as usual'.

At this stage of the war the British government's efforts were limited to improvising, making more use of existing research institutions and seeking ways to encourage cooperation between science and industry. Not until 1915 did they shape up into a plan for a central organisation for science. A few examples will serve here to illustrate the initial phase of improvisation. In 1914 the War Office sent questionnaires to all scientists asking them to register and describe any inventions of military significance.[9] The government extended its comprehensive search by supporting the Chemical Society in setting up an advisory body in the summer of 1915 'to consider, organise, and utilise all suggestions and inventions which may be communicated to it'.[10] By the end of August 1914 the Board of Trade, under an energetic Walter Runciman, had already appointed the Chemical Products Supply Committee, chaired by Lord Haldane and comprising scientists, civil servants and industrialists. Its function was 'to consider and advise as to the best means of obtaining for the use of British industries sufficient

---

7. William Osler, *Science and War* (Oxford, 1915), p. 14. Detailed information about the role played by British universities during the First World War can be found in Sanderson, *Universities*, esp. pp. 214–42. MacLeod and MacLeod, 'War and Economic Development' is interesting as a case study. See also the *Report of the Committee of the Privy Council for Scientific and Industrial Research for the Year 1915–16*, Cd. 8336 (London, 1916), pp. 31–4.

8. *The Times*, 9 February 1916. The article continues optimistically: 'When the war is over there will almost certainly be a rallying of the nation to the Universities which perhaps nothing but the cataclysm through which we are passing could have brought about.'

9. See Chaim Weizmann, *Trial and Error: The Autobiography*, 4th edn (London, 1950), p. 218.

10. *Nature*, 95 (1915), p. 524. After the war began the Royal Society also set up committees to examine new technical and scientific inventions for any possible military applications.

supplies of chemical products, colours and dyestuffs of kinds hitherto largely imported from countries with which we are at present at war'.[11] Thanks to the organisational work done by this committee, Britain's supply crisis, at least as far as dyes were concerned, was overcome.

Protection of German patents was lifted in Britain immediately after the war began. Trade with neutral Switzerland and its important chemical industry was stepped up. From autumn of 1914 the Board of Trade, which of all the ministries maintained the closest contacts with industry, negotiated with producers and consumers of aniline dyes about building up an efficient British dye industry capable of filling the gaps in supply. The Board of Trade clearly did the planning, but it willingly accepted suggestions from industry, especially the Midlands textile industry, and from Haldane's committee set up in August 1914.[12] The 'combined national effort on a scale which requires and justifies an exceptional measure of State encouragement',[13] resulted in the establishment of British Dyes Ltd, at the government's urging, in the spring of 1915. The state provided the majority of the starting capital of £2 million and had some say in the management of the company. Today the state's decision to involve itself in an entrepreneurial venture is justifiably seen as a turning point in the relations between government and industry in Britain.[14] The building up of a dye industry in Britain and the large-scale industrial application of nitrogen synthesis in Germany are regarded as outstanding achievements by the chemical industry during the war.

One of the small existing aniline dye companies, Read Holliday and Sons Ltd in Huddersfield provided the core of British Dyes Ltd. It was developed on the model, which had previously been studied in Britain, of the Bayer-Werke in Leverkusen, and built up by the purchase and addition of further companies. But the government's high expectations of the company's commercial success and of the quality of its products were only partially fulfilled during the war. It encountered numerous organisational and technical problems which militated against its success. Anticipating renewed

11. Armytage, *Social History*, p. 251; Haber, 'Government Intervention', p. 81; Marwick, *The Deluge*, p. 228; *Report of the Committee of the Privy Council 1915–16*, p. 8; *Nature*, 94 (1914–15), p. 30.
12. See Reader, *Imperial Chemical Industries*, vol. 1, p. 268.
13. Thus the Liberal (!) president of the Board of Trade, Runciman, in the House of Commons (Hansard, Parl. Deb., H. C. 5th Series, vol. 68, cols. 759–62, 23 November, 1914).
14. Reader, *Imperial Chemical Industries*, vol. 1, p. 270; Rees, *Trusts in British Industry*, p. 161.

German competition after the war, therefore, the Board of Trade began again in 1917 to plan a restructuring of the British dye industry. Finally, in November 1918, it forced a merger between British Dyes Ltd and the largest private British dye producer, Ivan Levinstein Ltd in Manchester,[15] by promising to supply further financial aid and to impose import restrictions on German chemical products. The chairman of the new company thus created, the British Dyestuffs Corporation, was Lord Moulton. At the time of its creation, this company controlled three-quarters of the growing British dyestuffs market and, as had been the case with British Dyes Ltd, the government had a controlling interest in the new company.[16]

The rather laborious and also expensive development of a British aniline dye industry, whose output increased fivefold between 1914 and 1919,[17] provides a textbook example of how German industry lost important export markets as a result of the war. But on the other hand, contemporaries had already noted with astonishment that after economic relations with Germany ceased in August 1914 and the price of dye spiralled upwards, British industry was slow to seize the opportunity thus offered to fill the gap in the market by increasing its output, either by its own efforts or with state support.[18] In February and March of 1915, during the House of Commons debate on British Dyes Ltd, Sir Alfred Mond was the only one among many representatives of industry to support wholeheartedly the development of the British dye industry and state intervention in a coordinating role. How can this remarkable reticence on the part of industry be explained?

Fear of making a bad investment in anticipation of German competition returning to the British market may have been one factor. Another important reason for industrialists' passivity in this respect is certainly to be found in British industry's comparatively low expenditure on, and lack of opportunity for, research — its

15. The founder of the company, Ivan Levinstein (1845–1916), was born in Charlottenburg, had studied in Berlin and came to Manchester as a young man (Kargon, *Science in Victorian Manchester*, p. 205. See also the obituary in *Nature*, 97, 1916, p. 89).
16. In 1926 the British Dyestuffs Corporation was taken over by the newly founded Imperial Chemical Industries (ICI), which was established in reaction to the foundation one year previously of IG Farbenindustrie in Germany. For details, see Reader, *Imperial Chemical Industries*, vol. 1, pp. 272–80, 317–450; Haber, 'Government Intervention', pp. 83–4; idem, *The Chemical Industry 1900–1930*, pp. 234–5 and 291–301; Miall, *British Chemical Industry*; Pollard, *British Economy*, pp. 42–4; Rees, *Trusts in British Industry*, pp. 161–2.
17. Haber, 'Government Intervention', p. 86. While before the First World War only the German Reich and Switzerland exported synthetic dyes, by the 1920s Britain was among the more important exporters.
18. See, for example, *Nature*, 94 (1914–15), p. 61.

scientific and technical backwardness, as already diagnosed before 1914. British industry's confidence in undisturbed trade relations with Germany had led it to neglect investment in research and technical development. Nothing in Britain could match the large chemical research laboratories which German industry had developed since the 1880s. Thus, despite the release of German patents, British industry was not equipped to produce the goods needed at the beginning of the war. Early in September 1914 Norman Lockyer, fully aware of the situation, wrote in *Nature*: 'Our manufacturers will have the greatest difficulty in carrying out the Government's intention precisely in those branches of industry in which technical instruction of the most advanced kind, with accompanying research, has been most lacking in Britain'.[19] Some time later, a similar observation was made in *Nature*: 'There is a general lack of appreciation by manufacturers of the advantages to be derived from the application of science to industry, and a tendency to avoid the employment of scientifically trained men'.[20] In May 1915 the president of the Board of Trade, Walter Runciman, blamed Britain's backwardness in industrial research on the 'traditional conservatism' of British entrepreneurs.[21] The attitude of British industrialists, often criticised, especially by scientists, as inappropriate for the times, was also attributed to the national character. It is presented in exaggerated form here, but could undoubtedly be observed often enough before the war: 'This partial famine in essential scientific materials and apparatus is not due to any real want of scientific ability on the part of British inventors or manufacturers. It is due to causes which are very deep-seated. For one thing, our easy-going national temperament has found it less trouble to buy from abroad than to make for ourselves'.[22] But the war forced hesitant entrepreneurs to display initiative, to invest and to innovate. Suddenly the whole question of the purpose and value of scientific research acquired a totally new significance.

By 1916 at the latest it was becoming obvious that, under the enormous pressures of war, the British public's appreciation of science was changing and that science and industry were cooperat-

19. 'The War and After', *Nature*, 94 (1914–15), p. 30.
20. *Nature*, 97 (1916), p. 92.
21. 'Report of the Deputation to the Presidents of the Boards of Education and Trade', 6 May 1915, Ed. 24/1579, PRO.
22. *Nature*, 96 (1915–16), p. 694. On this see also Reader, *Imperial Chemical Industries*, vol. 1, p. 270: 'Lack of enthusiasm on the private side was only too evident. The users of dyestuffs were not at all anxious to put their money into a dyestuffs company which, sooner rather than later, would have to compete with the Germans'.

ing more closely. The Board of Education noted that 'the War has brought the professor and the manufacturer together, with results which neither of them is likely to forget'.[23] Awareness of the far-reaching social and economic consequences of the war, and of the new nature of this war, was widespread among scientists and politicians. 'This war', declared the Nobel Prize-winning chemist Sir William Ramsay, 'in contradistinction to all previous wars, is a war in which pure and applied science plays a conspicuous part.'[24] The First World War was early seen as a military conflict 'in which victory will depend as much upon science and machinery as upon men'.[25] According to J. A. Fleming, professor of electrical engineering at University College London, there were grounds for hoping that 'there will be some increased appreciation in the minds of the politicians who govern us of the enormous influence of scientific research and discovery, even in its most abstruse forms, on the prosperity and safety of the Empire'. For it had been clearly shown that 'this war is a war quite as much of chemists and engineers as of soldiers and sailors'.[26] This comparison was repeatedly drawn by Fleming and other scientists during the war. In the middle term it had some effect on the attitude of Parliament and the British public to science and technology.

An explanation made to the House of Commons early in 1916 by Walter Runciman, president of the Board of Trade, provides interesting insights into the learning process which was taking place at that time among both politicians and high ministry officials. Looking back over the year and a half which had passed since the war began, Runciman, who had been president of the Board of Education from 1908 to 1911, recapitulated:

> At the Board of Trade, we had the idea strongly fixed in our minds that merely by assistance given in Government Departments it would be

23. *Report of the Board of Education for the Year 1915–1916*, Cd. 8594 (London, 1917), p. 70.
24. Speech to the ninth annual meeting of the British Science Guild, 1 July 1915 (*Nature*, 95, 1915, p. 521). See also Morris W. Travers, *A Life of Sir William Ramsay* (London, 1956).
25. 'Science in National Affairs', *Nature*, 96 (1915–16), p. 195.
26. J. A. Fleming, 'The Organisation of Scientific Research' (lecture delivered to the Society of Arts on 9 February 1916), *Nature*, 96, (1915–16), pp. 692–6, quotation on p. 692. Fleming had already used almost exactly the same expression in his Inaugural Lecture at University College on 6 October 1915 (printed as 'Science in the War and after the War', *Nature*, 96, 1915–16, pp. 180–5). In June 1915 Fleming had pointed out in *The Times* that in the ten months which had passed since the war began, he had never been asked 'to serve on any committee, cooperate in any experimental work, or place expert knowledge, which it has been the work of a lifetime to obtain, at the disposal of the forces of the Crown' (*The Times*, 15 June 1915).

possible for us to foster trade and industry. Let me say quite frankly that I have never been so foolish as to foster that notion. Government Departments can do a great deal, and I believe they ought to do more; but without the personal ability, without the training, skill and industry of the individual nothing can be done by Government Departments. I therefore put down, as one of the first necessities of this country if she is to hold her own during times of war and when war is over, that we must improve our research methods, the education of our people, and the training of our young men.[27]

Eighteen months later Christopher Addison, who had taken over the Ministry of Munitions in 1916, spoke of 'a serious neglect of research and scientific work as applied to industry' in Britain.[28] And four years after the end of the war a leader in *The Morning Post* commented: 'One of the results of the War, in which the scientific brains of this country were mobilised to such good purpose, is an appreciable increase of public interest in the achievements of science, whether theoretical or practical. In pre-war days neither the man in the street nor the man at the club window could have been persuaded to read articles about the Einstein v. Newton controversy'. The paper believed that this was now the case. It suggested that it was up to scientists to consolidate this newly awakened interest so that it would not prove to be a merely transitory phenomenon. 'Publicity is the only remedy', recommended the paper, 'and our scientists, as a body, are greatly to be blamed for not devising some system of propaganda which would keep the public informed of their discoveries and inventions. Some scheme is required which would do throughout the year what the British Association for the Advancement of Science does for a few days in September.'[29]

The military conflict which, in August 1914, had replaced the economic and political rivalry between the German empire and Britain stimulated the interest of government, the public and industry in scientific research. It also resulted in Germany losing the status of a model for science organisation which it had long enjoyed in Britain. After August 1914 it was no longer possible to draw upon the German science system, German universities or *Technische Hochschulen*; in fact they were discredited in many publications and speeches. Scientific communication between the countries of the Entente and the Central Powers almost came to a

27. Hansard, Parl. Deb., H. C., 5th Series, vol. 77, col. 1360 (10 January 1916).
28. Ibid., vol. 95, col. 587 (28 June 1917).
29. *Morning Post*, 27 September 1922 ('The Scientific Front').

complete halt. 'Many of us', wrote Norman Lockyer in *Nature* in September 1914,

> have been great admirers of Germany and German achievements along many lines, but we have now learned that her 'culture' and admirable organisation have not been acquired . . . for the purpose of advancing knowledge and civilisation, but, in continuation of a settled policy, they have been fostered and used in order that a military caste in Germany, with the Kaiser at its head, shall ride roughshod over Europe.[30]

In Britain it was suddenly denied that German science and science organisation possessed any originality, and its superiority, proclaimed for decades, was discounted as a general delusion.[31] The ability of German scientists was now seen as lying not in creative investigation of the unknown, but as limited to skilful appropriation and economic exploitation of discoveries made elsewhere. The history of synthetic dyes was held up as an instructive example.[32] Early in 1915 *The Morning Post* wrote bombastically that the war 'would pull down from its pedestal and shatter for ever the notion of the German super-man in science, literature, art, of ingenuity created by German self-assertion, and supported by the effusive adulation of a few professors of our own, proud of a smattering of second-rate Teutonic learning'.[33] A few days before the end of the war Virginia Woolf recorded in her diary a conversation with the historian H. A. L. Fisher, president of the Board of Education in Lloyd George's cabinet. It reflects the switch from admiration to aversion in the attitude towards Germany of many British intellectuals. 'I was a great admirer of the Germans in the beginning', said Fisher. 'I was educated there, and I've many friends there, but I've lost my belief in them. The proportion of brutes is greater with them than with us. They have been taught to be brutes.'[34] We can dispense with a more detailed discussion of the arguments, often fantastic and repetitive, which scientists and politicians developed after the war began in order to explain their

---

30. 'The War — and After', *Nature*, 94 (1914–15), p. 29.
31. Lord Haldane's political fate illustrates the change of mood in Britain. His admiration for the German system of education and science marked him as a friend of Germany after the outbreak of war. In 1915 he resigned his position as lord chancellor. On the hostility and defamation he was exposed to after August 1914, see Sommer, *Haldane of Cloan*, pp. 309–29; and Crowther, *Statesmen of Science*, pp. 271–300. On this in general, see also Badash, 'British and American Views', pp. 99–100.
32. Thus the zoologist E. Ray Lankester (*Nature*, 94, 1914–15, p. 486).
33. *Morning Post*, 21 January 1915 ('German Super-man Deposed').
34. H. A. L. Fisher on 15 October 1918 in Olivier Bell (ed.), *The Diary of Virginia*

rejection of the German scientific system. They have been comprehensively described by Lawrence Badash.[35]

The change in the assessment of science and its value for industry and society that took place in Britain during the war was thus accompanied by a change in the framework of reference used for promoting and organising science. While negative references to the German scientific system and the quality of German research became more and more frequent in British newspapers and scientific journals, the American system was increasingly held up as a model for British planning. As the 'pioneering center' for science,[36] and especially for its application in industry, the USA now came to the fore. How and why the USA became a 'world centre' for science does not fall within the scope of this study; this problem must be seen in the context of American history of science. Here we need only register that the First World War finally made British science planners aware of the high standard of science in the USA and of its exemplary organisational forms. *Nature* provides one indicator of the process by which the American model replaced the German model accepted for so long in Britain. During the course of the war *Nature* published numerous articles about American scientific institutions and methods of research.[37] The burgeoning popularity of American science organisation went hand in hand with a detailed examination, in public and in government circles, of its institutional preconditions. Documents from the Department of Scientific and Industrial Research show that British science continued to orientate itself by the American model throughout the 1920s. Within a few years international scientific relations, as they had emerged in the nineteenth century, had undergone a lasting change as a result of the First World War: Germany had forfeited its almost unchallenged supremacy in science. It no longer set the pace in the organisation of scientific research, as it had done since the first half of the nineteenth century. This change, which is of interest beyond the history of science, is reflected in the fact that German lost its position as the language of science because of the war and the international boycott of German scientists imposed after the war by the French in particular.[38] After 1918 German scientific

---

Woolf, vol. 1: *1915–1919* (London, 1977), p. 204.
35. Badash, 'British and American Views'.
36. Ben-David, *The Scientist's Role*, p. 176.
37. The following are instructive examples: 'Industrial Research in the United States', *Nature*, 97 (1916), pp. 270–2; 'The National Research Council of the United States', *Nature*, 97 (1916), pp. 464–6; 'Research Institutions in the United States', *Nature*, 99 (1917), pp. 274–6.
38. On the international boycott of German scientists during the 1920s, and on

journals never regained the outstanding significance which they had possessed for the international scientific community before the First World War.

The Organisation of Science by the State: The Founding of the Department of Scientific and Industrial Research

The various measures which had been introduced since August 1914 by the British government, by industry and the large scientific societies to adapt industrial output to the requirements of the war, on the whole fulfilled expectations. Initial shortages in the supply of certain chemical products, medicines and optical glass were overcome with the aid of scientists. But the scale of the war in Europe soon made it clear that a much closer relationship was needed between science, industry and politics than had been anticipated when improvised measures were introduced in the first few months of the war. They had not been able to eliminate the fundamental weaknesses of the British scientific system. In the autumn of 1915 *Nature* emphasised that although more than a year had passed since the war began, there were still a large number of scientists 'whose energies and expert knowledge are not being effectively used'.[39] Politicians with responsibility for these areas were coming around to a view which had been expressed for years by the British science lobby: that greater and more systematic support and more comprehensive organisation of science by the state could no longer be put off in Britain. Since at least the beginning of the war, the issue at stake had no longer been state guidance, but state coordination and support of research, and making it more beneficial to society. From 1915 the government began to tackle these problems.

After the outbreak of war, state interest in science was concentrated mainly on military research, in Britain as in other countries involved in the war. Even then, however, military research and civil research interlocked so closely that they could not always be

---

the German 'anti-boycott', see Daniel J. Kevles, '"Into Hostile Political Camps": The Reorganization of International Science in World War I', *Isis*, vol. 62 (1971), pp. 47–60; Brigitte Schroeder-Gudehus, *Les scientifiques et la paix. La communauté scientifique internationale au cours des années 20* (Montreal, 1978); idem, 'Challenge to Transnational Loyalties: International Scientific Organizations after the First World War', *Science Studies*, vol. 3 (1973), pp. 93–118; idem, 'Deutsche Wissenschaft und internationale Zusammenarbeit 1914–1928', PhD dissertation, University of Geneva, 1966.
39. *Nature*, 96 (1915–16), p. 195.

clearly distinguished from each other. Research for the military sector was undertaken primarily by the Admiralty and the War Office joined by the newly founded Ministry of Munitions and Air Ministry in mid 1915 and January 1918, respectively. From 1914 on they created a confusing array of new research agencies and bodies. These will not be described here; neither will the numerous support programmes and new forms of organising research which were often implemented in cooperation with universities, technical colleges, non-university research institutes and private industry. A comprehensive history of military research in Britain during the First World War, in so far as it can unequivocally be identified as such, still needs to be written.[40] The following discussion is concerned solely with the organisation of science in the civilian sector. Inevitably, under the conditions of war, however, it could also be used for military purposes.

Faced with the necessity of organising the country's scientific potential beyond the narrow field of the organic-chemical industry quickly and effectively to aid the war effort, the British government attempted to adopt an unconventional course. The science policy which took shape at this time broke with previous practice in two respects: it brought together responsibility for state support of science and for science organisation in a newly created central agency, and it endeavoured to place the promotion of science in a long-term context. From early in 1915 the government's plans assumed more definite shape. After a conversation with David Lloyd George, chancellor of the exchequer in Asquith's government, Christopher Addison, who at this time was still parliamentary secretary to the Board of Education, noted:

> Our recent experience of the enormous handicap the country is suffering from owing to the neglect of scientific training and its application to trade and industrial methods has deeply impressed both the Chancellor and his colleagues. There was no need to rub it in . . . He [Lloyd George] has also realized . . . that at any cost we should have to wake up in this respect when the war is over.[41]

---

40. Further references can be found in MacLeod and MacLeod, 'War and Economic Development'; idem, 'Social Relations of Science', pp. 308–9; Roy M. MacLeod and E. Kay Andrews, 'Scientific Advice in the War at Sea, 1915–1917: the Board of Invention and Research', *Journal of Contemporary History*, vol. 6, no. 2 (1971), pp. 3–40; Marwick, *The Deluge*, p. 230; Haber, *The Chemical Industry 1900–1930*, pp. 208–10, 231; *Report of the Committee of the Privy Council 1915–16*, p. 42; Ian Varcoe, 'Scientists, Government and Organised Research in Great Britain 1914–16: The Early History of the DSIR', *Minerva*, vol. 8 (1970), pp. 196–7.

41. Addison, *Four and a Half Years*, vol. 1, p. 56 (diary entry dated 19 January 1915).

Walter Runciman also indicated that the Liberal government was working on a fundamental revision of the state's traditional attitude towards science when, a few months after Addison's conversation with Lloyd George, he said in the House of Commons: 'The whole question of Government assistance to scientific research for the benefit of British industry is at present receiving careful consideration'.[42] At almost the same time the newly appointed president of the Board of Education, Arthur Henderson, wrote to Lord Rayleigh: 'It is obviously a matter of the first importance that a new departure in the relations of the State to scientific industry should take place under the most favourable auspices'.[43] The result of the government's deliberations was the creation of the Department of Scientific and Industrial Research (DSIR).

The concept behind the DSIR came from various sources.[44] Several of those who had played a leading part in innovations in science organisation before the shock of August 1914 were involved in its creation. Lloyd George, Lord Haldane and Christopher Addison head the list. Like other scientific institutions created in Britain since the turn of the century, the DSIR had several fathers, but the idea was clearly born and nurtured in the Board of Education. Strangely enough, industry, the Royal Society and the Board of Trade remained in the background while this attempt was made to encourage greater cooperation between research and industry. To repeated queries from the Treasury as to whether this issue did not primarily involve the Board of Trade, Runciman replied by expressly welcoming the leadership displayed by the Board of Education.[45] As a state institution with responsibility for coordinating all government measures to promote civilian research, the DSIR was an original innovation of British science policy. At that time neither the German Reich nor the USA, whose scientific system had come under close scrutiny by British science planners since the beginning of the war, possessed an institution which could have served as a model for the British foundation. In

42. Hansard Parl. Deb., H. C., 5th Series, vol. 71, col. 1475 (11 May 1915).
43. Arthur Henderson to Lord Rayleigh, 4 June 1915, Ed. 24/1580, PRO.
44. On the early history of the DSIR, see Roy M. MacLeod and E. Kay Andrews, 'The Origins of the D.S.I.R.: Reflections on Ideas and Men, 1915–16', *Public Administration*, vol. 48 (1970), pp. 23–48; Eric Hutchinson, 'Scientists as an Inferior Class: The Early Years of the DSIR', *Minerva*, vol. 8 (1970), pp. 396–411; idem, 'Government Laboratories and the Influence of Organized Scientists', *Science Studies*, vol. 1 (1971), pp. 331–56; Harry Melville, *The Department of Scientific and Industrial Research* (London and New York, 1962); Varcoe, 'Scientists'. An important source is *Report of the Committee of the Privy Council for Scientific and Industrial Research for the Year 1915–16*, Cd. 8336 (London, 1916).
45. See Ed. 24/1576 (28 June 1915), PRO.

many respects, however, the newly established Medical Research Committee provided a framework, and its organisational structure and working methods were frequently referred to during the period of the DSIR's development. In the same way that the Medical Research Committee had promoted medical research in the United Kingdom since 1913 by administering public funds, a central body was intended in future to support basic and industrial research by administering public funds, to coordinate research programmes and to promote cooperation between science and industry.

One view which early gained currency, even in official circles, was that the DSIR had been established largely in response to agitation by British scientists after the beginning of the war and pressure from the large specialist learned and scientific societies.[46] This widespread notion is only partially correct; it reflects no more than one phase in the institution's genesis. Public discussion of Britain's research deficit and the demonstration of the faults which its scientific system displayed by the second half of the nineteenth century at the latest undoubtedly influenced the government as much as did the various *démarches* undertaken by learned and scientific societies. Finally, in May 1915 — when the government's plans for the new state institution were already far advanced — the Royal Society and the Chemical Society together sent a deputation to ask the government to hasten the achievement of cooperation between industry and universities by making more public funds available.[47] In their study based on new sources, MacLeod and Andrews have shown, however, that plans for a central British organisation to promote pure and applied science go back to before 1914. Ever since 1909 a department had existed within the Board of Education whose function was to deal with ways of improving research at British universities and of making its results more accessible to industry.[48] Since that time various plans had been discussed in the ministry, but they suddenly acquired a completely new quality as a result of the outbreak of war.

The appointment in August 1914 of Christopher Addison as parliamentary secretary to the Board of Education introduced some

---

46. 'The action of the government in setting up the new machinery for the encouragement of research was accompanied, it not instigated, by vigorous discussion and debate in the public press and the learned societies' (*Report of the Committee of the Privy Council 1915–16*, p. 19).

47. *Nature*, 95 (1915), pp. 295–6; 'Report of the Deputation to the Presidents of the Boards of Education and Trade', 6 May 1915, Ed. 24/1579; Poole and Andrews (eds.), *Government of Science*, pp. 54–7; Varcoe, *Organizing for Science*, pp. 11–12; McLeod and Andrews, 'Origins of the D.S.I.R.', pp. 30–1.

48. Ibid., pp. 25–6.

urgency into these deliberations. Addison was a science organiser of unusual administrative talent, who had been a scientist himself prior to his election to the House of Commons in 1907. Since 1911 he had demonstrated his abilities during the establishment of the Medical Research Committee. Addison's friendship with Lloyd George also dates from this time. It now helped him in his efforts 'to link up science to industry better than we are doing'.[49] In preparing his bill of December 1914 to reform technical education in Britain and to create a coordinating 'research council', Addison could build on the preliminary work which had been done by the Board of Education. In their assessment of Addison's role in the development of the DSIR, MacLeod and Andrews point out that the 1914 bill, which provided the occasion for the discussion with Lloyd George referred to above, did not develop any new ideas.[50] Later Addison himself made it clear that his suggestions were based on ideas which had been circulating for some time in the Board of Education and in public.[51] In particular, the proposal to set up a 'Central Council of Commercial and Industrial Research' or a 'Central Advisory Committee on Research' — as it was called at this early stage — was made by Frank Heath, a colleague of Addison's who was in charge of matters relating to university affairs in the Board of Education at that time.[52] Undeniably, however, the credit for filtering out from among the most diverse ideas about state promotion of research those which had a good chance of realisation, in both organisational and political terms, must go to Addison. In addition, he deserves credit for having made his proposals to the responsible minister at an appropriate moment, and for pursuing them with great determination after 1914.

From January 1915 on Addison had the backing not only of Lloyd George, still chancellor of the exchequer, but also of the lord chancellor, Haldane, and of the president of the Board of Education, Sir Joseph A. Pease, Runciman's successor in this office in which ministers changed so frequently. On instruction from the three ministers, Addison became chairman of a ministerial committee which was to lay down, on the basis of his suggestions of

---

49. Addison, *Four and a Half Years*, vol. 1, p. 56; idem, *Politics from Within*, vol. 1, pp. 47–8. See also Morgan and Morgan, *Portrait of a Progressive*, pp. 22–3.
50. MacLeod and Andrews, 'Origins of the D.S.I.R.', p. 27.
51. Addison, *Four and a Half Years*, vol. 1, p. 57.
52. See MacLeod and Andrews, 'Origins of the D.S.I.R.', pp. 25–7. H. Frank Heath (1863–1946), professor of English literature at Bedford College, London (1890–96), civil servant in the Board of Education (1903–16), permanent secretary to the DSIR (1917–26).

December 1914, the organisational structure and the scope of the future government body's functions. The committee's deliberations issued in a result relatively quickly: in May 1915 Pease wrote a memorandum which gained cabinet approval for the planned foundation, and another memorandum, based on the first one and explaining the conception, functions and financing of the new organisation, was issued in July 1915.[53]

The Board of Education's short memorandum again refers to the 'strong consensus of opinion among persons engaged both in science and in industry that a special need exists at the present time for new machinery and for additional State assistance in order to promote and organise scientific research with a view especially to its application to trade and industry'.[54] It explicitly emphasises that establishing a state institution to promote research was not only a response to the immediate demands of the war, but that it was also intended to be a permanent solution after the war. The memorandum suggested that two committees, one political and the other scientific, be set up to implement the proposals which had been made. These two committees were created by an Order in Council immediately after the memorandum was published.

The Committee of the Privy Council was to be responsible to Parliament for the distribution of the new institution's grant-in-aid. The lord president of the Privy Council was appointed chairman of the nine-member committee. Its other *ex officio* members were the president of the Board of Education, who was deputy chairman, the chancellor of the exchequer, the president of the Board of Trade, the secretary for Scotland and the chief secretary for Ireland. Lord Haldane, Sir Joseph A. Pease, who had been succeeded at the Board of Education by the Labour MP Arthur Henderson, and A. H. D. Ackland, also a former president of the Board of Education (1892–5), were co-opted as independent members when the committee was constituted in the summer of 1915. In May 1916 the colonial secretary was added as another *ex officio* member. The committee for the 'promotion of industrial and scientific research' was allocated to the Privy Council rather than to the Board of

---

53. 'Proposals for a Scheme of Advanced Instruction and Research in Science, Technology and Commerce', Ed. 24/1581, printed in Poole and Andrews (eds.), *Government of Science*, pp. 49–53. See also Pease's note on a meeting with the prime minister, Asquith, and the chancellor of the exchequer, Lloyd George, on 3 May 1915: 'The Prime Minister, without waiting for me to develop my case, said that he approved the Scheme and thought . . . that Mr George might proceed to authorise the expenditure' (Ed. 24/1576, 5 May 1915). *Scheme for the Organization and Development of Scientific and Industrial Research*, Cd. 8005 (London, 1915).
54. *Scheme for the Organization*, p. 1.

Education largely because of the Board's limited authority under its constitution: it had responsibility only for England and Wales, whereas the Privy Council covered the whole of the United Kingdom. As the lord president was always a member of the House of Lords, however, the president of the Board of Education as deputy chairman had to answer for the committee's spending policy in the House of Commons.

To the Committee of the Privy Council was attached an Advisory Council to advise it in all research questions and matters connected with the founding and development of new scientific institutions which were intended to serve the needs of industry. It was conceived of as 'the body which effectively decides the Research policy'.[55] The Advisory Council therefore had the difficult job of defining areas of special interest in research and formulating a long-term public science policy. The responsibility for implementing it lay with the interdepartmental Committee of the Privy Council. The notion of a body of this sort was not new. The idea that a 'Council of Science' would be useful to advise a prospective minister for science had already come up during the Devonshire Commission's discussions in the 1870s. Norman Lockyer's speech to the British Association for the Advancement of Science in 1903 had also anticipated the thrust of the measures taken in 1915: he had suggested setting up a 'scientific council' under the jurisdiction of the Privy Council.[56] And at the founding of the British Science Guild in 1905, Haldane, referring back to the Devonshire Commission's discussions, had made a suggestion similar to Lockyer's — the establishment of a 'scientific corps under a permanent committee' to advise the government in all matters concerning science.[57] As Poole and Andrews have pointed out, the Advisory Council's creation represented 'the culmination of over fifty years of argument and persuasion on the part of both scientists and statesmen for the consistent support of scientific research on a significant scale'.[58]

The memorandum of July 1915 recommended that the Advisory Council 'should act in intimate cooperation with the Royal Society and the existing scientific or professional associations, societies and institutes, as well as with the Universities'. Its members were to be 'eminent scientific men and men actually engaged in industries dependent upon scientific research'.[59] These guidelines were fol-

55. R. L. Morant to the permanent secretary to the Board of Education, 21 June 1915, Ed. 24/1576, PRO.
56. See *Nature*, 68 (1903), p. 446.
57. See *Nature*, 73 (1905–6), p. 12.
58. Poole and Andrews (eds.), *Government of Science*, p. 27.
59. *Scheme for the Organization*, pp. 2–3.

lowed with the result that when the Advisory Council was constituted in August 1915, seven of its eight members were Fellows of the Royal Society. They included the famous physicist Lord Rayleigh, the chemists Raphael Meldola, Sir George T. Beilby and Sir Richard Threlfall, the engineer Bertram Hopkinson and the chairman of the Committee on Grants to University Colleges, Sir William S. McCormick. The Advisory Council's members were appointed by the lord president, acting for the government, after consultation with the president of the Royal Society. By 1915, therefore, the Royal Society had not yet fully given up its function as an unofficial advisory body for the government. But, as the permanent secretary to the Board of Education noted, the Royal Society did not express any desire that 'the work of promoting research should be entrusted to the Royal Society as such'.[60]

By creating the Committee of the Privy Council and the Advisory Council, the state had 'recognised the necessity for organising the national brain power in the interests of the nation at peace'.[61] This completed the first phase in the foundation history of the DSIR. No doubt the autonomous Advisory Council was intended to be more than a consultative body; it was to play a key part in formulating state science policy. The Order in Council of July 1915 emphasised this: 'The said Council may itself initiate such proposals and may advise the Committee on such matters whether general or particular, relating to the advancement of trade and industry by means of scientific research'.[62] Its recommendations would represent 'the progressive realization of a considered programme and policy'.[63] This comprehensive transfer of authority was based on the 'modern experience' that 'a properly composed body . . . can be allowed a sufficiently free hand by Parliament to be given the right to disburse the grant-in-aid, Parliament being content with no more than a Ministerial mouthpiece'.[64]

When the provisions laid down in 1915 were revised at the end of 1916, the Advisory Council retained its central position in the development of British science policy. The revision became necessary because the arrangement by which the Committee of the Privy Council was closely linked with the Board of Education, whose staff did its administrative work, soon proved to be unsatisfactory.

60. Note of June 1915, Ed. 24/1572, PRO.
61. *Report of the Committee of the Privy Council 1915–16*, p. 9.
62. Order in Council of 28 July 1915, printed in ibid., p. 45.
63. Fifth Draft Scheme for the Organization and Development of Scientific and Industrial Research, 13 June 1915, DSIR 17/1, PRO.
64. R. L. Morant to the permanent secretary to the Board of Education, 21 June 1915, Ed. 24/1576, PRO.

While the Board of Education had close connections with the universities, its officials were not experts in this new area, and it did not have the contacts with industry which were necessary to promote industrial research effectively. 'The scheme was conceived rightly enough', commented *Nature* in 1915,

> but when it passed into the hands of officials of the Board of Education much of its early promise was lost . . . The belief that the expert — whether scientific or industrial — has to be controlled or guided by permanent officials having no special knowledge of the particular subject in hand is typical of our executive system. While such a state of things exists, most of the advantages of enlisting men of science for national services must remain unfulfilled.[65]

In view of the fact that its functions were of a different nature, and that it had hitherto operated in other areas, the Board of Education accepted this criticism. As early as November 1916 the Board began negotiations with the Treasury aimed at transforming the Committee of the Privy Council into an independent body. They took place against the background of strong pressure from the science lobby, which in May 1916 had created a new forum for itself, the Neglect of Science Committee.[66] Within a few weeks the establishment of the Department of Scientific and Industrial Research was settled. The progress of the plans was hastened when Lloyd George became prime minister and the whole government administration was restructured. From time to time during these negotiations the creation of a Ministry of Science and Industry was considered,[67] something the British science lobby had been calling for since the war began. But in the end the government stopped short of implementing such a far-reaching change. Like the Committee of the Privy Council before it and the Medical Research Council after 1920, the Department of Scientific and Industrial Research, established in December of 1916, was responsible to the lord president. Formally, he was assisted by an interdepartmental committee, but in practice its significance diminished. The lord president was the cabinet member with responsibility for science, but as holders of this office after 1915 (with the exception of Arthur J. Balfour) showed little interest in the DSIR, administrative leadership in practice fell to the permanent secretary appointed to the

---

65. 'Science in National Affairs', *Nature*, 96 (1915–16), p. 195.
66. See above, p. 97 in this volume.
67. Thomson, *Medical Research*, vol. 1, p. 40; Hans Daalder, *Cabinet Reform in Britain, 1914–1963* (Stanford, Calif., 1963), pp. 37–53.

Department, while the Advisory Council made science policy decisions within the financial constraints laid down by Parliament.

British scientists acknowledged that the DSIR was an attempt by the state to respond to the qualitative and quantitative changes which had taken place in research since the mid nineteenth century. *Nature* called its establishment 'a most hopeful sign of the times',[68] and looking back to the war years, called them 'an epoch in the history of science of which it is impossible to exaggerate the significance and potentiality'.[69] At the 1921 annual meeting of the British Association for the Advancement of Science, held in Edinburgh, its president, the chemist Sir T. Edward Thorpe, observed with satisfaction that 'research has now become a national and State-aided object' and that 'for the first time in our history its pursuit with us has been organised by Government action'.[70] As a public body whose function was to formalise cooperation between government and industry in promoting research, the DSIR, like the Committee of the Privy Council before it, represented a new type of institution. Like the longer-established Medical Research Committee, it subsequently became a model for the reform of science organisation in other countries. Immediately after its creation, and partly on the instigation of the British government, the DSIR was copied when similar institutions were set up in the dominions — in South Africa, Australia, New Zealand, India and Canada. These parallel organisations were intended also to promote scientific cooperation within the empire. Even in the USA a great deal of interest was taken in the new British institution when similar government agencies were being planned.[71]

For almost fifty years the DSIR was Britain's foremost body for the promotion and coordination of civil research 'with a view to its application to trade and industry'.[72] In 1964 it was disbanded in the course of an extensive government reorganisation undertaken by Harold Wilson's new Labour cabinet; it was replaced by the Science Research Council (SRC) and the Ministry for Technology. During the years of its existence, the DSIR promoted research, excluding only medicine, agriculture, fisheries and forests, in three ways:

68. *Nature*, 96 (1915–16), p. 259.
69. *Nature*, 107 (1921), p. 802.
70. T. Edward Thorpe, 'Some Aspects and Problems of Post-war Science, Pure and Applied', *Nature*, 108 (1921), p. 48.
71. See the explanation given by William Hewins, under-secretary of state in the Colonial Office, on 29 November 1917 in the House of Commons (Hansard, Parl. Deb., H. C., 5th Series, vol. 77, cols. 2204–5). See also *Nature*, 97 (1916), pp. 464–5; *Nature*, 99 (1917), p. 186; *Nature*, 101 (1918), pp. 155–8.
72. *Report of the Machinery of Government Committee, 1918*, Cd. 9230 (London, 1918), p. 29.

(1) Like the Medical Research Committee, the DSIR provided non-recurring grants or long-term financial support for research projects by individuals or groups of scientists, learned and scientific societies and university departments. As a rule, basic and applied research were not explicitly differentiated, but from the start projects which seemed to promise quick industrial applications were favoured. Research which took place outside the universities was also clearly favoured. The Advisory Council obviously subscribed to the English tradition which saw universities primarily as educational and teaching institutions, and only secondarily as centres of research.

(2) In addition to promoting scientific work in any scientific institution in the United Kingdom, the DSIR also financed research in existing 'special institutions', which were placed under its direct control immediately after it was established. As in the case of the Medical Research Committee, there were few such institutions at first, but their number gradually increased over the years. In 1918 the DSIR took over the largely state-financed National Physical Laboratory after protracted negotiations, mostly about the extent to which the Laboratory should retain the scientific autonomy guaranteed at its foundation. Scientific responsibility for this large state-run laboratory, which by 1917 employed a scientific staff of more than 150 as its tasks multiplied during the war, remained with its Executive Committee, whose composition continued to be determined by the Royal Society.[73] In 1919 the DSIR took over the Geological Survey of Great Britain from the Board of Education. Later, the important Chemical Research Laboratory, renamed the National Chemical Laboratory in 1928, also came under the control of the DSIR.

(3) After mid 1917, the DSIR, together with industry and the Board of Trade, developed a form of science promotion which was new in Britain. In this case the idea can be traced back to Christopher Addison. 'If anything is to be done', he had written in his bill of December 1914, 'it must mean a courageous State-inspired scheme to deal with the big industrial groups who can define the chief needs of their industry and be associated with the scheme.'[74] It is difficult to assess how much these deliberations were influenced by A. P. M. Fleming's study, written in 1916 after he had visited the USA for the British government; his conclusions had included a recommendation that companies in Britain work together more

---

73. Hutchinson, 'Government Laboratories', p. 334; idem, 'Scientists and Civil Servants'; idem, 'Scientists as an Inferior Class'.
74. Addison, *Four and a Half Years*, vol. 1, p. 56.

closely by establishing common research laboratories.[75] Autonomous Research Associations were to be the new means of promoting industrial research in Britain. Half of their costs were borne by the Department, and the other half by branches of industry with related research problems.[76] Thus it was not individual companies, most of which had inadequate research facilities, who cooperated with the DSIR, but groups of companies or whole branches of industry that voluntarily agreed to work together on certain research projects. setting aside their competitiveness and own specific preferences. The financial basis for this new cooperation between state and industry was provided by the Imperial Trust for the Encouragement of Scientific and Industrial Research, with capital of £1,000,000 voted by Parliament for five years. Pease, president of the Board of Education, had already requested this amount in his cabinet memorandum dated May 1915. The extraordinarily generous endowment of this fund from public funds, going far beyond any previous allocation for science, alone shows how radically the British government's attitude towards science had changed since the end of the nineteenth century. But the grant was also based on the realisation that 'it would not be possible to develop systematic research on a large scale unless the Government were in the position to assist financially over an agreed period of years'.[77] Out of the Imperial Trust, the DSIR matched, on a pound-for-pound basis, money put forward by industry for cooperative research. This was obviously intended to stimulate industry's interest in research, and its financial involvement in research projects. Over the years the government's strategy on the whole proved successful. 'At any rate', reported the Committee of the Privy Council in 1917, 'the larger and more prosperous industries might be expected, after an initial impetus, to find themselves both willing and able to continue the work of research without direct assistance from the State.'[78] Tax concessions were also granted to encourage participation in the scheme; money put into research was no longer subject to taxation. Patents resulting from the work of Research

75. A. P. M. Fleming, *Industrial Research in the United States of America* (London, 1917), p. 52. See also Ronald S. Edwards, *Co-operative Industrial Research: A Study of the Economic Aspects of the Research Associations Grant-Aided by the Department of Scientific and Industrial Research* (London, 1950), pp. 36–7.

76. On the Research Associations see (apart from the study by Edwards) H. F. Heath and A. L. Hetherington, *Industrial Research and Development in the United Kingdom: A Survey* (London, 1946); and Rose and Rose, *Science and Society*, pp. 40–2.

77. *Report of the Committee of the Privy Council for Scientific and Industrial Research for the Year 1916–17*, Cd. 8718 (London, 1917), p. 4.

78. Ibid.

Associations could be used by all the companies involved.

By 1921 twenty-one Research Associations had been created on the initiative of industry, under very different initial conditions, and despite many reservations on the part of both industry and the universities.[79] There were, among others, Research Associations for photography, scientific instruments, engine manufacturing, iron, glass, fuel and radio technology, almost all 'new' industries. Established industries like coal, steel and shipbuilding did not participate. The chemical industry, too, which had undergone a process of concentration during the First World War, did not set up a Research Association, pointing to its existing research capacity. The fact that in the following years the applications for funds offered by the Imperial Trust fell far short of the total available suggests that industrial research was not pursued with great intensity in Britain. Nevertheless, by 1920 about half of British industry was represented in Research Associations. The most active and financially strongest of these, which were joined by the majority or all of the companies in their respective branches of industry, were the Glass Research Association, the British Scientific Instruments Research Association, the British Cotton Industry Research Association, and the British Electrical and Allied Industries Research Association. Their annual budget was, as a rule, in the region of £10,000 to £20,000. Other Research Associations had much smaller funds because industry was less interested, and their research was consequently much more limited.

Research Associations on the whole contributed to improving the institutional conditions for practically orientated research in Britain after the First World War. By creating more positions, they also improved the job opportunities available to the growing number of trained scientists.[80] The establishment of Research Associations thus in some ways reflects the development by which scientific work shed its character as an amateur pursuit and the scientific profession established itself outside the universities in Britain as it had done in Germany much earlier. The next chapter deals with this 'belated professionalisation' of science in Britain.

---

79. Sanderson, *Universities*, p. 234; Rose and Rose, *Science and Society*, pp. 41–3. Several Research Associations found that industry had little interest in them, and therefore stopped work after only a few years.

80. But high expectations were disappointed. In 1925 the Research Associations employed only 144 scientists (MacLeod and MacLeod, 'Social Relations of Science', p. 332).

5
Science as a Profession: From Amateur to Outsider?

Science in Victorian England

'Much less recognition by the State is given in Britain to original work than in foreign countries. All the greater is the need for the energetic impulse of private associations.'[1] This observation made in 1903 by Lord Reay, president of the British Academy for the Promotion of Historical, Philosophical and Philological Studies established in 1902, indicates the position which science occupied in public awareness at the time. It also points to the fact that, in light of this lack of awareness, the task of promoting science fell to the learned and scientific societies and academies. The state's caution and the lack of interest in science and research displayed by both Houses of Parliament simply reflected the low public appreciation of science in Britain until the First World War. Consequently, financial support for science from public funds was relatively limited and conditions for scientific work were unfavourable. Under these difficult circumstances, what did it mean to take up — and here we refer to Max Weber's famous lecture of 1919[2] — science as a profession in Britain in the nineteenth and early twentieth centuries? What can be said about the social status of scientists in Britain at this time? Hitherto, few aspects of the subject have been adequately scrutinised. The following general observations about conditions in Britain at the time are therefore incomplete in many respects. More detailed research on the social position of scientists in Britain since the early nineteenth century will undoubtedly refine many of these views.

In the second chapter of this book it was shown that the people

---

1. From Lord Reay's speech to the British Academy's first annual meeting on 26 June 1903, in *Proceedings of the British Academy 1905–1906* (London, n.d.), p. 9.
2. Max Weber, 'Wissenschaft als Beruf' in idem, *Gesammelte Aufsätze zur Wissenschaftslehre*, ed. by Johannes Winckelmann, 3rd edn (Tübingen, 1968), pp. 582–613.

and organisations who lamented the 'neglect' of science in Britain since the early nineteenth century made various attempts to encourage the British public and British politicians to change their attitude towards science. Their efforts were not free of exaggerations and distortions, used with an educative purpose. 'It is known to all the world', wrote *Nature* in 1873, complaining about the unsatisfactory conditions under which British scientists had to work, 'that science is all but dead in England.'[3] According to the biologist and zoologist T. H. Huxley, the natural sciences had to contend with public prejudices which separated them from the humanities by an unbridgeable gulf. C. P. Snow returned to this argument in a slightly modified form after the Second World War, when he developed his theory of the two cultures into which British society falls: that of the artists, writers and scholars, and that of the scientists and engineers.[4] 'How often have we not been told', Huxley had said at the opening of Mason College in Birmingham in 1880,

> that the study of physical science is incompetent to confer culture; that it touches none of the higher problems of life; and what is worse, that the continual devotion to scientific studies tends to generate a narrow and bigoted belief in the applicability of scientific methods to the search after truth of all kinds? How frequently one has reason to observe that no reply to a troublesome argument tells so well as calling its author a 'mere scientific specialist'. And, I am afraid it is not permissible to speak of this form of opposition to scientific education in the past tense.[5]

To the president of the Royal Society it was obvious that the educational system of the country must take the blame for the persistence of these prejudices:

> The evidence seems clear that the present inappreciative attitude of our public men, and of the influential classes of society generally, towards scientific knowledge and methods of thought must be attributed to the too close adherence of our older Universities, and through them of our public schools, and all other schools in the country downwards, to the traditional methods of teaching of medieval times.[6]

In similar vein, the Royal Society's Annual Report two years later suggested that the reason for the limited public support for science

---

3. *Nature*, 8 (1873), p. 21.
4. C. P. Snow, *The Two Cultures and the Scientific Revolution* (London, 1959). The second edition was published in 1964 as *The Two Cultures and a Second Look*.
5. Quoted in *Nature*, 60 (1899), p. 324.
6. 'Anniversary Meeting of the Royal Society', *Nature*, 67 (1902–3), p. 108.

in Britain had to be sought 'in the absence in the leaders of public opinion, and indeed throughout the more influential classes of society, of a sufficiently intelligent appreciation of the supreme importance of scientific knowledge and scientific methods in all industrial enterprises, and indeed in all national undertakings'.[7]

Advocates of greater support for science repeatedly declared, especially after the turn of the century, that this peculiar disregard for science was characteristic of British society. Lord Reay's remark quoted at the beginning of this chapter is one example. 'The English people do not manifest that interest in, and belief in the powers of science which is noticeable among the peoples of the Continent, or of America', observed Norman Lockyer in 1904 in the memorandum in which he recommended that the British Science Guild be founded. 'In spite of the efforts of many years, the scientific spirit, essential to all true progress, is still too rare, and, indeed, is often sadly lacking in some of those who are responsible for the proper conduct of many of the nation's activities.'[8] Haldane fully endorsed Lockyer's assessment when, at the British Science Guild's constituent assembly, he said that 'we lived in a country where science was not so much appreciated as it should be. Our people liked to see cash over the counter, and they did not like to wait for deferred payment'.[9] And in 1906, referring with admiration to Germany and the German science system, the committee chaired by Haldane which had been preparing the foundation of Imperial College of Science and Technology since 1904, wrote that 'in no country has more thought been given to the development of a University system than in Germany; the pride taken in University institutions and the distinction conferred on the leaders of learning by the popular estimation in which they are held are not surpassed elsewhere'.[10] In contrast to these supposedly ideal conditions in Germany, in Britain 'public interest in scientific research must still be considered to be on a low level — certainly lower here than in many other leading nations, and most decidedly lower than is desirable in the best interests of our country'.[11] Sensational scien-

---

7. *Nature*, 71 (1904–5), p. 108. Even a man as enlightened as Lyon Playfair, a student of Liebig's and one of Britain's leading chemists after the middle of the century, expressed the opinion in 1875 that 'the stronghold of literature should be built in the upper classes of society while the stronghold of science should be in the nation's middle class' (quoted in Ashby, *Technology and the Academics*, p. 32).

8. 'The British Science Guild', *Nature*, 70 (1904), p. 343.

9. We quote his address as reported in indirect speech in *The Times*, 31 October 1905.

10. *Final Report of the Departmental Committee on the Royal College of Science*, p. 11.

11. Raphael Meldola's presidential address to the annual meeting of the Chemical Society on 22 March 1907, *Nature*, 76 (1907), p. 231.

tific discoveries certainly aroused a passing interest among the public, the president of the Chemical Society in 1907 went on to explain, but 'the steady, plodding work which culminates in great discoveries is being carried on quite unheeded by the general public, and the workers themselves are practically unknown outside the ranks of science. Research as a "cult" is not understood; the national attitude towards the workers is one of "payment by results" in the very narrowest sense of the term'.

These examples could easily be multiplied. They all show, with unmistakably polemical undertones, that appreciation of science and the social standing of scientists were low in Britain in the nineteenth and early twentieth centuries. British scientists certainly did not enjoy the status of privileged 'mandarins' which Fritz K. Ringer ascribes to the professorial class in the German Kaiserreich.[12] The social and political elite in Britain admitted scientists into its ranks only to a limited extent, in sharp contrast to, say, France under the Third Republic. It is common knowledge that graduates of the Ecole Polytechnique in Paris filled high positions in public administration and private industry. British scientists did not occupy comparable positions in their country until the Second World War and even later. In the eyes of Britain's social and political ruling classes, scientists have, as a rule, remained rather eccentric and 'unworldly' outsiders to the present day. At the beginning of the twentieth century *The Times* expressed the hope that every British politician of the next generation might possess at least a basic knowledge of the natural sciences.[13] It has not yet been fulfilled. The chasm dividing C. P. Snow's two cultures often seems to be deeper than ever before in the Britain of today.

Public recognition of scientific achievement was rare in Britain throughout the period of this inquiry. *Nature* could write in 1873 of the 'comparative indignity' with which scientists were treated in Britain.[14] In 1904 the vice-president of the Society of Arts found that 'the lay official mind has, with some few exceptions, never fully grasped the importance of orderly and continued scientific investigation in order to increase national prosperity'.[15] A few years later *Nature* expressed similar sentiments: 'Neither the political nor the official mind in this country yet realises the power which science can give to the modern State; because classical and literary

---

12. Fritz K. Ringer, *The Decline of the German Mandarins: The German Academic Community, 1890–1933* (Cambridge, Mass., 1969).
13. *The Times*, 24 January 1900.
14. See above, pp. 185–90 in this volume.
15. William Abney, 'Science and the State', *Nature*, 71 (1904–5), p. 90.

studies still form the chief high-road to preferment in Parliament or in public offices'.[16] On the initiative of Sir Robert Peel, one of the few nineteenth-century British prime ministers who were interested in science and advocated greater support for it, the government instituted the Royal Medal in 1825 to be awarded annually for 'scientific work of exceptional merit'.[17] The medal, sponsored by the Royal Society, was associated with a sum of money. Also at Peel's instigation, the government had granted several outstanding scientists honoraria since the 1830s. Early beneficiaries of this form of financial support included the astronomer George B. Airy and the physicists Michael Faraday and David Brewster.

While knighthoods for scientists were relatively common in the nineteenth century, baronetcies and peerages were extremely rare. 'During the eleven years of the present reign', calculated Charles Babbage in the mid-nineteenth century,

> one solitary instance is to be found of a baronetcy given for science, and that too occurred only at a festival [i.e. the coronation of Queen Victoria in 1838] at which baronetages and peerages were showered upon those whose sole claim was founded on the mere support of party. During the same interval, about half a dozen of those who cultivate science, have been knighted . . . . When this is compared with the most successful prizes in the army, the navy, the church, or the bar, it shows at once the inferior position occupied by science.[18]

By the end of the nineteenth century the highest representatives of state had not fundamentally changed their assessment of scientists' worth. When the fiftieth anniversary of the discovery of the first aniline dye was celebrated in 1906, the experimental physicist Silvanus P. Thompson indignantly pointed out in *The Times* that William Henry Perkin (1838–1907), who had discovered mauveine and was still alive, 'has not been made a peer, nor a baronet, not even a knight-bachelor, not even a Privy Councillor nor a University professor'.[19] Although the First World War had brought scientists into a 'closer relationship with national affairs than ever

16. 'Science in National Affairs', *Nature*, 96 (1915–16), p. 195. See also Roy M. MacLeod, 'Science and the Civil List, 1824–1914', *Technology and Society*, vol. 6 (1970), p. 47.
17. Quoted in Roy M. MacLeod, 'Of Medals and Men: A Reward System in Victorian Science 1826–1914', *Notes and Records of the Royal Society of London*, vol. 26 (1971), pp. 81–105.
18. Charles Babbage, *The Exposition of 1851* (London, 1851), p. 193 (reprinted Farnborough, 1969). The baronetcy was given to John Herschel.
19. *The Times*, 3 March 1906.

before',[20] they did not believe that this was reflected in appropriate public recognition. 'Among the scores of names', commented *Nature* on the honours list published in the New Year of 1916, 'we do not find a single honour given specifically for scientific work. Several men of science engaged in Government departments, as well as leading surgeons and physicians, are selected for various honours, but outside what may be termed official circles, science is practically ignored.' The leader continued:

> The indifference thus shown to science, when all its resources are needed for the successful prosecution of the war in which we are engaged, and for the industrial conflict to follow it, makes us wonder whether our statesmen are capable of understanding what scientific work means to a nation.[21]

The Order of Merit, which was never as prestigious as the Pour le Mérite in Prussia or Germany, was not linked with elevation to the peerage. It had been instituted by Edward VII on his accession to the throne, following suggestions made by the prime minister, Lord Salisbury, in the 1880s. Originally it was to be awarded for 'particularly outstanding service in the armed services or exceptionally meritorious service towards the advancement of art, literature and science'.[22] Later, however, it was also conferred upon statesmen. Haldane received the Order of Merit in 1915, and Arthur J. Balfour one year later. On the whole, British scientists were adequately represented in the Order. The first twelve recipients, named by the King on the occasion of his coronation, included, in addition to three field marshals and two admirals, the physicists Lord Kelvin and Lord Rayleigh, Sir William Huggins, astronomer and president of the Royal Society from 1900 to 1905, and the surgeon Lord Lister.[23] Subsequently, all presidents of the Royal Society were awarded the Order of Merit. Recipients before 1918 were the botanist Sir Joseph Dalton Hooker (1817–1911), the zoologist and botanist Alfred Russell Wallace (1823–1913), the chemist Sir William Crookes (1832–1919), the physicist Sir Joseph John Thomson (1856–1940), and the geologist Sir Archibald Geikie (1835–1924). Between 1902 and 1918 thirty-six people received the

---

20. 'Merit and Reward', *Nature*, 96 (1915–16), p. 503.
21. Ibid., pp. 503–4.
22. Ivan de la Bere, *The Queen's Orders of Chivalry* (London, 1964), p. 168.
23. The field marshals were Lord Roberts, Lord Wolseley and Lord Kitchener of Khartoum; the admirals, Sir Henry Keppel and Sir Edward Seymour. The other recipients of the Order in 1902 were the scholar and politician John Morley, the historian W. E. H. Lecky, and the painter and sculptor George Frederic Watts.

Order of Merit, including eight scientists and one doctor. After 1918, too, the share of awards going to this group remained constant at exactly one quarter of the total number awarded. The twenty-eight recipients between 1919 and 1936 included six scientists and one doctor.[24]

Agitation by the science lobby since the early nineteenth century had obviously had little impact on the social status of scientists and on the attitude of the public towards science. On the eve of the First World War Raphael Meldola, one of the most active spokesmen for science at the time, could write in *The Morning Post* of the 'unscientific atmosphere' prevailing in Britain, without anyone writing to contradict him. He complained that 'public opinion has not yet been educated to the point of realising the enormously important part which scientific research has taken, is taking, and will continue to take in the development of civilisation'.[25] We need not emphasise that the majority of British scientists still shared this opinion during the war. A spokesman for the deputation sent by the Royal Society and the Chemical Society to urge the government to give more support to research, declared in May 1915 that 'in this country science and its applications suffer severely from the total ignorance of the nation at large, and from the neglect by the authorities and by the manufacturers of the applications of science'.[26] The writer H. G. Wells protested against the 'very small part we are still giving the scientific man and the small respect we are showing scientific methods in the conduct of the war'.[27] And the memorandum of 1916 that led to the foundation of the Neglect of Science Committee, stated baldly: 'Not only are our highest Ministers of State ignorant of science, but the same defect runs through almost all the public departments of the Civil Service. It is nearly universal in the House of Commons, and is shared by the general public, including a large proportion of those engaged in industrial and commercial enterprise'.[28]

It has already been pointed out that this undifferentiated assessment no longer adequately reflected the complexities of the real

---

24. These figures are based on a membership list of the Order of Merit, kindly placed at the author's disposal by the Central Chancery of the Orders of Knighthood, St James's Palace, London.
25. 'Science and the State', *Morning Post*, 27 May 1914. In 1907 Meldola had spoken of Britain as a country 'where the true position of scientific research is imperfectly understood' (*Nature*, 76, 1907, p. 232).
26. Minutes of the meeting between the deputation from the Royal Society and the Chemical Society, and the presidents of the Boards of Trade and Education on 6 May 1915, p. 8 (Ed. 24/1579, PRO).
27. *The Times*, 11 June 1915.
28. Printed in *The Times*, 2 February 1916.

situation. Especially during the First World War, British society's attitude towards science had changed to such an extent that a former president of the Institute of Chemistry could state that 'our nation has been led to realise the error of its ways, and chemistry can never again be relegated to the humble position it formerly occupied in the national esteem'.[29] But despite the experience of war, which had demonstrated the importance of science in modern warfare, science and scientists were not able to overcome their traditionally low status, even by the 1920s and 1930s.

One reason for the low position which science and the profession of scientist occupied in public awareness, and to a large extent still occupies today, lies in the fact that research and an interest in science were for a long time seen as a hobby of the aristocracy and the wealthy upper classes. Prominent examples from the late nineteenth century are the Duke of Devonshire, chairman of the Devonshire Commission in the 1870s,[30] and the prime ministers Lord Salisbury and Arthur J. Balfour. The Duke of Devonshire and Lord Salisbury both had private laboratories. In some respects Lord Rayleigh, also a member of the aristocracy, can still be considered one of the aristocratic amateur scientists. The amateurism which was an important aspect of many areas of British life — one suspects that it was deliberately cultivated — led the journal *The Nineteenth Century* to speak of Britain as 'a Nation of Amateurs'.[31] The constitutional lawyer Sidney Low went so far as to equate 'government in England' with 'government by amateurs'.[32] Amateurism, often blatantly vaunted, by no means precluded brilliant achievements being made by individuals. Rayleigh, who won the Nobel Prize for Physics in 1904 for his work on the inert gas argon, provides an impressive example. The physicist Lord Kelvin concluded, in 1898, that 'the best of experimental physics in this country has been undertaken by wealthy amateurs'.[33] As late as 1916 J. A. Fleming, professor of electrical engineering in London and a shrewd observer, wrote:

> In fact, the greater part of past British scientific research may be said to have been amateur work, not in the sense that it was lacking in the

---

29. A. Chaston Chapman, *The Growth of the Profession of Chemistry during the Past Half-Century (1877–1927)* (London, 1927), p. 21.
30. See above, pp. 80–1 in this volume, and Meadows, *Science and Controversy*, pp. 75–112.
31. *The Nineteenth Century*, vol. 48 (1900), p. 521. See also Coleman, 'Gentlemen and Players'.
32. Sidney Low, *The Governance of England* (London, 1904), p. 199.
33. *Report of the Committee Appointed by the Treasury to Consider the Desirability of*

highest qualities, but only in the sense that it was pursued in the sheer pleasure and interest of it by private individuals. It was done mostly at odd times, and nearly always at the worker's own expense.[34]

As soon as an interest in the natural sciences or the humanities lost its amateurish and playful aspect, it aroused astonishment. Prince Albert's persistent encouragement of science, and his insistence that it needed greater financial and institutional support, was greeted with an indulgent smile by the English aristocracy, who attributed his interest in this field to his German origins. For this reason, Justus von Liebig did not see Victorian England as 'das Land der Wissenschaft'. During his repeated visits to the United Kingdom since 1837 he had observed that what existed there was 'nur ein weitgetriebener Dilettantismus'.[35] Among large sections of the English middle classes, too, scientific research was not considered a 'profession' like medicine, law or the Church. It was seen as a hobby for educated and well-to-do gentlemen who possessed as much idealism as leisure. Many examples of this attitude can be found even at the beginning of the twentieth century. 'The general public', wrote a Fellow of the Royal Society in *Nature* in 1915, 'looks upon scientific investigation as a hobby.' Looking for reasons, he continued:

> The idea of scientific investigation as a hobby does not necessarily originate with the general public; it is indigenous in the older universities, where there are a large number of college officials intellectually competent to undertake researches, some of whom do and some do not. At Cambridge in my time scientific investigation was the occupation of the leisure of men whose maintenance was provided by the fees and emoluments of teaching. It was as much a hobby as chess or photography. There was no sense of collective responsibility for providing the nation with answers to its scientific questions . . . The idea of 'making a living' by scientific investigation never reached the surface, though the merit acquired by research might weigh in the appointment to a post for teaching or administration.[36]

Establishing a National Physical Laboratory, Minutes of Evidence, C. 8977 (London, 1898), p. 33.
34. J. A. Fleming, 'The Organisation of Scientific Research', *Nature*, 96 (1915–16), pp. 692–3.
35. Justus von Liebig to J. J. Berzelius, 26 November 1837, printed in Justus Carrière (ed.), *Berzelius und Liebig. Ihre Briefe von 1831–1845* (Munich and Leipzig, 1893), p. 134. See also Justus von Liebig, *Die Chemie und ihre Anwendung auf Agricultur und Physiologie*, vol. 1, 7th edn (Brunswick, 1862), p. 76 (see the section 'Der Zustand der Naturwissenschaft in England', pp. 74–86); Jakob Volhard, *Justus von Liebig* (Leipzig, 1909), vol. 1, p. 144, and vol. 2, p. 374.
36. *Nature*, 96 (1915–16), p. 453.

Sir William Crookes, president of the Royal Society, attributed the scant regard for science in Britain to the specific mentality of the upper and middle classes, shaped by public schools and Oxbridge:

> The nation's attitude towards science is, I think, largely due to the popular idea that science is a kind of hobby followed by a certain class of people, instead of the materialisation of the desire experienced in various degrees by every thinking person to learn something about innumerable natural phenomena still unsolved.

Crookes, who had studied at neither Oxford nor Cambridge, believed that only a protracted process of rethinking could effect a fundamental change in the assessment of science in Britain:

> I believe that the 'Hobby' attitude is due to our national character, and can only be rectified slowly, step by step. We cannot suddenly become a truly scientific nation, either now during the war, or immediately on its conclusion. We shall have to make many fundamental alterations in our ideas and almost to change our natures before such a change can be effected.[37]

Consequently, 'researcher' or 'scientist' was not, as a rule, a profession that provided a livelihood until well into the Victorian age. The idea that a scientist could draw a fixed salary — perhaps even from public money — for carrying out research was alien in the nineteenth century, at least in England. Career opportunities for scientists from the middle classes were accordingly limited, and Britain had substantially fewer middle-class scientists than comparable countries on the Continent until after the middle of the century. British scientists in the nineteenth century who did not possess a large fortune had to face the difficult question of what chance they had of earning a living by scientific work. In Germany a scientific career modelled on a career in the civil service was possible, but not in England. There were basically only three courses open to a young scientist in Victorian England: (1) he could try to find a post at a school or university college, with all the teaching duties this involved; (2) he could become a scientific adviser and in this way maintain close links with research; or (3) he could apply for one of the few positions for scientists in industry or the civil service.

To what extent these jobs, whose total number was small, did in fact offer opportunities for research depended on the individual

37. Ibid., p. 374.

circumstances in each case. We can assume that the state observatories and the Geological Survey of Great Britain offered more opportunities for research than did, for example, the office of the Government Chemist, which was set up in 1842 to do routine analyses. Roy MacLeod points out that even an institution as modern as University College London 'with its history of distinguished scientific appointments, offered limited opportunities for research'.[38] This is supported by the fact that fellowships at Oxford and Cambridge colleges were rarely used to enable a scientist to do research work. Until the 1870s fellowships were regarded only as periods of transition, limited as they were to unmarried scholars.[39] A life-long academic career was more usual at the Scottish universities and the new university colleges in the Midlands and the north of England, but even for professors at these institutions, as at Oxford and Cambridge, the main emphasis was on teaching, even after the First World War. But the fact that almost all well-known British scientists held professorships in the late nineteenth century speaks against MacLeod's argument: obviously their teaching duties left them sufficient time for research. As late as 1910 there was practically only one institution, the Royal Institution in London, whose scientists did not have to teach. Their one duty was 'to press forward into the unknown'.[40]

Outside the universities, a career as a scientist was 'little more than a phantom. The prospect of a post in any government institution, such as the British Museum, is, to say the most, scarcely a pittance'.[41] A survey conducted by the Institute of Chemistry among its just under 1,000 British and Irish members in the autumn of 1902 showed that 30 per cent worked on a free-lance basis as advisers or analysts, 27 per cent were employed in industry, 16 per cent taught at schools or universities, and 7.3 per cent worked in the civil service or for government.[42] In general, chemists and engineers had the best chances of finding a suitable job and being able to combine it with research in the nineteenth century. But at

---

38. MacLeod, 'Support of Victorian Science', p. 202.
39. This situation was sharply criticised by Mark Pattison: 'Hitherto, a university life has not been a life-profession. A fellowship and tutorship has been only held *in transitu* to a living. It is a principle of any reorganisation of the university . . . to erect teaching and learning, inseparably united, into a life-profession, for which a young man may regularly qualify himself, and look forward to it as a maintenance' (*Suggestions on Academical Organisation*, pp. 204–5); Engel, *From Clergyman to Don*, pp. 106–22.
40. The chemist Sir William A. Tilden, in *Nature*, 85 (1910–11), p. 29.
41. *Nature*, 9 (1873–4), p. 237.
42. Richard B. Pilcher, *The Institute of Chemistry of Great Britain and Ireland: History of the Institute 1877–1914* (London, 1914), p. 165.

the beginning of the twentieth century the president of the Chemical Society could still speak of the 'poor outlook for chemical research as a career'.[43] Biologists, physicists and astronomers had comparatively much worse chances of employment because of the tiny number of vacancies in their fields and the limited relevance of their subjects to industrial production processes. For this reason Charles Darwin remained a private scholar for most of his life, living on his own fortune and private support. The biologist T. H. Huxley and the physicist John Tyndall earned their livings by academic teaching and journalism, and the same was true of Norman Lockyer. Sir William Ramsay financed his research on the inert gas argon, for which he received the Nobel Prize with Lord Rayleigh in 1904, from his income as a chemical expert and his salary as professor of chemistry at University College London.[44] It has been estimated that there were only 300 positions for scientists at British universities in 1902. Added to this were about 250 positions in the civil service, and between 180 and 230 positions in industry.[45] In Germany, by contrast, 4,000 chemists were employed in 1897; the German chemical industry alone employed 3,000 chemists. The total number of German chemists in 1914 has been estimated at 9,000.[46]

Since the 1860s British scientists had been pointing to the increasingly scientific nature of industrial production processes being used abroad. A remarkable change was taking place, and the German chemical industry was only its most spectacular pioneer. British scientists had also drawn attention to the omissions of British industry. 'The supreme value of research in pure science for the success and progress of the national industries of a country', summed up the president of the Royal Society in 1902, 'can no longer be regarded as a question open to debate, since this principle has not only been accepted in theory, but put in practice on a large scale, at a great original cost, in a neighbouring country, with the most complete success.'[47] British industry had begun to make good its undeniable deficit in research by establishing its own laborato-

43. Raphael Meldola, 'The Position and Prospects of Chemical Research in Great Britain', *Nature*, 76 (1907), p. 232.
44. Roy M. MacLeod, 'Science in Grub Street', *Nature*, 224 (1969), pp. 423–7; Bernal, *Social Function of Science*, p. 28.
45. See MacLeod and McLeod, 'Social Relations of Science', p. 306.
46. *Report on Chemical Instruction in Germany*, p. 32; Varcoe, 'Scientists', p. 193; Lothar Burchardt, 'Professionalisierung oder Berufskonstruktion? Das Beispiel des Chemikers im wilhelminischen Deutschland', *Geschichte und Gesellschaft*, vol. 6 (1980), pp. 326–48.
47. 'Anniversary Meeting of the Royal Society', *Nature*, 67 (1902–3), p. 108; see also above, pp. 122–7 in this volume.

ries to monitor production and to undertake research, especially in the last ten years of the nineteenth century. It defended itself against what was considered unjustified criticism of its omissions by pointing out that British scientists were inadequately qualified for research work, referring to the large number of German and Swiss chemists in British industry at the end of the nineteenth century.[48] This argument does indeed appear to be valid. In 1910 *Nature* published an article in which the author conceded that twenty years ago 'the research chemist qualified for industrial work could scarcely be obtained from English laboratories. He had to be imported from Germany'.[49] The change which led to industrial laboratories being set up, however, was paralleled by a change in the attitude of industry to scientific research. At the end of 1914 *Nature* noted:

> There is little doubt that the value of employing numerous chemists in chemical works is becoming more and more appreciated in this country. The supply of well-trained chemical workers which is available for the coming struggle will be the most important factor in its decision.[50]

All available figures indicate that since the late nineteenth century the number of jobs for scientists in industry had not kept pace with the number of trained scientists and engineers. Especially between 1900 and 1914 there seems to have been a sudden spurt in the number of scientists in Britain, while the number of positions available more or less stagnated by comparison. It was estimated that there were 2,000 scientists in Britain in 1900, half of whom taught in schools. By 1914 their number was said to have risen to more than 7,000.[51] Thus for several years there was a discrepancy between supply and demand caused by the expansion of the university colleges, on the one hand, and the belated development of new science-dependent industries, on the other. It was reflected after the First World War in a persistently high unemployment rate among scientists.

The expansion of the state's functions in the nineteenth century

---

48. Sanderson, *Universities*, p. 18.
49. R. Blair, 'The Relation of Science to Industry and Commerce', *Nature*, 84 (1910), p. 347. See also Cardwell, *Organisation of Science*, pp. 204–6.
50. 'The Development of Chemical Industries in the British Isles', *Nature*, 94 (1914–15), p. 292.
51. MacLeod, 'Support of Victorian Science', p. 228. Before the First World War between 500 and 530 scientists and engineers took final examinations at British universities each year (*Report of the Committee of the Privy Council, 1915–16*, p. 34).

could not compensate for the small demand for scientists in industry. Since the early nineteenth century the state had employed geologists, astronomers, doctors and a few chemists. For physicists, the National Physical Laboratory provided the first opportunities for employment in an institution partially financed by the state in 1900. The state employed zoologists, botanists and biologists only in the few botanical gardens and in museums. The situation was aggravated by the fact that in Britain — in contrast to Germany, for example — university courses were not designed to qualify students for a clearly defined career in the civil service. In addition, compared with today, the civil service was very small at the turn of the century. In 1890 only 3.5 per cent of all employees, including the armed services, were employed by the state.[52]

The enormous expansion of the civil service did not begin in Britain until just before the First World War, and then continued throughout the war years. The number of scientists employed by the state correspondingly rose. The experience of the First World War had made politicians aware that Britain had become disastrously dependent on the scientific research of other countries. In the House of Commons the president of the Board of Education, Sir Joseph A. Pease declared:

> We have realised that it is essential, if we are going to maintain our position in the world, that we must make better use of our scientifically trained workers, that we must increase the number of those workers, we must endeavour to see that industry is closely associated with our scientific workers, and we must promote a proper system of encouragement of research workers, especially in our universities. The fault in the past, no doubt, has been partly due to the remissness on the part of the Government in failing to create careers for scientific men.[53]

Awareness that creating 'careers for scientific men' was an important condition for any more systematic attempt to promote research was becoming more widespread within the government. The permanent secretary of the Board of Education expressed a hope, along the same lines as the statement made by Pease just quoted, that 'the opening of better prospects of a career in Research would have the effect of retaining in the service of science young men who

---

52. In absolute figures, the state employed 412,000 people in 1891, 61 per cent of them in the armed forces, 22 per cent in central government and 17 per cent in local government (Moses Abramovitz and Vera F. Eliasberg, *The Growth of Public Employment in Great Britain*, Princeton, NJ, 1957, pp. 8, 16; R. K. Kelsall, *Higher Civil Servants in Britain from 1870 to the Present Day*, London, 1955).
53. Hansard, Parl. Deb., H. C., 5th Series, vol. 71, col. 1903 (13 May 1915).

for want of any reasonable prospect of a career drift into other professions of employment'.[54] When *Nature* surveyed the employment of scientists by the state in the spring of 1921 it declared the results satisfactory. While the number of scientists in state employment had been very small at the turn of the century, the situation had fundamentally changed after the war:

> The rapid growth of the public Services within the last fifteen years, the assimilation of public utility companies into the State system, the creation of entirely new Departments, and the realisation forced upon Ministers by the war of the necessity for scientific research in the nation's interest, have resulted in the employment of thousands of scientific and technical workers. Many of those engaged temporarily during the war have returned to the universities or other institutions from which they were recruited, but a large number remain and have been absorbed by various State establishments.[55]

In 1927 the number of chemists alone in the civil service was estimated to be between 500 and 600.[56] But the science lobby would not have been doing its job if it had been satisfied simply to register the success of its efforts. *Nature* therefore qualified its assessment: 'Prejudice dies hard, and there are still many men in high administrative positions in the Civil Service who hold science in contempt, and this feeling is reflected in their attitude towards scientific workers in their Departments'.[57]

The consolidation of science as a profession in Britain during the First World War was an intrinsic part of the wider social change which had taken place in Western Europe and the USA since the early nineteenth century, assuming various forms and with national variations in timing. One of the factors which greatly accelerated this process of professionalising science[58] was undoubtedly the new

---

54. Permanent secretary to the Board of Education to the Royal Society, 29 June 1915 (Ed. 24/1573, PRO).
55. 'Science and the Civil Service', *Nature*, 107 (1921), p. 1.
56. Chapman, *Profession of Chemistry*, p. 10.
57. *Nature*. 107 (1921), p. 1.
58. Of the voluminous literature, see Hans-Ulrich Wehler (ed.), 'Professionalisierung in historischer Perspektive', *Geschichte und Gesellschaft*, vol. 6 (1980), pp. 311–402; Joseph Ben-David, 'Akademische Berufe und die Professionalisierung' in D. V. Glass and René König (eds.), 'Soziale Schichtung und soziale Mobilität', *Kölner Zeitschrift für Soziologie und Sozialpsychologie*, Sonderheft 5 (1961), pp. 105–21; Mendelsohn, 'The Emergence of Science'; Armytage, *Rise of the Technocrats*; Cardwell, *Organisation of Science*, pp. 228–59; A. H. Halsey and Martin A. Trow, *The British Academics* (London, 1971); William Joseph Reader, *Professional Men: The Rise of the Professional Classes in Nineteenth-Century England* (New York, 1966); B. W. G. Holt, 'Social Aspects in the Emergence of Chemistry as an Exact Science: The

Science as a Profession

constellation emerging in the relationship between state and science. The gradual increase in the state's support for science throughout the nineteenth century, the allocation of public funds for scientific research, as well as the enormous increase in the amount of knowledge, all contributed substantially to the fundamental change in the social conditions for science, which Cardwell has called 'one of the remarkable social changes of the nineteenth century'.[59]

The indicators of the process of social and institutional change by which science lost its amateur character are well known. They include the differentiation of scientific disciplines as well as scientists' attempts, growing in urgency throughout the nineteenth century, to legitimise their work in social terms. As has been shown in the first chapter of this book, the accelerating process of differentiation between various scientific disciplines began in Britain early in the nineteenth century. It manifested itself externally in a growing number of specialist learned and scientific societies created on the initiative of private individuals or groups of individuals. The foundation of a specialist society generally indicated that a new scientific discipline had been formally recognised by the scientific community. In Britain the number of new foundations passed its peak between 1870 and 1890, while in Germany, for example, the peak was not reached until 1900.[60] After the middle of the century the progressive differentiation and specialisation of science was also expressed in an increase in the number of general and specialised scientific journals, most of which were published by the specialist societies.[61] The foundation of *Nature* in 1869 as a weekly science journal produced on a commercial basis is only the best-known example. In the forty years between 1860 and 1900 the number of specialised scientific journals in Britain increased from twelve to more than seventy; during the same period the number of technical and scientific journals for a more general market rose from forty to more than 130.[62] As early as 1894 the president of the Chemical Society complained at the Society's annual meeting, that

---

British Chemical Profession', *British Journal of Sociology*, vol. 21 (1970), pp. 181–99; G. S. R. Kitson Clark, *An Expanding Society: Britain 1830–1900* (Cambridge, 1967); Haines, *German Influence*, pp. 47–87 (ch. 3, 'The Professionalization of the Sciences').
  59. Cardwell, *Organisation of Science*, p. 249.
  60. See Appendix 1 below, p. 256 in this volume; and Pfetsch, *Wissenschaftspolitik in Deutschland*, p. 200.
  61. On this see Meadows, *Communication in Science*, pp. 66–86.
  62. Armytage, *Rise of the Technocrats*, p. 108.

chemical literature is fast becoming unmanageable and uncontrollable from its very vastness. Not only is the number of papers increasing from year to year, but new journals are constantly being established. Something must be done in order to assist chemists to remain in touch with their subject and to retain their hold on the literature generally.[63]

Greater differentiation in science was accompanied by a rise in the number of scientists, both in Britain and elsewhere. This development must be seen as a consequence of improved education and a slow but steady increase in the number of positions available in scientific institutes, the academic sector, the civil service and certain branches of industry. Growing membership figures in the learned and scientific societies, more and more of which were following the lead set by the Royal Society and making membership dependent on certain qualifications, graphically reflect the growth of the scientific community. The thirteen most important learned and scientific societies in London had a total of 5,000 members in 1850, 10,000 in 1870 and 20,000 in 1910.[64] Membership of the Chemical Society in London rose from 323 in 1860 to 2,292 in 1900, and to 3,721 in 1920.[65] The question of which social classes provided the growing number of British scientists in the nineteenth century requires more detailed investigation. Hitherto, only superficial information has on the whole been available. The statement that scientists came from the middle classes of the 'recently industrialized cities'[66] provides inital rough bearings, but it is too general to be satisfying. Pfetsch has suggested that in the nineteenth century, English scientists were recruited from a broader spectrum of classes than German scientists;[67] this also needs to be investigated further. British historiography still lacks a socio-historical analysis like Christian von Ferber's study of the social origins of German academics since the mid-nineteenth century.[68]

Since the middle of the nineteenth century, the aims, functions and significance of scientific work had increasingly been discussed

63. Quoted in Meadows, *Communication in Science*, p. 86. 'Special knowledge and special power are everywhere in demand. In law, in the arts, and more especially in science, the specialist is looked upon with a peculiarly favourable eye. Mankind gratefully acknowledges its obligations to the man who devotes his life to the attainment of perfect knowledge within a certain narrow sphere' (Bernhard H. Becker, *Scientific London*, London, 1874; reprinted Haarlem, 1968, p. 136).
64. See above, p. 17 in this volume.
65. Meadows, *Communication in Science*, p. 13. In general, see Chapman, *Profession of Chemistry*. See also the figures in Lyons, *The Royal Society*, pp. 228–9.
66. Mendelsohn, 'The Emergence of Science', p. 41.
67. Pfetsch, *Wissenschaftspolitik in Deutschland*, p. 317.
68. Christian von Ferber, *Die Entwicklung des Lehrkörpers der deutschen Universitäten und Hochschulen 1864–1954* (Göttingen, 1956), esp. pp. 163–86.

by British scientists. An oft-cited example of the growing self-awareness of scientists, boosted by the general improvement in communications, was the use of the word 'scientist' for all who worked on natural phenomena using generally accepted theoretical approaches and methods. This term, which quickly replaced those in normal usage at the time ('natural philosopher', 'naturalist', 'savant', 'cultivator of science' or 'man of science'), was coined in 1840 by the Cambridge mineralogist and philosopher William Whewell (1794–1866).[69] The growing self-confidence of scientists, who tended to have studied at London University or at one of the new university colleges rather than at Oxford or Cambridge, was also reflected in their efforts to achieve a better representation of the professional interests of science in society and to claim more public recognition for their work. Their pursuit of social legitimation and acceptance as a profession with economic and social privileges was expressed most clearly in the creation of large professional associations, encompassing all scientists, which functioned as pressure groups. They have been subsumed here under the term 'science lobby'. These associations, almost without exception, worked for better science education; they attempted to establish qualification criteria for courses, and they tried to gain public recognition for the specific interests of science. The best known of these associations, which existed separately from and mostly independently of the specialist societies, were the British Association for the Advancement of Science (founded in 1831) and the British Science Guild (established in 1905). This process received another boost during and shortly after the First World War, when scientists from related fields organised themselves in professional associations. Examples are the Association of University Teachers, the Institution of Professional Civil Servants, the British Association of Chemists, the National Union of Scientific Workers and the Institution of Chemical Engineers.

When the Institute of Physics was founded in 1921, more than forty years after a similar institution had been founded to represent the interests of chemists, *Nature* reflected upon the changes that had taken place in British science over the last fifty years:

Physics as a profession by which numbers of men would earn a

69. 'We need very much a name to describe a cultivator of science in general. I should incline to call him a Scientist' (William Whewell, *The Philosophy of the Inductive Sciences Founded upon their History*, vol. 1, London, 1840, p. CXIII). Sydney Ross, 'Scientist: The Story of a Word', *Annals of Science*, vol. 18 (1962), pp. 65–85; Raymond Williams, *Keywords. A Vocabulary of Culture and Society* (London, 1976), pp. 232–5 ('Science').

livelihood . . . never entered [James Clerk] Maxwell's thoughts. Contrast this . . . with the position at present — a university or technical school in almost every great town, each with its well-equipped physical laboratory, its keen professor and its enthusiastic students; laboratories in all the larger schools, with a staff of teachers numbering many hundreds. Fifty years ago the army of physicists was small in numbers; its generals were great men, but they had few of the rank and file to command.[70]

The enormous social changes which took place in the scientific community with the professionalisation of science have not yet been investigated in detail. But even a first assessment of contemporary sources shows clearly that the transition from 'natural philosopher' or 'cultivator of science' to 'scientist' or 'physicist' was not accepted totally and without reservations, especially by the older generation of scientists. They were expressed as early as 1881 by the physicist and mathematician William Spottiswoode (1825–83), a member of the influential X-Club,[71] during his period of office as president of the Royal Society (1878–83):

> It may be a matter of regret, although doubtless inevitable, that the same causes which have affected the social, the intellectual, the industrial and the political life of our generation, and have made them other than what they were, should affect also our scientific life; but, as a matter of fact, if science is pursued more generally and more ardently than in former times, its pursuit is attended with more haste, more bustle, and more display than was wont to be the case.

The whole structure of the science system had changed, and according to Spottiswoode, this gave rise to dangerous trends in science:

> Apart from other reasons, the difficulty . . . of ascertaining what is new in natural science; the liability at any moment of being anticipated by others, constantly present to the minds of those to whom priority is of serious importance; the desire to achieve something striking, either in principle or in mere illustration; all tend to disturb the even flow of scientific research. And it is perhaps not too much to say that an eagerness to outstrip others rather than to advance knowledge, and a struggle for relative rather than for absolute progress, are among the

---

70. 'Physics as a Profession', *Nature*, 107 (1921), p. 289. The word 'physicist' also goes back to Whewell: 'As we cannot use physician for a cultivator of physics, I have called him a *physicist*' (*Philosophy of the Inductive Sciences*, vol. 1, p. CXIII).
71. See above, pp. 83–4 in this volume.

dangerous tendencies peculiar to the period in which we live.[72]

As far as the state of professionalisation of science at the turn of the century is concerned, the difference between Britain and Germany, for example, lay in the lack in Britain of careers for scientists in which the interests of state and science coincided. The most important professional organisation of British chemists, the Institute of Chemistry, established in 1877,[73] found the time ripe early in 1919 'for taking steps to secure for the profession of chemistry a position corresponding with that occupied by other learned professions'. One of the main demands made of the state by chemists and listed by the Institute in a memorandum was that 'adequate and uniform conditions of appointment' be 'accorded to chemists directly engaged in the service of the State'.[74] Legal differentiation between chemists and pharmaceutical chemists was another area in which the state could assist the professionalisation of science. The fact that two different professions were called by the same name contributed to many confusions and ambiguities which long plagued the image of chemistry as a profession.[75] The Institute of Chemistry believed that chemists in public employment who had the strictly defined qualifications required should 'be graded as Civil Servants in the higher division, with status, emoluments, and pension comparable with those of the members of other technical and learned professions employed by the Government'.[76] The Institute also took the opportunity to suggest titles for the various stages in the career of chemists in the civil service, beginning with 'junior assistant chemist' and ending with 'chief chemist'. Richard B. Pilcher's book *The Profession of Chemistry* (1919), commissioned by the Institute of Chemistry, pursued the same purpose: to define chemistry as a profession and to place it on a par with other recognised 'professions'.

72. Quoted in Meadows, *Communication in Science*, p. 86.
73. 'The aims of the Institute include the elevation of the profession of chemistry and the maintenance of the efficiency, integrity and usefulness of persons practising the same, by compelling the observance of strict rules of membership and by setting up a high standard of scientific and practical proficiency' (Richard B. Pilcher, *The Profession of Chemistry*, 4th edn, London, 1938, p. 9). In 1878 the Institute had 225 members; fifty years later the number had risen to 5,186 (Chapman, *Profession of Chemistry*, p. 5). Colin A. Russell et al., *Chemists by Profession: The Origins and Rise of The Royal Institute of Chemistry* (Milton Keynes, 1977).
74. *Nature*, 103 (1919), p. 34.
75. In November 1837, after a visit to England, Justus von Liebig wrote: 'Die Chemiker schämen sich Chemiker zu heißen, weil die Apotheker, welche verachtet sind, diesen Namen an sich gezogen haben' (printed in Carrière, ed., *Berzelius und Liebig*, p. 134).
76. *Nature*, 103 (1919), p. 35.

Long after these and similar attempts were made to create a career structure for scientists in the civil service and universities, there were no precisely defined career patterns for scientists such as had long existed in Germany. The professional career of a British scientist in the nineteenth century might resemble that of a German academic. After studying at Oxford, Cambridge or London and subsequently at a foreign university, a scientist would usually start on his career at one of the new provincial university colleges. If he was lucky and could point to above-average scientific achievements, his career might lead back to Oxford, Cambridge or London. But careers could deviate considerably from this pattern. Two examples, both from the field of chemistry, will illustrate the broad range of scientists' social backgrounds in Britain. Alexander William Williamson (1824–1904), who at his death was considered one of the 'most notable of British chemists',[77] went abroad in 1840 to study under Leopold Gmelin in Heidelberg and Justus von Liebig in Giessen, taking his doctoral degree with the latter in 1846. Williamson never studied at a British university. After further study in Paris he was appointed professor of chemistry at University College London in 1849, at the age of 25, and he remained there until his retirement in 1887. His career included Fellowship of the Royal Society, and he also became president of the Chemical Society and the British Association for the Advancement of Science. In contrast to Williamson's straightforward and seemingly uncomplicated career, that of William Crookes (1832–1919), who was almost the same age as Williamson and an even more distinguished scientist, followed less secure channels. William Crookes, knighted in 1897, awarded the Order of Merit in 1910, and elected president of the Royal Society in 1913, was the son of a tailor from the north of England. After completing his studies with August Wilhelm Hofmann at the Royal College of Chemistry, Crookes did not find an academic position. He earned his living and financed his research by taking casual scientific jobs, initially at the Royal College and later at observatories and meteorological stations, by working extensively as an expert, and with the help of modest support from the Royal Society. In 1856 he turned to scientific journalism, first as editor of the *Liverpool Photographic Journal*, then in 1859 as editor of the *Chemical News*, and from 1864 as editor of the *Quarterly Journal of Science*. His finances were always strained;[78] he never became a professor. Until his scientific work,

---

77. See his obituary in *Nature*, 70 (1904), p. 32. See also *Dictionary of National Biography*, supplementary vol. 1901–11, pp. 678–80.
78. 'A stationary income', wrote Crookes in 1864, 'will not do with an increasing

pursued practically as a part-time occupation in his own laboratory, received public acclaim late in his life, Crooke's social position was typical of that of many British scientists of his time. Despite all the efforts of the science lobby, the situation of scientists who increasingly had to earn a living by their work improved only marginally in the course of the nineteenth century.

The Social Situation of British Scientists in the Nineteenth Century

The demands made with growing self-confidence by British scientists for more positions in the civil service and in industry brought in their wake a demand for appropriate payment. On the eve of the twentieth century the social position of scientists was not comparable with that of lawyers, senior civil servants, clergymen or doctors, and this was reflected in their pay. T. H. Huxley sardonically and succinctly summed up the financial aspect of being a 'scientist' in the second half of the nineteenth century from the point of view of one involved: 'Science in England does everything but pay. You may earn praise but not pudding!'[79]

The science lobby set about altering the widespread popular view of science as a 'private hobby' and the attendant consequences for the social situation, and the pay, of scientists. Complaints about totally insufficient pay for scientists and demands for an adequate income continued until after the First World War. In the 1860s and 1870s the Endowment of Research Movement, the Devonshire Commission, Alexander Strange and other science journalists all opposed the view that 'the man of science should work for love and die . . . in poverty'. It was pointed out that 'scientific discovery and research' were 'national work', and that this state of affairs, only too often denied, gave the community a financial responsibility for scientists.[80] The social problems of the scientific community could be solved only by state salaries for scientists, professors paid by the state, and 'scientific careers'. But before the turn of the century little was done to make a scientific career and its financial rewards more attractive. The physiologist Sir Michael Foster de-

---

family and domestic necessities are apt to make scientific men very mercenary' (E. E. Fournier d'Albe, *The Life of Sir William Crookes*, London, 1923, p. 90). See also *Dictionary of National Biography*, supplementary vol. 1912–21, pp. 136–7.

79. Quoted in Brock, 'Science Patronage', p. 173.

80. *Nature*, 4 (1871), p. 133; MacLeod, 'Support of Victorian Science', p. 206; idem, 'Resources of Science', p. 114.

scribed the disheartening situation, using as an example a doctor whose scientific interests had led him to specialise in pathology, a subject taught at almost all the new universities in Britain. In reply to the questions of what awaited the young scientist, and what difficulties he would have to overcome, Foster observed:

> The posts allotted to the science of pathology are very few, and most of them at least are ill paid. If he devotes his life to pathological inquiry, he must look forward to a continual struggle against a narrow income, not to say poverty, and even brilliant success will not bring him more than the salary of many a routine official.[81]

Other contemporary observers too referred to the low salaries of academics at universities and research institutes. The body in charge of distributing the grant-in-aid for British universities in 1905 noted the 'inadequacy of the remuneration of the professors and teachers in some of the Colleges' — meaning the new university colleges and universities in the Midlands and the north of England. This body believed that the consequences for scientific research in Britain were catastrophic, for

> a Professor or Assistant, whose salary is not sufficient to keep him in moderate comfort, is driven to extend his work either inside or outside the College until there is left to him neither time nor energy to do full justice to his students or to keep himself abreast of the latest developments of the branch of learning which he represents.[82]

Shortly thereafter, in his presidential address to the Chemical Society, Raphael Meldola expressed his amazement that

> so many men of ability can be found willing to take service in these newer institutions, the more especially as, apart from the absurdly inadequate remuneration often given to the chiefs of the chemical departments, the payment of the subordinate members of the staff is

---

81. Foster, 'State and Scientific Research', pp. 746–7. See also *Fifth Annual Report of the Medical Research Committee, 1918–19*, Cmd. 412 (London, 1919), pp. 9–10: 'In this country . . . this work [research in physiology and pathology] has been left almost wholly unpaid; it has been expected that the men capable of it shall pursue it, if at all, as a private hobby in the midst of duties often directly conflicting with it. This has long been the general national attitude towards research work in every science, but it can be shown that pathological science is under peculiar disadvantages in this respect. In all the sciences in the past the pursuit of research has been made possible for particular men only by the accident of private wealth or by some private endowment, given directly or by some institution.'
82. *University Colleges (Great Britain) (Grant in Aid)*, Cd. 2422 (London, 1905), p. 9.

generally on a scale which is nothing short of a scandal to the wealthiest of European nations.

He found it almost unbelievable that 'the average scale of remuneration should not exceed the wages earned by an artisan, and is often below that standard'.[83] This ensured that the 'best brains of the nation' were deterred from doing science, because 'unless the income of a professor is made in some degree commensurate with the earnings of a professional man who has succeeded in his profession, it is idle to suppose that the best brains will be attracted to the teaching profession'.[84] The demand for better payment for scientists was treated with more understanding by the British government during the war, but it did not yet evoke a positive response. 'We have not merely to make the best use of the scientific men we now have', said a Board of Education memorandum of 1915, 'but to provide a fuller supply in the future. The deficiency in the supply is mainly due to the slight prospect which applied science offers of a useful and remunerative career, and this in turn is due to the fact that the leaders of industry do not appreciate the service which science might render to them.'[85] The only practicable way to set up a productive university, explained H. A. L. Fisher, president of the Board of Education, two years later, was to appoint distinguished scholars and researchers, and 'not a bad way to secure these was to pay them well'.[86]

Although spokesmen for science were still complaining during the First World War that 'British scientific men, including engineers, have formed a habit of rendering the nation gratuitous services of the greatest intrinsic value',[87] isolated voices were raised in warning against the appointment and payment of scientists by the state. In an essay on 'The State and Scientific Research', Sir Michael Foster opposed the scientist in the civil service, whom he described as the 'paid scientific servant of the State, who for hire is working out the answer'. This sort of scientist was not allowed to deviate from the prescribed purpose of his research. Foster therefore praised the 'independent man of science', the 'private worker in any inquiry', because only he could work in a 'spirit of perfect freedom', autonomously determining the aim and method of

83. Meldola. 'Positions and Prospects', *Nature*, 76 (1907), p. 233.
84. Sir William Ramsay's presidential address to the annual meeting of the British Association for the Advancement of Science, printed in *Nature*, 87 (1911), p. 283.
85. Memorandum on a National Scheme of Advanced Instruction and Research in Science, Technology and Commerce, 1915 (DSIR 17/1, PRO).
86. *Nature*, 100 (1917–18), p. 122.
87. *Nature*, 97 (1916), p. 13.

his research. According to Foster, experience had shown that 'absolute freedom to follow wherever Nature leads is the one thing needful to make an inquiry a truly fruitful one'.[88] But Foster's plea for the 'independent inquirer', who could be 'reckless in his research' evoked little response. Taking its values from a world which had long since disappeared, this opinion was too far removed from the realities of the day. But it reveals to what a large extent the ideal of the financially independent researcher still existed, even within the scientific community, long after it had ceased to be relevant to the actual social position of scientists.

In conclusion, we must ask what the social situation of scientists was really like in late-nineteenth- and early-twentieth-century Britain. Did the perpetual complaints made by the science lobby about the inadequate income of scientists, undoubtedly determined by powerful interests, have a basis in reality? Available statistics do not permit an exact analysis of the income situation of the time, but they provide a few points of reference on which to hang an initial investigation of the complaints which were expressed.

Firstly, some basic data: it is generally accepted that around the turn of the century, a period which was on the whole characterised by relatively stable prices, an annual income of between £200 and £1,000 would have been enough to maintain a 'middle class' existence.[89] These figures represent a comparatively wide range. There are many views and definitions of what constituted a 'middle class' existence, and another factor which makes a great difference in calculations of this sort is whether the income had to support life in the country, in a small town or in a city such as London. For example, £600 was considered a good income for doctors in East Anglia in 1905 — 'enough for comfort, but hardly for affluence'.[90] For London the sum would certainly have been larger. According to William J. Reader, an annual income of £1.000 was regarded as 'considerable worldly success'. This becomes comprehensible if we take into account that for a white-collar worker in industry or in the civil service, £200 was a good salary, and a managing director of a factory was considered successful if he had a salary of between £500 and £600.[91] Guy Routh has calculated that the average profes-

88. Foster, 'State and Scientific Research', pp. 742–43.
89. 'In late Victorian or early Edwardian England, the income necessary to maintain a middle-class life-style, i.e. to buy education and to have the bare decency of a domestic staff, was between £200 and £1,000 per annum' (Rose and Rose, *Science and Society*, p. 33).
90. Reader, *Professional Men*, p. 192.
91. Ibid., p. 202. 'On an income of £500 a year, soon after 1900, the editors of *Mrs Beeton's Every Day Cookery* allowed "two women servants only" for a family of four' (ibid.).

sional income was considerably below £500: the average income of solicitors was £478, dentists £368, doctors £395, chemists £314, engineers £292 and the clergy £206.[92] Industrial workers in the same period earned an average of £277 per annum for a 54-hour working week, and primary school teachers £154.[93] The top of the income scale, which in 1914 covered a very wide range, was open. Successful lawyers like Richard Haldane, for example, who on his own admission gave up an annual income of between £15,000 and £20,000 to become secretary of state for war in 1905, must be placed at the top of the range.[94] The attorney-general, Sir Rufus Isaacs, earned almost £17,000 in 1912/13.[95] Even a permanent secretary earned between £2,000 and £2,500 before the First World War. At the same time the prime minister's salary was about £5,000,[96] while Haldane, as a minister, would have received at least £4,000 after 1905.

If we take these figures as a framework of reference, three factors relating to the income of British scientists around the turn of the century are particularly striking:

(1) The enormous differences in income in British society as a whole before 1914, hardly reduced by direct taxation, were reflected in the incomes of scientists at universities and research institutes, and in state employment. All available statistics point to a large gradation in salaries. Scientists' incomes were related to age and experience, but they depended equally on the particular field and the employer. The president of the Chemical Society had addressed this problem in 1907, when, as quoted above, he had pointed to the 'absurdly inadequate remuneration' of professors of chemistry at some of the new English universities, and had called the payment of 'subordinate members of the staff' of chemistry departments nothing 'short of a scandal'.[97] In practice, this meant that while the two professors of chemistry at Manchester, for instance, had an unusually high salary of £1,200 in 1910, assistants and demonstrators received only £100, and senior assistants between £100 and £180.[98] At Glasgow, the professor of chemistry's salary in 1905/6 was as high as £1,300; a lecturer in the same

92. Guy Routh, *Occupation and Pay in Britain, 1906–60* (Cambridge, 1965), p. 64.
93. Hobsbawm, *Industry and Empire*, vol. 1, p. 140; Routh, *Occupation and Pay in Britain*, p. 69.
94. See Ashby and Anderson, *Portrait of Haldane*, p. 87.
95. 'Merit and Reward', *Nature*, 96 (1915–16), p. 503.
96. See John L. M. Morrison, 'An Engineer Looks at British Society', *Interdisciplinary Science Reviews*, vol. 4 (1979), p. 22; D. Butler and J. Freeman, *British Political Facts 1900–1916* (London, 1963), p. 172.
97. See above, pp. 236–7 in this volume.
98. Roy M. MacLeod and E. Kay Andrews, *Selected Science Statistics Relating to*

department received £400.[99] In 1905 the following salaries were proposed for the physics department of the planned Imperial College of Science and Technology in London: £800 for the director, £400 for the assistant professor, £250 for the demonstrator, £150 for the electrician and £78 for the technicians.[100] When the National Physical Laboratory opened in 1900, the director's salary was fixed at £1,200, that of the superintendents at £400, while assistants received between £200 and £250, and junior assistants between £100 and £150.[101] In 1904 the Treasury expressly imposed on the NPL the condition that 'no member of the Staff, below the Director and the Superintendent of the Engineering Department, shall receive a higher salary than £400 per annum'.[102] Leading scientists on the Medical Research Committee received between £1,000 and £1,250 after 1913, while younger scientists earned between £300 and £500.[103] In 1913/14 the office of the Government Chemist employed forty-eight scientists, whose salaries ranged between £120 and £1,500; the majority received between £120 and £500.[104]

Different subjects also commanded different salaries at universities. At Imperial College of Science and Technology, for example, the professors of mathematics and mechanics, and of geology received salaries of £1,000 in mid-1914, the two professors of physics received £1,075, the professors of botany and zoology £700, and of engineering £1,200.[105] Even for the same subject, salaries varied between universities in the United Kingdom. Imperial College carried out a confidential survey of the salaries of mathematics professors in 1913, prior to appointing a professor itself, with the surprising result that salaries ranged from £350 to

---

Research Endowment and Higher Education, 1850–1914, Mimeograph, Science Policy Research Unit (University of Sussex, 1967), p. 35; Haber, *The Chemical Industry 1900–1930*, p. 55.

99. Haber. *The Chemical Industry 1900–1930*, p. 55.

100. Memorandum enclosed with letter from Sir Arthur Rücker to Robert L. Morant, 16 March 1905 (Ed. 24/530).

101. 'National Physical Laboratory: Minutes of the Executive Committee', vol. 1, p. 7 (9 June 1899). See also 'The National Physical Laboratory: Memorandum on the Future Organization and Expenditure', 19 February 1904 (Royal Society Archives, MS 538). On the salaries of junior assistants, Moseley writes: 'As though this figure was not a large enough deterrent, those who found themselves in this last category were not even considered — despite their undoubted academic qualifications — as holding permanent appointments but were merely accepted on a form of probation' ('National Physical Laboratory', p. 237).

102. Treasury to the Royal Society, 26 December 1904 (Royal Society Archives, MS 538).

103. Figures of 14 February 1916 (Medical Research Council Archives, P. F. 2/1).

104. *Nature*, 111 (1923), p. 447.

105. *Seventh Annual Report of the Imperial College of Science and Technology 1913–1914*, Cd. 7765 (London, 1915), p. 58.

£1,450. The lowest salary was paid by University College Galway, one of the colleges of the National University of Ireland; the highest salary was paid by Manchester. Oxford paid £900, Cambridge £850, University College London £1,000, Edinburgh £1,200 and Leeds £700.[106] A similarly large range existed for other subjects too. In 1916 the highest average wage for scientists of all subjects was paid at Manchester (£888), followed by Liverpool, (£853), while Southampton (£325) and Aberystwyth (£320) paid the lowest.[107] According to contemporary observers, the average annual salary of professors in the sciences was £628 on the eve of the First World War.[108] Haldane, who was extremely knowledgeable on this subject, pointed out that the earnings of few professors in the United Kingdom exceeded the range £700 to £1,100.[109]

(2) With few exceptions, the salaries paid to scientists in industry do not appear to have been higher than those of university professors before 1914. The paucity of statistics, however, does not allow generalisations to be made. Salaries of young chemists in industry seem to have started at between £100 and £150, and, as a rule, doubled only after considerable experience had been gained. Schools and technical colleges paid chemists at the same rates.[110] In 1915 the president of the Board of Trade, Walter Runciman, met a deputation from the Royal Society and the Chemical Society, and expressed his surprise at coming across 'manufacturers who were dependent for their success upon chemical knowledge, still regarding, I will not say £120, but certainly £250 and £300, as ample remuneration for the man on whom their whole success in the

---

106. Questionnaire of 20 January 1913, Minutes and Reports of Imperial College Appointments Board, 1912–13 (Mathematics Joint Committee, Institution of Department of Mathematics for the whole College, file no. 413). Imperial College set the salary of the newly appointed professor at £1,000, following the example of University College London.
107. *Nature*, 98 (1916–17), p. 342.
108. See *The Morning Post* of 5 June 1914 (E. N. da C. Andrade, 'Science and the State'). 'By then', Chaim Weizmann wrote in 1911, 'we were solidly settled, my income at the university [Manchester] had risen to six hundred a year, my wife [a doctor] was making three hundred and fifty, and, with some other earnings, we had about a thousand a year between us, a considerable sum in those days' (*Trial and Error*, p. 148). At the beginning of the twentieth century salaries of German professors were apparently lower than those of their British counterparts. Berlin professors earned £325, at other Prussian universities £250, and outside Prussia between £200 and £225 (Haber, *The Chemical Industry 1900–1930*, p. 45). Ringer estimates that around 1900 the average income of a full professor in Prussia was £600 (Fritz K. Ringer, 'Higher Education in Germany in the Nineteenth Century, *Journal of Contemporary History*, vol. 2, no. 3, 1967, p. 127). See also Charles E. McClelland, *State, Society, and University in Germany 1700–1914* (Cambridge, 1980), pp. 269–73, 309–12.
109. See Ashby and Anderson, *Portrait of Haldane*, p. 24.
110. Haber, *The Chemical Industry 1900–1930*, p. 59; H. A. Roberts, *Careers for University Men* (Cambridge, 1914), pp. 11, 14–15, 21.

future must depend'.[111] It is understandable that under such unfavourable conditions, the career of an industrial chemist in England 'has hitherto been not altogether alluring'.[112] Low salaries were one reason why there were considerably fewer industrial chemists in Britain than in Germany. One cause may lie in the structure of the British chemical industry, especially in its poor research capacity before 1914. Richard B. Pilcher attributed the inadequacy of the salaries paid to British chemists to the fact that in many cases their work 'consisted largely of routine analysis which could be carried out sufficiently well by youths who had a short training in a limited variety of analytical processes and who were willing to accept positions affording them experience'.[113]

(3) Salaries quoted so far of scientists at universities and research institutes seem at first glance to contradict the standard complaint made by the British science lobby since the early nineteenth century. In this period scientists could obviously achieve incomes which would adequately support a 'middle class' life-style and which compared with those of other professions. A few scientists in senior administrative posts even had earnings which were far above the average. The director of the Royal Naval College at Greenwich, for example, had a salary of £1,200 as early as 1873.[114] The position of director-general of munitions design in the Ministry of Munitions commanded a salary of £2,000 in 1916; the departmental head of research, however, received only £850.[115] The Treasury was of the opinion that the vice-chancellor of the University of London had to be 'a person of such a position and dignity' that his 'salary, if given at all, can hardly be less than £2,000 a year'.[116]

If, however, we disregard these few prominent positions, the income situation of British scientists before 1914 appears much less favourable. It was the 'subordinate members of the staff' above all, as the president of the Chemical Society had correctly observed in 1907, who were particularly affected by the general undervaluation of science in Britain. While according to the source quoted above, the average annual salary of university professors in the sciences was £628 in 1914, for all other 'teachers of science in a modern University, including assistant professors', it was only £137.[117]

111. Minutes of the meeting in the Board of Trade on 6 May 1915 (Ed. 24/1579).
112. 'The English Charlottenburg', *Illustrated London News*, 17 February 1906.
113. Pilcher, *Profession of Chemistry*, p. 38.
114. Brock, 'Science Patronage', p. 198.
115. *Nature*, 97 (1916), p. 144.
116. Permanent secretary to the Treasury to the chancellor of the Exchequer, 17 February 1900 (T1/9653B/3328/1901, PRO).
117. *Morning Post*, 5 June 1914.

Science as a Profession

Thus it was more or less the same as that of young chemists in industry. If we take this group of scientists, which had been growing rapidly in size since the end of the nineteenth century, it is true that 'scientific salaries compared poorly with those of other "professional" men such as lawyers and doctors'.[118] Before the First World War it was the young scientists at universities, research institutes and in industry who represented the real social problem which the science lobby repeatedly drew attention to with increasing bitterness.

The situation of these young scientists was not fundamentally improved by the war. In 1915 the Royal Arsenal at Woolwich, where important military research was done, still only paid graduate chemists a salary of £150.[119] Given the conditions of the time, this was justifiably described as 'miserably inadequate'.[120] At the same time the National Physical Laboratory paid its junior assistants, the level at which scientists started, a salary of £175. This led to a demand, made by the Laboratory's staff, for a minimum salary of £200.[121] By this time the salary of senior assistants at the NPL had reached £500 to £600. In general, however, during and after the First World War scientists in Britain were paid less than comparable officials in the civil service who did administrative work. In the war this inequality increasingly came under fire. The problem was raised in 1917 during Question Time in the House of Commons, but this did not change the situation.[122] The provost of Worcester College, Oxford, and a leader-writer in *Nature* in 1921 both pointed out the anomaly that ' a permanent head of a Government Department receives £3,000 or more per annum, a headmistress of a council secondary school may rise to £777 or £800 a year, whereas an Oxford tutor or a professor in one of our modern universities receives on the average a salary of about £850 a year'.[123] Also in 1921 *Nature* wrote of the 'glaring anomalies', as 'in no case do the status, pay, and prospects of promotion of scientific workers compare favourably with those which obtain in the higher clerical grades'.[124] The example cited was salaries paid by the Ministry of Agriculture. Equality of pay between scientists and higher officials was not achieved until after the Second World War. In 1945 the government decided that, in view of the contribution made by

118. MacLeod and MacLeod, 'Social Relations of Science', pp. 305–6.
119. *Nature*, 95 (1915), p. 119.
120. *Nature*, 97 (1916), p. 145.
121. Werskey, *The Visible College*, p. 51.
122. See Hansard, Parl. Deb., H.C., 5th Series, vol. 96, col. 1904 (31 July 1917).
123. 'University and Civil Service Salaries', *Nature*, 107 (1921), p. 802.
124. 'Science in the Civil Service', ibid., p. 2.

science 'towards the winning of the war', 'the salaries of the most highly qualified members of the Scientific Service are to be brought into relationship with those of the Administrative Class'.[125]

We must conclude, therefore, that science did not offer a particularly attractive career between 1850 and 1920. On the contrary, low social esteem for science, limited employment opportunities and bad pay all encouraged students at this time to consider occupations that had nothing to do with science. The social origins of students also contributed to the reservations they had about science. Before 1914 the majority of students still came from the upper and middle classes, who traditionally looked to a career in trade and in the City, in the civil service or the Royal Navy, in the colonial service, the Anglican Church or in politics.[126] Consequently, courses at Oxford and Cambridge were more or less strongly orientated towards the requirements of these occupations, and secondary education, too, was deeply influenced by the same considerations. Hilary and Steven Rose consider that from the 1830s onwards the educational system became 'geared to the production of gentleman administrators and officers, not scientists'. Those who did become scientists and engineers 'did so almost despite the prevailing ethos of the society'.[127]

In Germany science was a profession which could provide a means of social advancement for the bourgeoisie; in England this was rarely the case. Britain's imperial position, 'the success of Britain as an imperial power in the nineteenth century',[128] extensively influenced the thinking and behaviour of its social elite. It allowed a specific hierarchy of social values to evolve. But at the beginning of the twentieth century few perceived that a dangerous gap was opening between an imperial mentality and the demands of a modern industrial economy. Only in retrospect was it clear that until well into the twentieth century Britain was engaged in 'one of the least valuable national exercises'; that is, 'the frantic, at times hysterical, pursuit of an Empire that was increasingly meaningless'.[129] While Britain's economic and political rivals concentrated primarily on advancing their industrial development on

125. *The Scientific Civil Service: Reorganisation and Recruitment during the Reconstruction Period*, Cmd. 6679 (London, 1945), pp. 2–3.
126. See A. H. Halsey, 'British Universities and Intellectual Life' in idem et al. (eds.), *Education*, p. 508; Levine, *Industrial Retardation*, p. 73.
127. Rose and Rose, *Science and Society*, p. 24. See also Wiener, *English Culture*, pp. 127–37, and above, pp. 215–16, 222–3 in this volume. An overview is given by Harold Perkin, 'Die Rekrutierung der Eliten in der britischen Gesellschaft seit 1800', *Geschichte und Gesellschaft*, vol. 3 (1977), pp. 485–502.
128. Halsey, 'British Universities', p. 507.
129. Cardwell, *Technology*, p. 195. See also idem, *Organisation of Science*, p. 191.

the basis of technical and scientific innovation, Britain was seeking 'prestige in planting union flags over miles of jungle, desert and swamp; and glory in such names as those of Rhodes, Jameson and Milner'.[130] The dissipation of productive energies in the creation of a far-flung but transient colonial empire in which British industry enjoyed preferential tariffs ultimately allowed British science and industry to stagnate. The consequences for British economic history in the twentieth century are well known.

---

130. Cardwell, *Technology*, p. 195.

Conclusion
Science in a Liberal Industrial State

In 1919 *Nature*, the British science lobby's most important and influential journal, celebrated its fiftieth anniversary. Sir Archibald Geikie, geologist and former president of the Royal Society, took this opportunity to reflect upon one of the central themes to have occupied *Nature* in the preceding decades: the relationship between science and the state in Britain. 'That one of the great duties of a nation', wrote Geikie proudly, 'is to promote the cultivation of science by appropriating funds not only in aid of education in theory and practice, but also in support of research and experiment, never began to be realised until within living memory. British science has attained its greatness without State aid.' But one year after the end of the First World War, after fifty years of lobbying by scientists, a remarkable change could be noticed in Britain: 'The day of parsimony in regard to the prosecution of scientific inquiry and its applications is now gone beyond the power of any Government to revive', because the war 'has brought the economic value of science before the world on a colossal scale of demonstration.' Geikie was therefore optimistic for the future: 'If, now, we cast our eyes towards the future, the prospect for British science is eminently encouraging. The opportunities for research and experiment were never before so ample, the cooperation of the State never so cordial, the ranks of the investigators never so full, and the joy and enthusiasm for investigation never more ardent'.[1] Did Geikie's observations adequately reflect the real situation?

In broad terms, the 'intellectual climate' was more favourable to science in Germany than it was in Britain during the nineteenth century. In Germany, for reasons which are beyond the scope of this study, the state had had an exceptionally strong interest in promoting science since the beginning of the nineteenth century. In Britain, by contrast, scientists complained until the eve of the First

---

1. Archibald Geikie, 'Retrospect and Prospect', *Nature*, 104 (1919–20), pp. 195–7.

Conclusion

World War that the state's attitude towards science was determined by the economic principle of *laissez-faire*, which had contributed significantly to the decline of science in this country. Historians of science have only too often uncritically accepted the British science lobby's assessment and condemnation of the British state. 'The denial of State aid during the crucial period 1850–80', writes Cardwell, for example, 'was the final reason why applied science was later in making its appearance in England than in Germany.'[2]

The science lobby's claims were, of course, not always free of distortions, exaggerations and controversial arguments. A detailed examination of the relationship between science and the state in Britain during the nineteenth century shows that it was in fact far more differentiated than suggested by spokesmen for science who had an eye on conditions in Germany. At least since Oliver MacDonagh's seminal article 'The Nineteenth-Century Revolution in Government',[3] and the ensuing debate about the state's conception of its functions in the nineteenth century, it has been obvious that this relationship, then as now, cannot be reduced to the simplified alternatives 'private enterprise' or 'public enterprise'. This sort of black-and-white picture, which allows the state only a nightwatchman function in Victorian society, is a misrepresentation of reality. Even in the liberal Britain of the nineteenth century, *laissez-faire* did not mean that the state refrained on principle from intervention in society and economy under all circumstances. Fundamentally, this principle meant only that in practice the state left certain economic and social functions primarily to private initiative, without itself becoming involved. This attitude, which should not be equated with indifference towards the problems to be solved, by no means excluded state intervention if, under constantly changing circumstances and conditions, private enterprise could not cope and intervention seemed necessary. Thus even in Britain, the nineteenth century was only to a limited extent the 'classical' century of *laissez-faire*. Administrative historians of nineteenth-century Britain, for example – as Arthur J. Taylor has pointed out[4] — would not call the nineteenth century an era of doctrinaire *laissez-faire*; it was a period when the scope of administration increased and it became much more differentiated. Britain's massive intervention in

---

2. Cardwell, *Organisation of Science*, p. 244.
3. Oliver MacDonagh, 'The Nineteenth-Century Revolution in Government: A Reappraisal', *Historical Journal*, vol. 1 (1958), pp. 52–67. On this see also François Bédarida, 'L'Angleterre victorienne paradigme du laissez-faire? A propos d'une controverse', *Revue Historique*, vol. 261 (1979), pp. 79–98.
4. Taylor, *Laissez-faire and State Intervention in Nineteenth-Century Britain*, p. 54.

Conclusion

Ireland shows that the same is true of economic policy at this time.[5] As Henry Parris has written, '*laissez-faire* and state intervention were equally characteristic of the middle quarters of the nineteenth century and it is not necessary to assume that they were in contradiction to one another'.[6]

In the first chapter of this book, we tried to show that this characteristic mixture of *laissez-faire* and state intervention also typified the relationship between science and the state during the nineteenth century in Britain. The government revealed its priorities to the extent that, when asked, it was more willing to support scientific research and projects which had practical application than to support basic research, whose benefits for the state, administration and the general public were less immediately obvious. Not until the end of the nineteenth century did a gradual change take place in the government's attitude, despite opposition from the Treasury, and Parliament's continuing indifference towards scientific problems. In 1902 the Conservative prime minister, Arthur Balfour, candidly, and almost apologetically, admitted in reply to a question in the House of Commons that 'the amount of Government money given to experiments in the country is not very large'.[7] Late in 1904 the president of the Royal Society suggested that the growing number of scientific assignments which the Society was being given by the government showed 'fuller recognition by the Government and the public of the need for scientific advice and direction in connection with many matters of national concern'.[8] A government white paper, with an introduction by the president of the Board of Education, Walter Runciman, in 1910 established a 'marked and growing public interest in University Education during recent years'.[9] In 1907 Edward VII, as chancellor of the University of Wales, wrote: 'We must look ahead and endeavour to be ready to meet all the requirements of scientific and intellectual progress. The imperative necessity for higher education and research is becoming more and more recognised'.[10] And although the Liberal prime minister Asquith maintained, at the Royal Society's 250th anniversary, that is was not right that 'science should be a mendicant for State endowment',[11] the govern-

---

5. See R. D. Collison Black, *Economic Thought and the Irish Question, 1817–1870* (Cambridge, 1960).
6. Parris, *Constitutional Bureaucracy*, p. 270.
7. Hansard, Parl. Deb., H.C., 4th Series, vol. 101, col. 800 (24 January 1902).
8. *Nature*, 71 (1904–5), p. 107.
9. *Reports from those Universities*, p. 111.
10. Quoted in ibid., p. IV.
11. *The Times*, 17 July 1912.

Conclusion

ment was at the same time considering making 'the maintenance of Universities and of an adequate number of institutions for specialised instruction of the most advanced character as well as the provision of funds for the conduct of researches of national importance a first charge upon the national expenditure for education'.[12] The slowly emerging change in the government's and, more broadly, the state's attitude towards science was occasionally noted with approval by spokesmen for science. In 1910 the British Science Guild noticed that 'the present Government has shown itself more anxious to promote scientific inquiry than any of its predecessors'.[13]

If we look back to the relationship between science and the state in Britain between 1900 and 1920, the years on which this study has concentrated, we must conclude that the state did not yet have a long-term public science policy; *ad hoc* decisions continued to govern state aid granted to science. But, at the same time, relations between science and the state became much stronger after the turn of the century, and greater cooperation between science and the state meant that institutional conditions for science improved markedly. In Chapter 3 of this book, three case studies allowed us to show that the science lobby's persistent references to the economic and political significance of education and science had met with greater understanding from politicians since the end of the nineteenth century. The National Physical Laboratory and Imperial College of Science and Technology owed their origins to private initiative, but the subsequent development of the two institutions shows that the state was far more ready to join in private initiatives and offer financial support than it had been in the preceding decades. The establishment and funding of new provincial universities also illustrates this tendency. By complying with suggestions brought to it 'from outside', the state gradually stepped beyond its traditional domain and ventured into areas which had previously been left mainly to private initiative, organisation and finance. The state had never been concerned with matters relating to scientific research, maintained *Nature* late in 1917:

> They were subjects to be left to private enterprise and individual effort. But the circumstances of the time have changed much in our time-honoured and traditional view of the mutual relations of the individual and the State. Public opinion, under the hustling influence of the

---

12. A Policy of National Education (typewritten memorandum, 1912, no exact date, H. H. Asquith, Private Papers, vol. 93, fol. 1, Bodleian Library.)
13. Annual meeting of the British Science Guild (*Nature*, 83, 1910, p. 100).

moment, now compels the State to accept responsibilities and exercise initiative to an extent hitherto undreamt of.[14]

The establishment of the Medical Research Committee in 1913, and the spurt in science organisation during the First World War culminating in the founding of the Department of Scientific and Industrial Research, are examples from the science sector which clearly illustrate the movement of the state towards a regulative and initiatory role in social areas which it had previously avoided. They make clear that within a few years, the state's relationship with science had assumed a new quality: instead of merely reacting to the pace set by science, the state became active in its own right and attempted to improve the conditions under which scientific work was done in Britain. The initiative for innovations in science organisation, some of which broke new ground by international standards, more and more often originated from the state. Although Bentley B. Gilbert writes of a 'revolution in popular attitudes towards State activity that have been manifest in British politics since the First World War',[15] in science, and even more strongly in the areas of national insurance, education and the labour market, this revolution had been on the horizon at least since the turn of the century. Hans Medick has shown that in the economy, too, despite the failure of Chamberlain's campaign for tariff reform, an essentially pragmatic, tentative interventionism was characteristic of Liberal governments since 1906.[16] In this view, the period of transition lasted from the end of the nineteenth century to the end of the First World War, making the year 1914 much less of a caesura than many historians of British science policy have suggested. A completely new chapter in the relationship between science and the state did not begin in Britain in August 1914; rather, developments that had begun some time before 1914 accelerated and intensified after this date. Thus 1914 represents not a qualitative but only a quantitative leap towards a 'new and more conscious appreciation of the function of science in a modern industrial state'.[17]

If the period from 1900 to 1920 can be called a period of

14. *Nature*, 100 (1917–18), p. 266.
15. Bentley B. Gilbert, *British Social Policy, 1914–1939* (London, 1970, reprinted 1973), pp. 31–2.
16. Medick, 'Anfänge und Voraussetzungen des Organisierten Kapitalismus', p. 74. See also S. J. Hurwitz, *State Intervention in Great Britain: A Study of Economic Control and Social Response, 1914–1919* (New York, 1949; reprinted London, 1968); H. V. Emy, *Liberals, Radicals and Social Politics, 1892–1914* (Cambridge, 1973).
17. Bernal, *Social Function of Science*, p. 31.

transition during which the foundations of a modern public science policy emerged, then the question arises as to why the state began to promote science systematically so late in Britain as compared with Germany. Any answer to this question can only be speculative. But it seems that we cannot totally discount the idea that the relative delay with which the state began to organise the science sector was related to Britain's political system in the nineteenth century. The liberal political and social theories of the age obviously placed larger obstacles in the way of intervention and state initiative, not only in science but also in economic and social policy, than was the case in other European countries at the same time. With the exception of the examples we have mentioned in which science had direct practical application to the state's expanding functions, or held out the promise of such benefits in the near future, liberal theory left the science sector, including the universities, more or less to itself. Basic research in particular received only little support from the state.

The state's reticence as far as science was concerned is connected with the fact that in Britain private patronage of science was so prominent in the nineteenth century. It lost its significance only during the First World War and in the economically difficult years after 1918. Until this time innovations in science organisation were made possible by private patronage which made available for science the large amounts of money the state was not prepared – or did not need – to give. One conclusion to be drawn from this study is that the existence of generous private patronage in Victorian Britain contributed significantly to delaying any involvement by the state in promoting science on a large scale. As late as 1904 Austen Chamberlain could point to private patronage when, in his capacity as chancellor of the exchequer, he advocated financial support for the universities, 'with the object of stimulating private benevolence in the locality'.[18] It seems that the state's attempts to avoid expenditure on science and to hide behind private benefactors is symptomatic of the era. This study has suggested that private patronage placed no obstacles in the way of the state in pursuing this strategy.

Confronted with other problems and influenced by thinking directed primarily towards expanding the empire, British politicians were long satisfied to retain the status quo in science organisation. They left science promotion and the financing of research largely to private initiative, even when the inadequacy of this

18. See above, p. 35 in this volume.

situation became obvious during the second half of the nineteenth century. Not until the backwardness of sections of British industry (especially in the development of new science-dependent industries) in comparison with Germany and the USA emerged clearly at the end of the nineteenth century, were the arguments put forward by the British science lobby taken more seriously. It attributed the unfavourable development of British industry to the lack of support for scientific research. And only when the symptoms of economic crisis became unmistakable around the turn of the century, did doubts arise about the capacity and adaptability of the existing political system. 'Under a democratic constitution,' wrote *Nature* as late as autumn 1915, 'it is perhaps too much to expect that Parliament will pay much attention to scientific men or methods.'[19] Only a few weeks later it pointed out how paradoxical it was that 'we live in a scientific age, yet we are governed by men who belong to a century ago'.[20] During the crisis of confidence which afflicted late-Victorian Britain under the growing pressure of international economic competition and rapidly rising costs for scientific research, the nineteenth-century British model of science promotion briefly outlined in the first chapter of this book was increasingly put into question by a modernising elite in science and politics. A greater willingness to tolerate state intervention in the science sector among others thus emerged. This attitude was consciously influenced by the German model of promoting science, which conceded the state a dominant influence. Its success in the nineteenth century made it generally regarded as worthy of emulation. In Britain state intervention was intended primarily to provide financial aid and to create the institutional conditions necessary for scientific research. Thus after the turn of the century a policy for science was *de facto* introduced. The state no longer hesitated to intervene in the organisation of the university system, and traditional forms of science promotion, whether by learned and scientific societies or by private patrons, were pushed into the background. In other words, the British model of science promotion as it developed during the nineteenth century went through a process of transformation after the turn of the century, by which support for scientists was placed on a new footing and the relative significance of the various 'promoters of science' was established anew.

Britain's liberal economic and political climate and the existence

19. 'Science in National Affairs', *Nature*, 96 (1915–16), p. 195.
20. Ibid., p. 504.

Conclusion

of a broad stratum of private patrons among the wealthy industrial and commercial bourgeoisie are possible explanations for the state's neglect of science organisation and promotion in the nineteenth century, but Britain's specific experience during the Industrial Revolution was another important factor. In England the Industrial Revolution had taken place totally under the sway of economic *laissez-faire*, free from state intervention and aid, and independently of the universities, if not of the sciences. With few exceptions, the universities, academic researchers and technicians, or engineers with a theoretical training had played no part in it. Practical experience alone had led to the development of new technologies and production processes. As a result, British industry, with few exceptions – and many examples could be cited to illustrate this – remained bound by this historical experience. Even at the end of the nineteenth century, British industry, and especially the chemical industry, did not recognise the necessity of scientific research to the same extent as did industry in other European countries and the USA. 'The Channel of commerce is changing . . . ', noted the American consul in Bradford as early as 1879. 'Instead of changing with the stream, English manufacturers continue to sit upon the bank of the old watercourse and argue from plausible but unprofitable scientific and commercial premises that it had no right to alter.'[21] 'The great cause of German success', maintained Ernest E. Williams shortly before the turn of the century, 'is an alert progressiveness, contrasting brilliantly, with the conservative stupor of ourselves . . . . The mass of English manufacturers and traders clings to the faith of its fathers.'[22] 'Fifty years of industrial preeminence', wrote Derek Aldcroft in 1964, 'had bred contempt for change and had established industrial traditions, in which the basic ingredients of economic progress, science and research, were notably absent.'[23] By the end of the nineteenth century, Britain's pioneering role in the process of industrialisation had obviously become an obstacle to industrial development and success in economic competition with the new industrial nations of Europe and North America, which were beginning to challenge Britain's position as the dominant industrial power.

Despite an enormous expansion in the state's functions since the end of the nineteenth century, its traditional *laissez-faire* attitude was still felt in that British science policy attempted to guarantee

---

21. Quoted in Coleman, 'Gentlemen and Players', p. 93.
22. Williams, *'Made in Germany'*, p. 163.
23. Aldcroft, 'The Entrepreneur', p. 133. See also Wiener, *English Culture*, p. 139.

Conclusion

science as much autonomy as possible in determining its own long-term research aims and projects. Austen Chamberlain's maxim that 'State aid must always be accompanied by State control',[24] proved to be inappropriate for Britain. Modern British science policy, whose guiding ideas were formulated after the turn of the century, ensured that even after public funds were made available for science, concerns inherent to science itself provided the criteria by which decisions were made on scientific questions, and that the majority on the bodies which took these decisions was made up of scientists. Direct dependence and control by the state, which many scientists early this century still saw as an inevitable consequence of accepting financial support from the state,[25] was deliberately avoided in Britain, even under the exceptional conditions of the war. After the state began to support science on a larger scale, it could still exercise only indirect control by granting or withdrawing funds. From an organisational point of view, the autonomy of science in the distribution of public funds and in making decisions about research aims and areas of special interest was guaranteed by the esblishment or development of independent advisory, management, or controlling bodies, which mediated between the government and scientists. The Executive Committee of the National Physical Laboratory, the Governing Body of Imperial College of Science and Technology, the Medical Research Committee, the Advisory Council of the Department of Scientific and Industrial Research, and even the Research Associations are notable examples of the principle of divided responsibility in action. While this principle is generally accepted today in Western industrial nations, it was not taken for granted in the early twentieth century as a structural feature of public support and organisation of science.

The Machinery of Government Committee, appointed in 1917 by Lloyd George's government and chaired by Lord Haldane,[26]

---

24. Austen Chamberlain expressed this opinion in July 1904 when receiving a deputation of scientists and university representatives, which asked the government to increase its subsidies to universities and university colleges. On this occasion Chamberlain said that it was 'dangerous for the higher education of the country that it should have to conform itself, for the purpose of obtaining grants, to rules and regulations laid down by the Treasury' (Howarth, *The British Association*, p. 236).

25. In 1904 Sir Michael Foster, for example, maintained: 'We are placed in this dilemma: inquiry needs money, but has even greater need of freedom; the State can offer money, but it gives money at the cost of freedom' (Foster, 'The State and Scientific Research', p. 744).

26. The Haldane Committee was composed of three MPs: Beatrice Webb, Sir Robert L. Morant and Sir George Murray, former permanent secretary to the Treasury (1903–11). Haldane called it 'about the best committee over which I have ever presided and we see our way to producing a real plan for the Government of

Conclusion

explicitly supported the claim of scientists for research autonomy *vis-à-vis* not only politics but also industry. The Committee's report, the major parts of which were written by Haldane, also endorsed the claim made by politicians that the allocation of public funds for science must be subject to parliamentary control. That science should determine its own research aims and methods and be free from any direct control by the executive, known as the 'Haldane Principle', became the guiding principle of modern British science policy. Its identification with Haldane's name is an appropriate tribute to the outstanding achievements of the leading British spokesman for science and organiser of science in the first half of the twentieth century.

---

the nation in the days ahead' (Haldane to his mother, mid-July 1917, printed in Sommer, *Haldane of Cloan*, p. 346). Roseveare calls the committee's report 'a major landmark in British administrative history' (*The Treasury*, p. 244); *Report of the Machinery of Government Committee, 1918*, Cd. 9230 (London, 1918). On this see Daalder, *Cabinet Reform*, pp. 266–77; Charles H. Wilson, *Haldane and the Machinery of Government* (London, 1956).

Appendices

Appendix 1: Foundation of learned and scientific societies in all fields in the United Kingdom during the nineteenth century

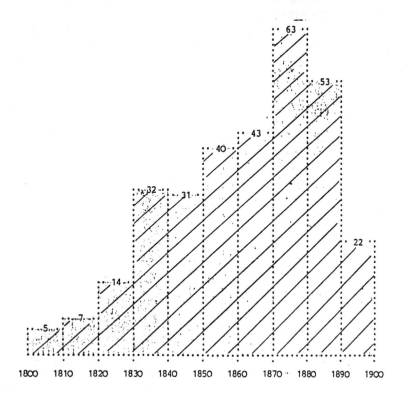

Source: Based on data from the *Year-Book of the Scientific and Learned Societies of Great Britain and Ireland* (London, 1900). It lists 463 learned and scientific societies and institutions. Not included here are ten government institutions, such as, for example, the Royal Observatory at Greenwich, the Meteorological Office and the Geological Survey of Great Britain. In 125 cases the foundation date could not be established; eighteen societies were founded before 1800.

Appendices

Appendix 2: Scientific societies and institutions in London, 1883

	Year of foundation	Membership figures
1. *General Sciences*		
The Royal Society of London for Improving Natural Knowledge	1660	471
The (Royal) Society for the Encouragement of Arts, Commerce, and Manufactures (The Society of Arts)	1754	3,445
The Royal Institution of Great Britain for the Promotion, Diffusion and Extension of Science and Useful Knowledge	1799	—
The British Association for the Advancement of Science	1831	c.4,000
The Lambeth Field Club and Scientific Society	1872	39
The Balloon Society of Great Britain	1880	c.10,000
2. *Mathematics and Physics*		
The Royal Astronomical Society	1820	591
The Royal Meteorological Society	1850	c.570
The Mathematical Society of London	1865	170
The Mathematical Association	1871	—
The Association for the Improvement of Geometrical Teaching	1871	—
The Physical Society of London	1874	c.330
3. *Chemistry and Photography*		
The Chemical Society of London	1841	c.1,300
The Pharmaceutical Society of Great Britain	1841	c.6,000
The Royal Photographic Society of Great Britain	1853	363
The Institute of Chemistry [of Great Britain and Ireland]	1877	440
The Society of Chemical Industry	1881	c.1,700
The London and Provincial Photographic Association	—	—
The Photographic Club	—	—
The Postal Photographical Society	—	—
The South London Photographical Society	—	—

continued on p. 258

Appendix 2: *continued*

	Year of foundation	Membership figures
4. *Geology and Geography*		
The Geological Society of London	1807	1,440
The Royal Geographical Society	1830	3,370
The Palaeontographical Society	1847	—
The Geologists' Association	1858	c.500
The Mineralogical Society	1876	—
5. *Biology, Microscopy, Anthropology*		
The Linnean Society of London	1877	c.715
The Zoological Society of London	1829	c.3,300
The Entomological Society of London	1833	238
The Royal Botanical Society of London	1836	c.2,500
The Royal Microscopical Society	1839	c.650
The Ray Society	1844	—
The City of London Entomological and Natural History Society	1858	—
The British Ornithologists' Union	1859	145
The Quekett Microscopical Club	1865	—
The (Royal) Anthropological Institute of Great Britain and Ireland	1871	c.500
The South London Microscopical and Natural History Club	1871	c.130
The South London Entomological and Natural History Society	1872	48
The Postal Microscopic Club	1873	170
The Physiological Society of London	1876	50
The Hackney Microscopical and Natural History Society	1877	111
6. *Engineering*		
The Institution of Civil Engineers	1818	1,381
The Institution of Mechanical Engineers	1847	1,440
The Society of Engineers	1854	386
The Civil and Mechanical Engineers' Society	1859	—
The Aeronautical Society of Great Britain	1866	c.100
The Surveyors' Institution	1868	c.1,200
The Iron and Steel Institute	1869	—

continued on p. 259

Appendix 2: *continued*

	Year of foundation	Membership figures
The Institution of Electrical Engineers	1871	—
The Society of Telegraph Engineers and Electricians	1871	—
7. Agricultural Science		
The Royal Horticultural Society	1804	*c.*2,500
The Royal Agricultural Society of England	1838	8,352
The Bedford Park Natural History and Gardening Society	1883	—

Source: Year-Book of the Scientific and Learned Societies, 1st edn (1884) and 17th edn (1900); also Roy M. MacLeod and E. Kay Andrews, *Selected Science Statistics Relating to Research Endowment and Higher Education, 1850–1914*, Mimeograph, Science Policy Research Unit (University of Sussex, 1967), p. 19 (whose membership figures sometimes differ from those given in the *Year-Book*) and monographs on individual specialist societies.

Appendix 3: Membership figures of the Royal Society of London, 1660–1935

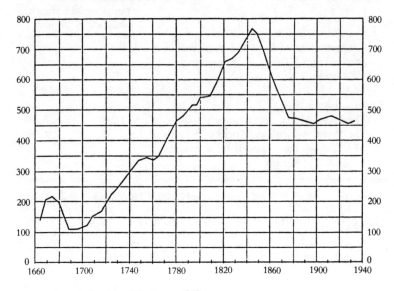

Source: Record of the Royal Society, p. 568.

Bibliography

Manuscript Sources

Bibliothek der Physikalisch-Technischen Bundesanstalt, Institut Berlin
  Tätigkeitsberichte der Physikalisch-Technischen Reichsanstalt 1899–1906
National Library of Scotland, Edinburgh
  Richard B. Haldane, Private Papers
  Philip Archibald Primrose, Fifth Earl of Rosebery, Private Papers
University Archives, Liverpool
  John T. Brunner, Private Papers
  H. A. Ormerod, 'History of University College, Liverpool' (typescript, n.d.)
British Library, London
  Henry Campbell-Bannerman, Private Papers
  Herbert Gladstone, Private Papers
British Library of Political and Economic Science, London
  Beatrice and Sidney Webb, Private Papers (Passfield Papers)
House of Lords Record Office, London
  David Lloyd George, Private Papers
Imperial College of Science and Technology Archives, London
  Minutes and Reports of Imperial College Appointments Board, 1912–13
  ABC. Board of Education. Departmental Committee of the Royal College of Science and Royal School of Mines, 1902–1909
Medical Research Council, London
  Harold P. Himsworth, 'The Support of Medical Research' (typescript, 1958)
  MRC 1472
  MRC, P.F. 1–3, General Correspondence 1914–16
Public Record Office, London
  Department of Education and Science, Private Office Papers, Series I, 1851–1935
  Department of Scientific and Industrial Research (DSIR)
  Treasury, Finance Files 1901–20
  Cabinet Papers 1895–1906
The Royal Society of London, London
  MS 538, Some letters and papers about the early days of the National Physical Laboratory
  M.C. 16, 1893–6, Miscellaneous Correspondence
  Minutes of Council, vols. 8–11 (1898–1920)
Bodleian Library, Oxford
  Herbert Henry Asquith, Private Papers
Broadlands Archives, Romsey, Hampshire

Bibliography

Ernest J. Cassel, Private Papers
National Physical Laboratory, Teddington
Minutes of the Executive Committee, vols. 1–4 (May 1899 to March 1914)

Printed Sources

1. Documents, official publications and annual reports

Annual Reports of the British Science Guild (London, 1907–20)
Annual Reports of the Governing Body of the Imperial College of Science and Technology 1908–1917 (London, 1909–18)
Annual Reports of the Medical Research Committee, 1914–1919 (London, 1915–19)
Annual Reports of the Meetings of the British Association for the Advancement of Science (London, 1892–1900)
Education in Britain, ed. by the Central Office of Information, Reference Pamphlet No. 7 (London, 1974)
Final Report of the Departmental Committee on the Royal College of Science, Etc., vol. 1, *Final Report with Appendix I*, Cd. 2872; vol. 2, *Minutes of Evidence*, Cd. 2956 (London, 1906)
Final Report of the Departmental Committee on Tuberculosis, vol. 1, Cd. 6641 (London, 1913)
Final Report of the Royal Commission Appointed to Inquire into the Relations of Human and Animal Tuberculosis, Cd. 5761 (London, 1911)
Hansard, House of Commons Debates, Official Report, 4th Series 1892–1908 and 5th Series, 1909– (to be completed)
The Imperial College of Science and Technology: Charter of Incorporation (London, 1907)
Magnus, Philip et al., *Report on a Visit to Germany, with a View of Ascertaining the Recent Progress of Technical Education in that Country, Being a Letter to his Grace the Duke of Devonshire, Lord President of the Council*, C. 8301 (London, 1896)
The National Physical Laboratory: Annual Reports 1901–1912 (London, 1901–13)
The Neglect of Science: Report of the Proceedings at a Conference at Burlington House, 3rd May 1916 (London, 1916)
Poole, J. B. and E. Kay Andrews (eds.), *The Government of Science in Britain* (London, 1972)
Preliminary Report of the Departmental Committee on the Royal College of Science, Cd. 2610 (London, 1905)
Proceedings of the Royal Society (London, 1901–21)
Proceedings of the British Academy 1905–1906 (London, no date)
Report of the Advisory Committee for the Tropical Diseases Research Fund for the year 1914, Cd. 7796 (London, 1915)
Report of the Board of Education for the Year 1908–1909, Cd. 5130 (London, 1910)
Report of the Board of Education for the Year 1915–1916, Cd. 8594 (London, 1917)

Bibliography

Report on Chemical Instruction in Germany and the Growth and Present Condition of the German Chemical Industries (Diplomatic and Consular Reports, Misc. Series), Cd. 430 (London, 1901)
Report of the Committee Appointed by the Treasury to Consider the Desirability of Establishing a National Physical Laboratory, 2 vols., C. 8976, and C. 8977 (London, 1898)
Report of the Committee of the Privy Council for Scientific and Industrial Research for the Year 1915–16, Cd. 8336 (London, 1916)
Report of the Committee of the Privy Council for Scientific and Industrial Research for the Year 1916–17, Cd. 8718 (London, 1917)
Report of the Machinery of Government Committee, 1918, Cd. 9230 (London, 1918)
Report of the Schools Inquiry Commission, vol. 6 (London, 1868)
Reports from those Universities and University Colleges in Great Britain which Participated in the Parliamentary Grant for University Colleges in the Year 1908–09, Cd. 5246 (London, 1910)
Reports of the Royal Commission on Scientific Instruction and the Advancement of Science, C. 318, C. 536, C. 868, C. 884, C. 1087, C. 1279, C. 1297, C. 1298 (London, 1871–5)
Royal Commission on University Education in London, Final Report, C. 6717 (London, 1913)
Scheme for the Organization and Development of Scientific and Industrial Research, Cd. 8005 (London, 1915)
The Scientific Civil Service: Reorganisation and Recruitment during the Reconstruction Period, Cmd. 6679 (London, 1945)
Scientific Societies (Government Grants): Return to an Order of the Honourable The House of Commons, dated 25 July 1906 (London, 1906)
University Colleges (Great Britain) (Grant in Aid), Cd. 2422 (London, 1905)
The Year-Book of the Royal Society of London 1902 (London, 1902)

2. Newpapers and journals

The Daily Mail
The Daily Telegraph
The Economist
The English Mechanic
The Illustrated London News
The Morning Post
Nature: A Weekly Illustrated Journal of Science
The Nineteenth Century and After
The Spectator
The Times
Zeitschrift für Instrumentenkunde

3. Diaries, letters, memoirs, autobiographies, and contemporary accounts

Addison, Christopher, *Politics from Within, 1911–1918*, 2 vols. (London,

1924)
—— *Four and a Half Years: A Personal Diary from June 1914 to January 1919*, 2 vols. (London, 1934)
Appleton, Charles E. (ed.), *Essays on the Endowment of Research by Various Writers* (London, 1876)
Babbage, Charles, *Reflections on the Decline of Science in England and on Some of its Causes* (London, 1830; reprinted Farnborough, 1969)
—— *The Exposition of 1851; or, Views of the Industry, the Science, and the Government of England* (London, 1851; reprinted Farnborough, 1969)
Becker, Bernhard H., *Scientific London* (London, 1874; reprinted Haarlem, 1968)
Bell, Olivier (ed.), *The Diary of Virginia Woolf*, vol. 1, *1915–1919* (London, 1977)
Carrière, Justus (ed.), *Berzelius und Liebig. Ihre Briefe von 1831–1845* (Munich and Leipzig, 1893)
Curtius, Ludwig, *Deutsche und antike Welt. Lebenserinnerungen*, 2nd edn (Stuttgart, 1958)
Fitzroy, Almeric, *Memoirs*, 2 vols. (London, 1925)
Fleming, A. P. M., *Industrial Research in the United States of America* (London, 1917)
Foster, Michael, 'The State and Scientific Research', *The Nineteenth Century and After*, vol. 55 (1904), pp. 741–51
Geikie, Archibald, *A Long Life's Work: An Autobiography* (London, 1924)
Glazebrook, Richard T., *Early Days at the National Physical Laboratory* (Teddington, 1933)
Haldane, Richard B., *An Autobiography* (London, 1929)
Harnack, Adolf, *Aus Wissenschaft und Leben*, vol. 1 (Giessen, 1911)
Howard, E. D., *The Cause and Extent of the Recent Industrial Progress of Germany* (London, 1907)
Lecky, W. E. H., *Democracy and Liberty*, 2 vols. (London, 1896)
Lockyer, J. Norman, 'The Influence of Brain-power on History', *Nature*, 68 (1903), pp. 439–47
—— *Education and National Progress: Essays and Addresses, 1870–1905* (London, 1906)
Lodge, Oliver, *Past Years: An Autobiography* (London, 1931)
Low, Sidney, *The Governance of England* (London, 1904)
Mackenzie, F. A., *The American Invaders* (London, 1902)
Mackenzie, Norman (ed.), *The Letters of Sidney and Beatrice Webb*, 3 vols. (Cambridge, 1978)
Osler, William, *Science and War* (Oxford, 1915)
Pattison, Mark, *Suggestions on Academical Organisation with Especial Reference to Oxford* (Edinburgh, 1868)
Pearson, Karl, *National Life from the Standpoint of Science* (London, 1901)
Proctor, Richard A., *The Wages and Wants of Science-Workers* (London, 1876, reprinted 1970)
Roberts, H. A., *Careers for University Men* (Cambridge, 1914)
Russell, Bertrand, *The Autobiography of Bertrand Russell*, vol. 1, *1872–1914* (London, 1967)
Shadwell, Arthur, *Industrial Efficiency: A Comparative Study of Industrial Life*

in England, Germany and America, 2 vols. (London, 1906)
Stead, W. T., *The Americanisation of the World* (London, 1902)
Taine, Hippolyte, *Notes on England*, trans. and ed. Edward Hyams (London, 1957)
Webb, Beatrice, *Our Partnership* (London, 1948, reprinted 1975)
―― *Diaries 1912–1924*, ed. Margaret I. Cole (London, 1952)
Webb, Sidney, *A Policy of National Efficiency* (London, 1901)
―― 'London University: A Policy and a Forecast', *The Nineteenth Century and After*, vol. 51 (1902), pp. 914–31
―― 'The London Charlottenburg', *The University Review* (October 1906), pp. 13–24
Weizmann, Chaim, *Trial and Error: The Autobiography of Chaim Weizmann*, 4th edn (London, 1950)
Whewell, William, *The Philosophy of the Inductive Sciences Founded upon their History*, 2 vols. (London, 1840)
Williams, Ernest E., *'Made in Germany'* (London, 1896)
Williams, L. Pearce (ed.), *The Selected Correspondence of Michael Faraday*, 2 vols. (Cambridge, 1971)

4. Reference works

Buchloh, Paul G. and Walter T. Rix (eds.), *American Colony of Göttingen: Historical and Other Data Collected between the Years 1855 and 1888* (Göttingen, 1976)
Butler, D. and J. Freeman, *British Political Facts 1900–1916* (London, 1963)
The Dictionary of National Biography, 66 vols. (London, 1885–1900), 6 supplementary volumes (1912–71)
Historical Statistics of the United States: Colonial Times to 1970 (Washington, DC, 1975)
MacLeod, Roy M. and E. Kay Andrews, *Selected Science Statistics Relating to Research Endowment and Higher Education, 1850–1914*, Mimeograph, Science Policy Research Unit (University of Sussex, 1967)
Mitchell, Brian R., *European Historical Statistics 1750–1970* (London, 1975)
Williams, Raymond, *Keywords: A Vocabulary of Culture and Society* (London, 1976)
Year-Book of the Scientific and Learned Societies of Great Britain and Ireland: Giving an Account of their Origin, Constitution, and Working, 1st edn (London, 1884)
Year-Book of the Scientific and Learned Societies of Great Britain and Ireland, 17th edn (London, 1900)

Books, Articles and Theses

Abramovitz, Moses and Vera F. Eliasberg, *The Growth of Public Employment in Great Britain* (Princeton, NJ, 1957)
Adler, Cyrus, *Jacob H. Schiff: His Life and Letters*, 2 vols. (London, 1929)
Aldcroft, D. H., 'The Entrepreneur and the British Economy, 1870–1914',

Economic History Review, vol. 17 (1964–5), pp. 113–34
―――― (ed.), *The Development of British Industry and Foreign Competition, 1875–1914* (London, 1968)
―――― and Harry W. Richardson, *The British Economy 1870–1939* (London, 1969)
Alford, B. W. E., *W. D. & H. O. Wills and the Development of the U. K. Tobacco Industry 1786–1965* (London, 1973)
Allen, Bernard M., *Sir Robert Morant: A Great Public Servant* (London, 1934)
Allen, David Elliston, *The Naturalist in Britain: A Social History* (Harmondsworth, 1978)
―――― 'The Women Members of the Botanical Society of London, 1836–1856', *British Journal for the History of Science*, vol. 13 (1980), pp. 240–54
Alter, Peter, 'Staat und Wissenschaft in Großbritannien vor 1914' in Helmut Berding et al. (eds.), *Vom Staat des Ancien Régime zum modernen Parteienstaat. Festschrift für Theodor Schieder zu seinem 70. Geburtstag* (Munich and Vienna, 1978), pp. 369–83
―――― 'The Royal Society and the International Association of Academies 1897–1919', *Notes and Records of the Royal Society of London*, vol. 34 (1979–80), pp. 241–64
Andrade, E. N. da Costa, *A Brief History of the Royal Society* (London, 1960)
Argles, Michael, *South Kensington to Robbins: An Account of English Technical and Scientific Education since 1851* (London, 1964)
Aris, Stephen, *The Jews in Business* (London, 1970)
Armytage, W. H. G., *Sir Richard Gregory: His Life and Work* (London, 1957)
―――― *The German Influence on English Education* (London, 1969)
―――― *The Rise of the Technocrats: A Social History* (London and Toronto, 1965, reprinted 1969)
―――― *A Social History of Engineering*, 4th edn (London, 1976)
Ashby, Eric, 'On Universities and the Scientific Revolution' in A. H. Halsey et al. (eds.), *Education, Economy and Society* (New York, 1961), pp. 466–76
―――― *Technology and the Academics: An Essay on Universities and the Scientific Revolution* (London, 1958; reprinted London and New York, 1966)
―――― and Mary Anderson, *Portrait of Haldane at Work on Education* (London, 1974)
Badash, Lawrence, 'British and American Views of the German Menace in World War I', *Notes and Records of the Royal Society of London*, vol. 34 (1979–80), pp. 91–121
Bailey, Edward, *The Geological Survey of Great Britain* (London, 1952)
Basalla, George (ed.), *The Rise of Modern Science: External or Internal Factors?* (Lexington, Mass., 1968)
Beales, H. L., 'The Great Depression in Industry and Trade', *Economic History Review*, vol. 5 (1934), pp. 65–75
Bédarida, François, 'L'Angleterre victorienne paradigme du *laissez-faire*? A propos d'une controverse', *Revue Historique*, vol. 261 (1979), pp. 79–98
Beer, John J., *The Emergence of the German Dye Industry* (Urbana, Ill., 1959)

Bibliography

Bellot, H. Hale, *University College, London, 1826–1926* (London, 1929)
Ben-David, Joseph, 'Akademische Berufe und die Professionalisierung' in D. V. Glass and René König (eds.), *Soziale Schichtung und soziale Mobilität*, Kölner Zeitschrift für Soziologie und Sozialpsychologie, Sonderheft 5 (1961), pp. 105–21
―― 'Scientific Growth: A Sociological View', *Minerva: A Review of Science, Learning and Policy*, vol. 2 (1964), pp. 455–76
―― 'The Rise and Decline of France as a Scientific Centre', *Minerva: A Review of Science, Learning and Policy*, vol. 8 (1970), pp. 160–79
―― *The Scientist's Role in Society: A Comparative Study* (Englewood Cliffs, NJ, 1971)
―― *Centers of Learning: Britain, France, Germany, United States* (New York, 1977)
Bere, Ivan de la, *The Queen's Orders of Chivalry* (London, 1964)
Berman, Morris, *Social Change and Scientific Organization: The Royal Institution, 1799–1844* (London, 1978)
Bernal, John D., *The Social Function of Science* (London, 1939; reprinted Cambridge, Mass., and London, 1967)
―― *Science in History*, 2nd edn (London, 1957)
Beveridge, Janet, *An Epic of Clare Market: Birth and Early Days of the London School of Economics* (London, 1960)
Bibby, Cyril, *T.H. Huxley: Scientist, Humanist and Educator* (London, 1959)
Birks, J. B. (ed.), *Rutherford at Manchester* (London, 1962)
Black, R. D. Collison, *Economic Thought and the Irish Question, 1817–1870* (Cambridge, 1960)
Brandt, Peter, 'Wiederaufbau und Reform. Die Technische Universität Berlin 1945–1950' in Reinhard Rürup (ed.), *Wissenschaft und Gesellschaft. Beiträge zur Geschichte der Technischen Universität Berlin 1879–1979*, vol. 1 (Berlin, Heidelberg and New York, 1979), pp. 495–522
Brennan, E. J. T. (ed.), *Education for National Efficiency: The Contribution of Sidney and Beatrice Webb* (London, 1975)
Briggs, Asa, *Victorian Cities* (Harmondsworth, 1977)
Brocke, Bernhard vom, 'Hochschul- und Wissenschaftspolitik in Preußen und im Deutschen Kaiserreich 1882–1907: das "System Althoff"' in Peter Baumgart (ed.), *Bildungspolitik in Preußen zur Zeit des Kaiserreichs* (Stuttgart, 1980), pp. 9–118
―― 'Der deutsch-amerikanische Professorenaustausch. Preußische Wissenschaftspolitik, internationale Wissenschaftsbeziehungen und die Anfänge einer deutschen auswärtigen Kulturpolitik vor dem Ersten Weltkrieg', *Zeitschrift für Kulturaustausch*, vol. 31 (1981), pp. 128–82
Brown, E. J., *The Private Donor in the History of the University of Leeds* (Leeds, 1953)
Bühl, Walter L., *Einführung in die Wissenschaftssoziologie* (Munich, 1974)
Bunge, Mario and William R. Shea (eds.), *Rutherford and Physics at the Turn of the Century* (New York and Folkstone, 1979)
Burchardt, Lothar, 'Deutsche Wissenschaftspolitik an der Jahrhundertwende. Versuch einer Zwischenbilanz', *Geschichte in Wissenschaft und Unterricht*, vol. 26 (1975), pp. 271–89
―― *Wissenschaftspolitik im Wilhelminischen Deutschland. Vorgeschichte, Grün-*

Bibliography

dung und Aufbau der Kaiser-Wilhelm-Gesellschaft zur Förderung der Wissenschaften (Göttingen, 1975)
—— 'Die Ausbildung des Chemikers im Kaiserreich', *Zeitschrift für Unternehmensgeschichte*, vol. 23 (1978), pp. 31–53
Burn, Duncan L., *The Economic History of Steelmaking, 1867–1939: A Study in Competition* (Cambridge, 1940, reprinted 1961)
Burstall, H. F. W. and C. G. Burton, *Souvenir History of the Foundation and Development of the Mason Science College and of the University of Birmingham, 1880–1930* (Birmingham, 1930)
Caine, Sydney, *The History of the Foundation of the London School of Economics and Political Science* (London, 1963)
Camplin, Jamie, *The Rise of the Plutocrats: Wealth and Power in Edwardian England* (London, 1978)
Cannon, Susan Faye, *Science in Culture: The Early Victorian Period* (Folkestone and New York, 1978)
Cardwell, D. S. L., *The Organisation of Science in England*, 2nd edn (London, 1972, reprinted 1980)
—— *Technology, Science and History* (London, 1972)
Caroe, G. M., *William Henry Bragg 1862–1942: Man and Scientist* (Cambridge, 1978)
Carter, C. F. and B. R. Williams, *Industry and Technical Progress: Factors Governing the Speed of Application of Science* (London, 1957)
Cecil, Lamar, 'The Creation of Nobles in Prussia, 1871–1918', *American Historical Review*, vol. 75 (1970), pp. 757–95
Chapman, A. Chaston, *The Growth of the Profession of Chemistry during the Past Half-Century (1877–1927)* (London, 1927)
Charles, John, *Research and Public Health* (London, 1961)
Charlton, H. B., *Portrait of a University 1851–1951 to Commemorate the Centenary of Manchester University* (Manchester, 1951)
Chester, Norman, *The English Administrative System 1780–1870* (Oxford, 1981)
Cipolla, Carlo M. (ed.), *The Fontana Economic History of Europe: The Twentieth Century* (Glasgow, 1976)
Clapham, J. H., *Economic Development of France and Germany 1815–1914*, 4th edn (Cambridge, 1963)
Clapp, B. W., *John Owens: Manchester Merchant* (Manchester, 1965)
Cohen, John Michael, *The Life of Ludwig Mond* (London, 1956)
Coleman, D. C., 'Gentlemen and Players', *Economic History Review*, vol. 26 (1973), pp. 92–116
Connell, Brian, *Manifest Destiny: A Study in Five Profiles of the Rise and Influence of the Mountbatten Family* (London, 1953)
Coppock, D. J., 'British Industrial Growth during the "Great Depression" (1873–96): A Pessimist's View', *Economic History Review*, vol. 17 (1964–5), pp. 389–96
Cottle, Basil and J. W. Sherborne, *The Life of a University*, 2nd edn (Bristol, 1959)
Crafts, N.F.R., *British Economic Growth during the Industrial Revolution* (Oxford, 1985)
Crequer, Ngaio, 'Rising from the Ruins: Liverpool's Success Story', *Times*

Higher Education Supplement, 1 February 1980, pp. 8–9
Crowther, J. G., *Statesmen of Science* (London, 1965)
Daalder, Hans, *Cabinet Reform in Britain, 1914–1963* (Stanford, Calif., 1963)
Dale, Henry H., 'Fifty Years of Medical Research', *British Medical Journal*, vol. 94 (1963), Part 2, pp. 1279–81 and 1287–94
Davies, P. N., *Sir Alfred Jones: Shipping Entrepreneur par Excellence* (London, 1978)
Deane, P. and W. A. Cole, *British Economic Growth 1688–1959*, 2nd edn (Cambridge, 1967)
Diehl, Carl, *Americans and German Scholarship 1770–1870* (New Haven and London, 1978)
Donnan, F. G., *Ludwig Mond, F. R. S.: 1839–1909* (London, 1939)
Dreyer, J. L. E. and H. H. Turner, *History of the Royal Astronomical Society, 1820–1920* (London, 1923)
Dugdale, Blanche E. C., *Arthur James Balfour*, 2 vols. (London, 1936, reprinted Westport, Conn., 1970)
Dumbell, Stanley, *The University of Liverpool 1903–1953: A Jubilee Book* (Liverpool, [1953])
Dunstan, Wyndham R. (ed.), *Imperial Institute: Technical Reports and Scientific Papers* (London, 1903)
Edwards, Ronald S., *Co-operative Industrial Research: A Study of the Economic Aspects of the Research Associations Grant-Aided by the Department of Scientific and Industrial Research* (London, 1950)
Egremont, Max, *Balfour: A Life of Arthur James Balfour* (London, 1980)
Emy, H. V., *Liberals, Radicals and Social Politics, 1892–1914* (Cambridge, 1973)
Engel, A. J., *From Clergyman to Don: The Rise of the Academic Profession in Nineteenth-Century Oxford* (Oxford, 1983)
Ensor, R. C. K., *England 1870–1914* (Oxford, 1936, reprinted 1968)
Evans, D. Emrys, *The University of Wales: A Historical Sketch* (Cardiff, 1953)
Farber, Eduard, *The Evolution of Chemistry* (New York, 1952)
Ferber, Christian von, *Die Entwicklung des Lehrkörpers der deutschen Universitäten und Hochschulen 1864–1954* (Göttingen, 1956)
Fiddes, Edward, *Chapters in the History of Owens College and of Manchester University, 1851–1914* (Manchester, 1937)
Flett, John Smith, *The First Hundred Years of the Geological Survey of Great Britain* (London, 1937)
Flexner, Abraham, *Universities: American, English, German* (New York, 1930; reprinted London, 1968)
Floud, Roderick and Donald McCloskey (eds.), *The Economic History of Britain since 1700*, vol. 2, *1860 to the 1970s* (Cambridge, 1981)
Foden, Frank, *Philip Magnus: Victorian Educational Pioneer* (London, 1970)
Fort, G. Seymour, *Alfred Beit: A Study of the Man and his Work* (London, 1932)
Fournier d'Albe, E. E., *The Life of Sir William Crookes* (London, 1923)
Gage, A. T., *A History of the Linnean Society of London* (London, 1938)
Geison, Gerald L., *Michael Foster and the Cambridge School of Physiology:*

The Scientific Enterprise in Late Victorian Society (Princeton, NJ, 1978)

Gerschenkron, Alexander, 'Economic Backwardness in Historical Perspective' in idem, *Economic Backwardness in Historical Perspective: A Book of Essays* (Cambridge, Mass., 1962) pp. 5–30

Gilbert, Bentley B., *The Evolution of National Insurance in Great Britain: The Origins of the Welfare State*, 2nd edn (London, 1973)

───── *British Social Policy, 1914–1939* (London, 1970, reprinted 1973)

Gizycki, Rainald von, 'Centre and Periphery in the International Scientific Community: Germany, France and Great Britain in the Nineteenth Century', *Minerva*, vol. 11 (1973), pp. 474–94

───── 'The Association for the Advancement of Science: An International Comparative Study', *Zeitschrift für Soziologie*, vol. 8 (1979), pp. 28–49

Goldsmith, Maurice, *Sage: A Life of J. D. Bernal* (London, 1980)

Gowing, Margaret, 'Science, Technology and Education: England in 1870', The Wilkins Lecture, 1976, *Notes and Records of the Royal Society of London*, vol. 32 (1977–8), pp. 71–90

Green, V. H. H., *Religion at Oxford and Cambridge* (London, 1964)

Griewank, Karl, *Staat und Wissenschaft im Deutschen Reich. Zur Geschichte und Organisation der Wissenschaftspflege in Deutschland* (Freiburg, 1927)

Grove, J. W., *Government and Industry in Britain* (London, 1962)

Guppy, Henry, *The John Rylands Library Manchester* (Manchester, 1935)

Habakkuk, H. J., *American and British Technology in the Nineteenth Century. The Search for Labour-Saving Inventions* (Cambridge, 1962)

Haber, Lutz F., *The Chemical Industry during the Nineteenth Century in Europe and North America* (Oxford, 1958, reprinted 1969)

───── *The Chemical Industry 1900–1930: International Growth and Technological Change* (Oxford, 1971)

───── 'Government Intervention at the Frontiers of Science: British Dyestuffs and Synthetic Organic Chemicals 1914–39', *Minerva*, vol. 11 (1973), pp. 79–94

Habermas, Jürgen, *Technik und Wissenschaft als 'Ideologie'*, 7th edn (Frankfurt, 1974)

Hagstrom, Warren O., *The Scientific Community* (New York and London, 1965)

Haines, George, *Essays on German Influence upon English Education and Science, 1850–1919* (Hamden, Conn., 1969)

Hall, A. Rupert, *Science for Industry: A Short History of the Imperial College of Science and Technology and its Antecedents* (London, 1982)

───── and Marie Boas Hall, 'The Intellectual Origins of the Royal Society — London and Oxford', *Notes and Records of the Royal Society of London*, vol. 23 (1968), pp. pp. 157–68

Halsey, A. H., 'British Universities and Intellectual Life' in idem, Jean E. Floud and C. Arnold Anderson (eds.), *Education, Economy and Society* (New York, 1961), pp. 502–12

───── 'The Changing Functions of Universities' in idem, Jean E. Floud and C. Arnold Anderson (eds.), *Education, Economy and Society* (New York, 1961), pp. 456–65

───── (ed.), *Trends in British Society since 1900: A Guide to the Changing Social Structure of Britain* (London, 1972)

Bibliography

―― and Martin A. Trow, *The British Academics* (London, 1971)
Haltern, Utz, *Die Londoner Weltausstellung von 1851, Ein Beitrag zur Geschichte der bürgerlich-industriellen Gesellschaft im 19. Jahrhundert* (Münster, 1971)
―― 'Die "Welt als Schaustellung". Zur Funktion und Bedeutung der internationalen Industrieausstellung im 19. und 20. Jahrhundert', *Vierteljahrschrift für Sozial- und Wirtschaftsgeschichte*, vol. 60 (1973), pp. 1–40
Harrison, Brian, 'Philanthropy and the Victorians', *Victorian Studies*, vol. 9 (1965–6), pp. 353–74
Hartmann, Fritz and Rudolf Vierhaus (eds.), *Der Akademiegedanke im 17. und 18. Jahrhundert* (Bremen and Wolfenbüttel, 1977)
Hartog, P. J. (ed.), *The Owens College, Manchester: A Brief History of the College and Description of its Various Departments* (Manchester, 1900)
Hartwell, R. M., *The Industrial Revolution and Economic Growth* (London, 1971)
Hearnshaw, F. J. C., *The Centenary History of King's College, London 1828–1928* (London, 1929)
Heath, H. F. and A. L. Heatherington, *Industrial Research and Development in the United Kingdom: A Survey* (London, 1946)
Heisenberg, Werner, *Tradition in der Wissenschaft. Reden und Aufsätze* (Munich, 1977)
Henriques, U. R. Q., 'The Jewish Emancipation Controversy in Nineteenth-Century Britain', *Past and Present*, vol. 40 (1968), pp. 126–46
Heuss, Alfred, 'Das Problem des "Fortschritts" in den historischen Wissenschaften', *Zeitschrift für Religions- und Geistesgeschichte*, vol. 31 (1979), pp. 132–46
Hill, Christopher, *Intellectual Origins of the English Revolution* (Oxford, 1965)
―― 'The Intellectual Origins of the Royal Society – London or Oxford?', *Notes and Records of the Royal Society of London*, vol. 23 (1968), pp. 144–56
Himsworth, Harold P., *The Development and Organization of Scientific Knowledge* (London, 1970)
Hobsbawm, Eric J., *Industry and Empire: An Economic History of Britain since 1750*, 4th edn (London, 1973)
―― *The Age of Capital 1848–1875* (London, 1975)
Hoffmann, Ross J. S., *Great Britain and the German Trade Rivalry, 1875–1914* (Philadelphia, Penn., and London, 1933)
Hoffmann, W. G., *British Industry, 1700–1950* (Oxford, 1955)
Hogg, Q. M., *Science and Politics* (London, 1963)
Hollenberg, Günter, *Englisches Interesse am Kaiserreich. Die Attraktivität Preußen-Deutschlands für konservative und liberate Kreise in Großbritannien 1860–1914* (Wiesbaden, 1974)
Hollinger, David A., 'T. S. Kuhn's Theory of Science and Its Implications for History', *American Historical Review*, vol. 78 (1973), pp. 370–93
Holt, B. W. G., 'Social Aspects in the Emergence of Chemistry as an Exact Science: The British Chemical Profession', *British Journal of Sociology*, vol. 21 (1970), pp. 181–99
Howard, Michael, *The Franco-Prussian War: The German Invasion of France,*

Bibliography

1870–1871 (London, 1961)

Howarth, O. J. R., *The British Association for the Advancement of Science: A Retrospect, 1831–1931* (London, 1931)

Huber, Ernst Rudolf, *Deutsche Verfassungsgeschichte seit 1789*, vol. 4, *Struktur und Krisen des Kaiserreichs* (Stuttgart etc., 1969)

Humberstone, Thomas Lloyd, *University Reform in London* (London, 1926)

Hurwitz, S. J., *State Intervention in Great Britain: A Study of Economic Control and Social Response, 1914–1919* (New York, 1949; reprinted London, 1968)

Husemann, Harald, 'Zu den deutsch-englischen Universitätsbeziehungen während der letzten hundert Jahre' in Hartmut Boockmann et al. (eds.), *Geschichte und Gegenwart. Festschrift für Karl Dietrich Erdmann* (Neumünster, 1980), pp. 459–90

Hutchinson, Eric, 'Scientists and Civil Servants: The Struggle over the National Physical Laboratory in 1918', *Minerva*, vol. 7 (1969), pp. 373–98

—— 'Scientists as an Inferior Class: The Early Years of the DSIR', *Minerva*, vol. 8 (1970), pp. 396–411

—— 'Government Laboratories and the Influence of Organized Scientists', *Science Studies*, vol. 1 (1971), pp. 331–56

Hutchison, Terence W., *On Revolutions and Progress in Economic Knowledge* (Cambridge, 1978)

Inkster, Ian and Jack Morrell (eds.), *Metropolis and Science: Science in British Culture, 1780–1850* (London, 1983)

Jaher, Frederic Cople, 'The Gilded Elite: American Multimillionaires, 1865 to the Present' in W. D. Rubinstein (ed.), *Wealth and the Wealthy in the Modern World* (London, 1980), pp. 189–276

Jordan, W. K., *Philanthropy in England, 1480–1660* (London, 1959)

Judson, Horace F., *The Eighth Day of Creation: Makers of the Revolution in Biology* (London, 1979)

Kargon, Robert H., *Science in Victorian Manchester: Enterprise and Expertise* (Manchester, 1977)

Kelsall, R. K., *Higher Civil Servants in Britain from 1870 to the Present Day* (London, 1955)

Kennedy, Paul M., *The Rise of the Anglo-German Antagonism 1860–1914* (London, 1980)

Kenyon, Frederic G., *The British Academy: The First Fifty Years* (London, 1952)

Kevles, Daniel J., '"Into Hostile Political Camps": The Reorganization of International Science in World War I', *Isis*, vol. 62 (1971), pp. 47–60

Kiker, B. F. (ed.), *Investment in Human Capital* (Columbia, S.C., 1971)

Kindleberger, Charles P., 'Germany's Overtaking of England, 1806 to 1914' in idem, *Economic Response: Comparative Studies in Trade, Finance and Growth* (Cambridge, Mass., and London, 1978), pp. 185–237

Kirby, M.W., *The Decline of British Economic Power since 1870* (London 1981)

Kitson Clark, G. S. R., *An Expanding Society: Britain 1830–1900* (Cambridge, 1967)

Kluke, Paul, *Die Stiftungsuniversität Frankfurt am Main 1914–1932* (Frank-

furt, 1972)
Knight, David, *The Age of Science* (Oxford, 1986)
Koch, Hansjoachim W., 'Social Darwinism as a Factor in the "New Imperialism"' in idem (ed.), *The Origins of the First World War: Great Power Rivalry and German War Aims* (London and Basingstoke, 1972), pp. 329–54
Koss, Stephen E., *Lord Haldane: Scapegoat for Liberalism* (New York and London, 1969)
____ *Sir John Brunner: Radical Plutocrat, 1842–1919* (Cambridge, 1970)
Kroker, Evelyn, *Die Weltausstellungen im 19. Jahrhundert. Industrieller Leistungsnachweis, Konkurrenzverhalten und Kommunikationsfunktion unter Berücksichtigung der Montanindustrie des Ruhrgebiets zwischen 1851 und 1880* (Göttingen, 1975)
Kuhn, Thomas S., *The Structure of Scientific Revolutions*, 2nd edn (Chicago, 1970)
____ 'The History of Science' in *International Encyclopedia of the Social Sciences*, vol. 14 (New York, 1968), pp. 74–83
____ *The Essential Tension: Selected Studies in Scientific Tradition and Change* (Chicago and London, 1977)
Landes, David S., *The Unbound Prometheus: Technological Change and Industrial Development in Western Europe from 1750 to the Present* (Cambridge, 1969)
Levine, A. L., *Industrial Retardation in Britain 1880–1914* (London, 1967)
Liebig, Justus von, *Die Chemie und ihre Anwendung auf Agricultur und Physiologie*, 2 vols., 7th edn (Brunswick, 1862)
Lindert, Peter H. and Keith Trace, 'Yardsticks for Victorian Entrepreneurs' in Donald N. McCloskey (ed.), *Essays on a Mature Economy: Britain after 1840* (London, 1971), pp. 239–74
Lipman, Vivian D., *Social History of the Jews in England 1850-1950* (London, 1954)
Locke, Robert R., 'Industrialisierung und Erziehungssystem in Frankreich und Deutschland vor dem 1. Weltkrieg', *Historische Zeitschrift*, vol. 225 (1977), pp. 265–96
____ *The End of Practical Man: Entrepreneurship and Higher Education in Germany, France, and Great Britain, 1880–1940* (Greenwich, Conn., 1984)
Lockyer, T. Mary and Winifred L. Lockyer, *Life and Work of Sir Norman Lockyer* (London, 1928)
Lübbe, Hermann, *Wissenschaftspolitik. Planung, Politisierung, Relevanz* (Zurich, 1977)
Lyons, Henry, *The Royal Society 1660–1940: A History of its Administration under its Charters* (Cambridge, 1944)
McClelland, Charles E., *State, Society, and University in Germany 1700 to 1914* (Cambridge, 1980)
McCloskey, Donald N., *Economic Maturity and Entrepreneurial Decline: British Iron and Steel, 1870–1913* (Cambridge, Mass., 1973)
____ 'Did Victorian Britain Fail?', *Economic History Review*, vol. 23 (1970), pp. 446–59
____ (ed.), *Essays on a Mature Economy: Britain after 1840* (London, 1971)
MacDonagh, Oliver, 'The Nineteenth-Century Revolution in Govern-

ment: A Reappraisal', *Historical Journal*, vol. 1 (1958), pp. 52–67
McDowell, R. B. and D. A. Webb, *Trinity College Dublin 1592–1952: An Academic History* (Cambridge, 1982)
Mackay, Ruddock, F., *Balfour: Intellectual Statesman* (Oxford, 1985)
McKie, Douglas, 'The Origins and Foundation of the Royal Society of London', *Notes and Records of the Royal Society of London*, vol. 15 (1960), pp. 1–37
—— 'Science and Technology' in *The New Cambridge Modern History*, vol. 12, *The Shifting Balance of World Forces 1898–1945*, ed. by C. L. Mowat (Cambridge, 1968), pp. 87–111
—— 'Scientific Societies to the End of the Eighteenth Century' in Allan Ferguson (ed.), *Natural Philosophy through the 18th Century*, 2nd edn (London, 1972), pp. 133–43
MacLeod, Roy M., 'The Alkali Acts Administration, 1863–1884: The Emergence of the Civil Scientist', *Victorian Studies*, vol. 9 (1965), pp. 85–112
—— 'Science and Government in Victorian England: Lighthouse Illumination and the Board of Trade, 1866–1886', *Isis*, vol. 60 (1969), pp. 5–38
—— 'Into the Twentieth Century', *Nature*, 224 (1969), pp. 457–61
—— 'Science in Grub Street', *Nature*, 224 (1969), pp. 423–7
—— 'The X-Club. A Social Network of Science in Late-Victorian England', *Notes and Records of the Royal Society of London*, vol. 24 (1969–70), pp. 305–22
—— 'Science and the Civil List, 1824–1914', *Technology and Society*, vol. 6 (1970), pp. 47–55
—— 'Of Medals and Men: A Reward System in Victorian Science 1826–1914', *Notes and Records of the Royal Society of London*, vol. 26 (1971), pp. 81–105
—— 'The Support of Victorian Science: The Endowment of Research Movement in Great Britain, 1868–1900', *Minerva*, vol. 4 (1971), pp. 197–230
—— 'The Royal Society and the Government Grant: Notes on the Administration of Scientific Research, 1849–1914', *Historical Journal*, vol. 14 (1971), pp. 323–58
—— 'Resources of Science in Victorian England: The Endowment of Science Movement, 1868–1900' in Peter Mathias (ed.), *Science and Society, 1600–1900* (Cambridge, 1972), pp. 111–66
—— 'Statesmen Undisguised', *American Historical Review*, vol. 78 (1973) pp. 1,386–405
—— 'The Ayrton Incident: A Commentary on the Relations of Science and Government in England, 1870–1873' in Arnold Thackray and Everett Mendelsohn (eds.), *Science and Values* (New York, 1974), pp. 45–78
—— 'Scientific Advice for British India: Imperial Perceptions and Administrative Goals, 1898–1923', *Modern Asian Studies*, vol. 9 (1975), pp. 343–84
—— and E. Kay Andrews, 'Scientific Careers of 1851 Exhibition Scholars', *Nature*, 218 (1968), pp. 1011–16
—— 'The Origins of the D.S.I.R.: Reflections on Ideas and Men, 1915–

1916', *Public Administration*, vol. 48 (1970), pp. 23–48
____ 'Scientific Advice in the War at Sea, 1915–1917: The Board of Invention and Research', *Journal of Contemporary History*, vol. 6, no. 2 (1971), pp. 3–40
____ and Peter Collins (eds.), *The Parliament of Science: The British Association for the Advancement of Science, 1831–1981* (Northwood, 1981)
____, J. R. Friday and Carol Gregor, *The Corresponding Societies of the British Association for the Advancement of Science 1883–1929: A Survey of Historical Records, Archives and Publications* (London, 1975)
____ and E. Kay MacLeod, 'War and Economic Development: Government and the Optical Industry in Britain, 1914–1918' in J. M. Winter (ed.), *War and Economic Development: Essays in Memory of David Joslin* (Cambridge and London, 1975), pp. 165–203
____ 'The Social Relations of Science and Technology 1914–1939' in Carlo M. Cipolla (ed.), *The Fontana Economic History of Europe: The Twentieth Century* (Glasgow, 1976), pp. 301–63
Macrosty, H. W., *The Trust Movement in British Industry: A Study of Business Organisation* (London, 1907)
Martin, Thomas, *The Royal Institution*, 2nd edn (London, New York and Toronto, 1948)
Marwick, Arthur, *The Deluge: British Society and the First World War* (London and Basingstoke, 1973)
____ 'The Impact of the First World War on British Society', *Journal of Contemporary History*, vol. 3, no. 1 (1968), pp. 51–68
Mathias, Peter, *The First Industrial Nation: An Economic History of Britain, 1700–1914*, 4th edn (London, 1972)
____ 'Who Unbound Prometheus? Science and Technical Change, 1600–1800' in idem (ed.), *Science and Society, 1600–1900* (Cambridge, 1972), pp. 54–80
____ (ed.), *Science and Society, 1600–1900* (Cambridge, 1972)
Matthew, H. C. G., *The Liberal Imperialists: The Ideas and Politics of a Post-Gladstonian Elite* (London, 1973)
Maurice, Frederick B., *Haldane*, 2 vols. (London, 1937–9; reprinted Westport, Conn., 1970)
Meadows, A. J., *Science and Controversy: A Biography of Sir Norman Lockyer* (Cambridge, Mass., 1972)
____ *Communication in Science* (London, 1974)
Medick, Hans, 'Anfänge und Voraussetzungen des Organisierten Kapitalismus in Großbritannien 1873–1914' in Heinrich August Winkler (ed.), *Organisierter Kapitalismus. Voraussetzungen und Anfänge* (Göttingen, 1974), pp. 58–83
Mehrtens, Herbert and Steffen Richter (eds.), *Naturwissenschaft, Technik und NS-Ideologie. Beiträge zur Wissenschaftsgeschichte des Dritten Reiches* (Frankfurt, 1980)
Melville, Harry, *The Department of Scientific and Industrial Research* (London and New York, 1962)
Mendelsohn, Everett, 'The Emergence of Science as a Profession in Nineteenth-Century Europe' in Karl Hill (ed.), *The Management of Scientists* (Boston, Mass., 1964), pp. 3–48

Bibliography

Merton, Robert K., *Science, Technology and Society in Seventeenth Century England* (Bruges, 1938; reprinted New York, 1970)
Miall, Stephen, *A History of the British Chemical Industry* (London, 1931)
Minihan, Janet, *The Nationalization of Culture: The Development of State Subsidies to the Arts in Great Britain* (London, 1977)
Minney, R. J., *Viscount Addison: Leader of the Lords* (London, 1958)
Mitchell, P. Chalmers, *Centenary History of the Zoological Society of London* (London, 1929)
Mommsen, Wolfgang J., *Das Zeitalter des Imperialismus* (Frankfurt, 1969)
Moody, T. W., 'The Irish University Question of the Nineteenth Century', *History*, vol. 43 (1958), pp. 90–109
—— and J. C. Beckett, *Queen's, Belfast 1845–1949: The History of a University*, 2 vols. (London, 1959)
Morgan, Kenneth O., *Rebirth of a Nation: Wales 1880–1980* (Oxford, 1981)
—— and Jane Morgan, *Portrait of a Progressive: The Political Career of Christopher, Viscount Addison* (Oxford, 1980)
Morrell, J. B., 'The University of Edinburgh in the Late Eighteenth Century: Its Scientific Eminence and Academic Structure', *Isis*, vol. 62 (1971), pp. 158–71
—— 'The Patronage of Mid-Victorian Science in the University of Edinburgh', *Science Studies*, vol. 3 (1973), pp. 353–88.
Morrison, John L. M., 'An Engineer Looks at British Society', *Interdisciplinary Science Reviews*, vol. 4 (1979), pp. 17–26
Moseley, Russell, 'The Origins and Early Years of the National Physical Laboratory: A Chapter in the Pre-history of British Science Policy', *Minerva*, vol. 16 (1978), pp. 222–50
—— 'Government Science and the Royal Society: The Control of the National Physical Laboratory in the Inter-War Years', *Notes and Records of the Royal Society of London*, vol. 35 (1980–1), pp. 167–93
Moulton, H. Fletcher, *The Life of Lord Moulton* (London, 1922)
Mountford, James, *British Universities* (London, New York and Toronto, 1966)
Musson, A. E., 'British Industrial Growth, 1873–96: A Balanced View', *Economic History Review*, vol. 17 (1964–5), pp. 397–403
—— (ed.), *Wissenschaft, Technik and Wirtschaftswachstum im 18. Jahrhundert* (Frankfurt, 1972)
—— and Eric Robinson, *Science and Technology in the Industrial Revolution* (Manchester, 1969)
Neave, S. A. and F. J. Griffin, *The History of the Entomological Society of London, 1833–1933* (London, 1933)
Olby, Robert C., *The Path to the Double Helix* (London, 1974)
Orange, A. D., 'The British Association for the Advancement of Science: The Provincial Background', *Science Studies*, vol. 1 (1971), pp. 315–30
Owen, David, *English Philanthropy, 1660–1960* (Cambridge, Mass., 1964)
Parris, Henry, 'The Nineteenth-Century Revolution in Government: A Reappraisal Reappraised', *Historical Journal*, vol. 3 (1960), pp. 17–37
—— *Constitutional Bureaucracy: The Development of British Central Administration since the Eighteenth Century* (London, 1969)
Pelling, Henry, 'The American Economy and the Foundation of the

British Labour Party', *Economic History Review*, vol. 8 (1955), pp. 1–17
Perkin, Harold, *Key Profession: The History of the Association of University Teachers* (London, 1969)
―― 'Die Rekrutierung der Eliten in der britischen Gesellschaft seit 1800', *Geschichte und Gesellschaft*, vol. 3 (1977), pp. 485–502
Pfetsch, Frank R., 'Wissenschaft als autonomes und integriertes System. Tendenzen in der neueren Literatur zur Wissenschaftspolitik und -soziologie', *Neue Politische Literatur*, vol. 17 (1972), pp. 15–28
―― *Zur Entwicklung der Wissenschaftspolitik in Deutschland 1750–1914* (Berlin, 1974)
―― and Avraham Zloczower, *Innovation und Widerstände in der Wissenschaft. Beiträge zur Geschichte der deutschen Medizin* (Düsseldorf, 1973)
Pilcher, Richard B., *The Institute of Chemistry of Great Britain and Ireland: History of the Institute, 1877–1914* (London, 1914)
―― *The Profession of Chemistry*, 4th edn (London, 1938)
Pohrt, Wolfgang (ed.), *Wissenschaftspolitik – von wem, für wen, wie?* (Munich, n.d. [1977])
Pollard, Sidney, *The Development of the British Economy, 1914–1967*, 2nd edn (London, 1969)
Porter, Roy, 'Gentlemen and Geology: The Emergence of a Scientific Career, 1660–1920', *Historical Journal*, vol. 21 (1978), pp. 809–36
Price, Derek J. de Solla, *Little Science, Big Science* (New York and London, 1963)
Purver, Margery, *The Royal Society: Concept and Creation* (London, 1967)
Pyatt, Edward, *The National Physical Laboratory: A History* (Bristol, 1983)
Rattansi, P. M., 'The Intellectual Origins of the Royal Society', *Notes and Records of the Royal Society of London*, vol. 23 (1968), pp. 129–43
Read, Donald, *England 1868–1914: The Age of Urban Democracy* (London and New York, 1979)
Reader, William Joseph, *Professional Men: The Rise of the Professional Classes in Nineteenth-Century England* (New York, 1966)
―― *Imperial Chemical Industries: A History*, 2 vols. (London, 1970, 1975)
The Record of the Royal Society of London for the Promotion of Natural Knowledge, 4th edn (London, 1940)
Rees, J. Morgan, *Trusts in British Industry 1914–1921: A Study in Recent Developments in Business Organisation* (London, 1922)
Richardson, Harry W., 'The Development of the British Dyestuffs Industry before 1939', *Scottish Journal of Political Economy*, vol. 9 (1962), pp. 110–29
―― 'Retardation in Britain's Industrial Growth, 1870–1913', *Scottish Journal of Political Economy*, vol. 12 (1965)
Ringer, Fritz K., 'Higher Education in Germany in the Nineteenth Century', *Journal of Contemporary History*, vol. 2, no. 3 (1967), pp. 123–47
―― *The Decline of the German Mandarins: The German Academic Community, 1890–1933* (Cambridge, Mass., 1969)
Ritter, Gerhard A. (ed.), *Vom Wohlfahrtsausschuß zum Wohlfahrtsstaat. Der Staat in der modernen Industriegesellschaft* (Cologne, 1973)
Roberts, David, *Victorian Origins of the British Welfare State* (New Haven, Conn., 1960)

Bibliography

Roderick, Gordon W., and Michael D. Stephens, *Education and Industry in the Nineteenth Century: The English Disease?* (London and New York, 1978)

Rohe, Karl, 'Ursachen und Bedingungen des modernen britischen Imperialismus von 1914' in Wolfgang J. Mommsen (ed.), *Der moderne Imperialismus* (Stuttgart, 1971), pp. 60–84

Rose, Hilary, and Steven Rose, *Science and Society* (Harmondsworth, 1971)

Roseveare, Henry, *The Treasury: The Evolution of a British Institution* (London, 1969)

―― *The Treasury 1660–1870: The Foundations of Control* (London and New York, 1973)

Ross, Sydney, 'Scientist: The Story of a Word', *Annals of Science*, vol. 18 (1962), pp. 65–85

Rostow, Walt W., *The Stages of Economic Growth* (New York, 1962)

―― *Politics and the Stages of Growth* (Cambridge, 1971)

―― (ed.), *The Economics of Take-Off into Sustained Growth*, 2nd edn (London, 1965)

Routh, Guy, *Occupations and Pay in Great Britain, 1902–60* (Cambridge, 1965)

The Royal Charter of the Imperial College of Science and Technology: Jubilee, 1907–1957 (Watford, 1957)

Rubinstein, W. D., 'Wealth, Elites and the Class Structure of Modern Britain', *Past and Present*, vol. 76 (1977), pp. 99–126

―― 'Modern Britain' in idem (ed.), *Wealth and the Wealthy in the Modern World* (London, 1980), pp. 46–89

Rürup, Reinhard (ed.), *Wissenschaft und Gesellschaft. Beiträge zur Geschichte der Technischen Universität Berlin 1879–1979*, 2 vols. (Berlin, Heidelberg and New York, 1979)

―― 'Die Technische Universität Berlin 1879–1979: Grundzüge und Probleme ihrer Geschichte' in ibid., vol. 1, pp. 3–47

Ruske, Walter, 'Außeruniversitäre technisch-naturwissenschaftliche Forschungsanstalten in Berlin bis 1945' in ibid., vol. 1, pp. 231–63

Russell, Colin A. et al., *Chemists by Profession: The Origins and Rise of The Royal Institute of Chemistry* (Milton Keynes, 1977)

St Aubyn, Giles, *Edward VII: Prince and King* (London, 1979)

Salomon, Jean-Jacques, *Science and Politics* (London, 1973)

Sampson, Anthony, *The New Anatomy of Britain* (London, 1971)

Sanderson, Michael, *The Universities and British Industry 1850–1970* (London, 1972)

―― 'Research and the Firm in British Industry, 1919–1939', *Science Studies*, vol. 2, (1972), pp. 107–52

―― 'The University of London and Industrial Progress 1880–1914', *Journal of Contemporary History*, vol. 7, nos. 3 and 4 (1972), pp. 243–62

Saul, S. B., *The Myth of the Great Depression, 1873–1896* (London, 1969)

―― *Industrialisation and De-Industrialisation? The Interaction of the German and British Economies before the First World War* (London, n.d. [1980])

Schieder, Theodor, 'Europa im Zeitalter der Nationalstaaten und europäische Weltpolitik bis zum I. Weltkrieg (1870–1918)' in idem (ed.), *Handbuch der Europäischen Geschichte*, vol. 6: *Europa im Zeitalter der*

Nationalstaaten und europäische Weltpolitik bis zum Ersten Weltkrieg (Stuttgart, 1968), pp. 1–196

—— 'Kultur, Wissenschaft und Wissenschaftspolitik im Deutschen Kaiserreich' in Gunter Mann and Rolf Winau (eds.), *Medizin, Naturwissenschaft, Technik und das Zweite Kaiserreich* (Göttingen, 1977), pp. 9–34

Schofield, Robert E., *The Lunar Society of Birmingham: A Social History of Provincial Science and Industry in Eighteenth-Century England* (Oxford, 1963)

—— 'Histories of Scientific Societies: Needs and Opportunities for Research', *History of Science: An Annual Review of Literature, Research and Teaching*, vol. 2 (1963), pp. 70–83

—— 'Die Orientierung der Wissenschaft auf die Industrie in der Lunar Society von Birmingham' in A. E. Musson (ed.), *Wissenschaft, Technik und Wirtschaftswachstum im 18. Jahrhundert* (Frankfurt, 1972), pp. 153–64

Schramm, Percy Ernst, 'Englands Verhältnis zur deutschen Kultur zwischen der Reichsgründung und der Jahrhundertwende' in Werner Conze (ed.), *Deutschland und Europa. Historische Studien zur Völker- und Staatsordnung des Abendlandes. Festschrift für Hans Rothfels* (Düsseldorf, 1951), pp. 135–75

Schroeder-Gudehus, Brigitte, 'Deutsche Wissenschaft und internationale Zusammenarbeit 1914–1928. Ein Beitrag zum Studium kultureller Beziehungen in politischen Krisenzeiten', PhD thesis, University of Geneva, 1966

—— 'Challenge to Transnational Loyalties: International Scientific Organizations after the First World War', *Science Studies*, vol. 3 (1973), pp. 93–118

—— *Les scientifiques et la paix. La communauté scientifique internationale au cours des années 20* (Montreal, 1978)

Searle, G. R., *The Quest for National Efficiency: A Study in British Politics and British Political Thought, 1899–1914* (Berkeley and Los Angeles, 1971)

Semmel, Bernard, *Imperialism and Social Reform: English Social-Imperial Thought 1895–1914* (London, 1960)

Shapin, Steven, 'Property, Patronage and the Politics of Science: The Founding of the Royal Society of Edinburgh', *British Journal for the History of Science*, vol. 7 (1974), pp. 1–41

Shils, Edward, 'The Order of Learning in the United States from 1865 to 1920: The Ascendancy of the Universities', *Minerva*, vol. 16 (1978), pp. 159–95

Shimmin, A. N., *The University of Leeds: The First Half-Century* (Cambridge, 1954)

Sinai, Robert, 'Was uns krank macht. Die Zivilisation ist am Ende', *Der Monat*, vol. 31, no. 3 (1979), pp. 7–18

Skeat, W. O., *King's College London Engineering Society 1847–1957* (London, 1957)

Snow, C. P., *The Two Cultures and the Scientific Revolution* (London, 1959), 2nd edn, *The Two Cultures and a Second Look* (Cambridge, 1964)

Sommer, Dudley, *Haldane of Cloan: His Life and Times 1856–1928* (London, 1960)

Späth, Manfred, 'Die Technische Hochschule Berlin-Charlottenburg und die internationale Diskussion des technischen Hochschulwesens 1900–1914' in Reinhard Rürup (ed.), *Wissenschaft und Gesellschaft. Beiträge zur Geschichte der Technischen Universität Berlin 1879–1979*, vol. 1 (Berlin, Heidelberg and New York, 1979), pp. 189–208

Spiegel-Rösing, Ina and Derek J. de Solla Price (eds.), *Science, Technology and Society: A Cross-Disciplinary Perspective* (London and Beverly Hills, 1977)

Stadelmayer, Peter, 'Das schwierige Geschäft, Gutes zu tun. Von Theorie und Praxis der Stiftungen' in Rolf Hauer et al. (eds.), *Deutsches Stiftungswesen 1966–1976. Wissenschaft und Praxis* (Tübingen, 1977)

Stiftungen in Europa. Eine vergleichende Übersicht (Baden-Baden, 1971)

Stimson, Dorothy, *Scientists and Amateurs: A History of the Royal Society* (New York, 1948)

Strutt, Robert J., *The Life of John William Strutt, Third Baron Rayleigh* (London, 1924), 2nd edn (Madison, Wisc., 1968)

―― *Lord Balfour in his Relation to Science* (Cambridge, 1930)

Sutherland, Gillian (ed.), *Studies in the Growth of Nineteenth-Century Government* (London, 1972)

Syfret, R. H., 'The Origins of the Royal Society', *Notes and Records of the Royal Society of London*, vol. 5 (1948), pp. 75–137

Taylor, Arthur J., *Laissez-faire and State Intervention in Nineteenth-Century Britain* (London and Basingstoke, 1972)

Temin, Peter, 'The Relative Decline of the British Steel Industry, 1880–1913' in Henry Rosovsky (ed.), *Industrialization in Two Systems: Essays in Honor of Alexander Gerschenkron* (New York, 1966), pp. 140–55

Thackray, Arnold and Jack Morrell, *Gentlemen of Science: The Origins and Early Victorian Years of the British Association for the Advancement of Science* (London, 1981)

Thomas, J. A., *The House of Commons 1906–1911: An Analysis of its Economic and Social Character* (Cardiff, 1958)

Thompson, Joseph, *The Owens College: Its Foundation and Growth* (Manchester, 1886)

Thomson, A. Landsborough, 'Origin and Development of the Medical Research Council', *British Medical Journal* (1963), Part 2, pp. 1290–2

―― *Half a Century of Medical Research*, vol. 1, *Origins and Policy of the Medical Research Council (UK)* (London, 1973); vol. 2, *The Programme of the Medical Research Council (UK)* (London, 1975)

Travers, Morris W., *A Life of Sir William Ramsay* (London, 1956)

Trebilcock, Clive, 'War and the Failure of Industrial Mobilisation: 1899 and 1914' in J. M. Winter (ed.), *War and Economic Development: Essays in Memory of David Joslin* (Cambridge and London, 1975), pp. 139–64

Turner, G. L'E. (ed.), *The Patronage of Science in the Nineteenth Century* (Leyden, 1976)

Vagts, Alfred, 'Die Juden im englisch-deutschen imperialistischen Konflikt vor 1914' in Joachim Radkau and Imanuel Geiss (eds.), *Imperialismus im 20. Jahrhundert. Gedenkschrift für George W. F. Hallgarten* (Munich, 1976), pp. 113–43

Varcoe, Ian, 'Scientists, Government and Organised Research in Great

Britain 1914–16: The Early History of the DSIR', *Minerva*, vol. 8 (1970), pp. 192–216

―― *Organizing for Science in Britain: A Case-Study* (London, 1974)

Vig, Norman J., *Science and Technology in British Politics* (Oxford, 1968)

Vincent, Eric W. and Percival Hinton, *The University of Birmingham: Its History and Significance* (Birmingham, 1947)

Volhard, Jakob, *Justus von Liebig*, 2 vols. (Leipzig, 1909)

Voss, Jürgen, 'Die Akademien als Organisationsträger der Wissenschaften im 18. Jahrhundert', *Historische Zeitschrift*, vol. 231 (1980), pp. 43–74

Ward, David, 'The Public Schools and Industry in Britain after 1870', *Journal of Contemporary History*, vol. 2, no. 3 (1967), pp. 37–52

Warwick, Paul, 'Did Britain Change? An Inquiry into the Causes of National Decline', *Journal of Contemporary History*, vol. 20 (1985), pp. 99–133

Watson, James D., *The Double Helix: A Personal Account of the Discovery of the Structure of DNA* (London, 1968; reprinted Harmondsworth, 1976)

Weber, Max, 'Wissenschaft als Beruf' in *Gesammelte Aufsätze zur Wissenschaftslehre*, ed. by Johannes Winckelmann, 3rd edn (Tübingen, 1968), pp. 582–613

Wehler, Hans-Ulrich, *Modernisierungstheorie und Geschichte* (Göttingen, 1975)

―― 'Sozialdarwinismus im expandierenden Industriestaat' in idem, *Krisenherde des Kaiserreichs 1871–1918*, 2nd edn (Göttingen, 1979), pp. 281–9

Weingart, Peter, *Die amerikanische Wissenschaftslobby. Zum sozialen und politischen Wandel des Wissenschaftssystems im Prozess der Forschungsplanung* (Düsseldorf, 1970)

―― *Wissensproduktion und soziale Struktur* (Frankfurt, 1976)

Werskey, Paul Gary, 'British Scientists and "Outsider" Politics, 1931–1945', *Science Studies*, vol. 1 (1971), pp. 67–83

―― 'The Perennial Dilemma of Science Policy', *Nature*, 223 (1971), pp. 529–32

―― *The Visible College: A Collective Biography of British Scientists and Socialists of the 1930s* (London, 1978)

Whitaker, Ben, *The Foundations: An Anatomy of Philanthropy and Society* (London, 1974)

Wiener, Martin J., *English Culture and the Decline of the Industrial Spirit, 1850–1980* (Cambridge, 1981)

Wiese, Leopold von, *Die Funktion des Mäzens im gesellschaftlichen Leben* (Cologne, 1929)

Williams, L. Pearce, 'The Royal Society and the Founding of the British Association for the Advancement of Science', *Notes and Records of the Royal Society of London*, vol. 16 (1961), pp. 221–33

Wilson, Charles H., *Haldane and the Machinery of Government* (London, 1956)

―― 'Economy and Society in Late Victorian Britain', *Economic History Review*, vol. 18 (1965), pp. 183–98

Winter, J. M. (ed.), *War and Economic Development: Essays in Memory of David Joslin* (Cambridge and London, 1975)

Bibliography

Wood, Alexander, *The Cavendish Laboratory* (Cambridge, 1946)
Wright, Maurice, *Treasury Control of the Civil Service, 1854–1874* (London, 1969)
—— 'Treasury Control 1854–1914' in Gillian Sutherland (ed.), *Studies in the Growth of Nineteenth-Century Government* (London, 1972), pp. 195–226
Young, Kenneth, *Arthur James Balfour: The Happy Life of the Politician, Prime Minister, Statesman and Philosopher 1848–1930* (London, 1963)
Zebel, Sydney H., *Balfour: A Political Biography* (London, 1973)
Zilsel, Edgar, 'Die Entstehung des Begriffs des wissenschaftlichen Fortschritts' in idem, *Die sozialen Ursprünge der neuzeitlichen Wissenschaft* (Frankfurt, 1976) pp. 127–50
Zunkel, Friedrich, *Der Rheinisch-Westfälische Unternehmer 1834–1879. Ein Beitrag zur Geschichte des deutschen Bürgertums im 19. Jahrhundert* (Cologne and Opladen, 1962)

Index

Aberdeen, 87
Académie des Sciences, 119
academies, 43, 214
Ackland, A.H.D., 206
Addison, Christopher, 175–6, 180, 198, 202–5, 211
Admiralty, 64, 202
Advisory Council for Research, 175–6, 207–8, 210–11, 254
Africa, 55, 152
African Entomological Research Committee, 65, 173
aircraft, 185–6
airships, 185–6
Airy, George B., 80, 218
Akademie der Schönen Künste, 48
Albert, Prince, 87, 159, 222
Albert Hall, 127
Aldcroft, Derek H., 115, 253
amateurism, 221
Amsterdam, 48
Anderson, Mary, 160
Andrews, E. Kay, 7, 204–5, 207
aniline dye industry, 109, 114, 122–3, 186–7, 194–5
aniline dyes, 100, 109, 194, 218
Antarctic expeditions, 66–7
anthropology, 90
Arctic Expedition (1875), 66
argon, 221, 225
Armytage, W.H.G., 8
Arnold, Matthew, 22
Ashby, Eric, 8, 23, 160
Ashton, Thomas G., 37
Asquith, Herbert H., 35, 97, 148, 182, 191, 202, 206, 248
Association of Scientific Workers, 96, 190
Association of University Teachers, 231
Astor, Waldorf, 173, 175
Astronomer Royal, 80
astronomy, 21, 61
Attlee, Clement, 190
Australia, 94, 152–3, 210

Austria, 100
aviation, 185

Babbage, Charles, 77–9, 86, 134, 218
Bacon, Francis, 119, 126
Badash, Lawrence, 200
Balfour, Arthur J., 46, 89, 96, 129, 143, 157–9, 180–2, 189, 209, 219, 221, 248
Balliol College, Oxford, 40
Bayer-Werke, 194
Bedford College, London, 205
Beilby, George T., 208
Beit, Alfred, 46–9, 51, 154, 165–6
Beit, Sir Otto John, 49
Beit Trust, 49
Belgium, 100, 112
Bentham, Jeremy, 25, 117
Berlin, 49–50, 86, 101, 140, 153, 171, 195
Berman, Morris, 7
Bernal, John D., 8
Bessemer Memorial Fund, 166
Beyer, Charles Frederick, 37, 59
Beyer, Peacock & Co., 37
Bibby, Cyril, 84
biochemistry, 24
biological research, 172–3
biology, 2, 21, 69
Birmingham, 14, 23, 27–9, 40, 44–5, 54, 181, 215
Birrell, Augustine, 159
Board of Education, 6, 11, 65, 110, 150, 158–60, 163–5, 167–9, 182, 197, 202–9, 211, 220, 227, 237
Board of Trade, 6, 64, 66, 70, 97, 143, 145, 148, 193–5, 203, 206, 211, 220, 241
Bodleian Library, 49
Boer War, 1, 6, 11, 27, 99, 103–4, 111, 122, 129
Bohr, Niels, 30
botanical gardens, 64, 227
Bradford, 253
Bragg, Sir William H., 42

Index

Brewster, Sir David, 86, 218
Briggs, Asa, 54
Bristol, 45
British Academy for the Promotion of Historical, Philosophical and Philological Studies, 18, 42–3, 47–8, 57, 59, 89, 181, 214
British Association for the Advancement of Science (BAAS), 16, 66, 79–80, 82, 84–92, 94, 96, 101–2, 114, 117, 120, 126, 129, 131–2, 139–42, 179–80, 188, 198, 207, 210, 231, 234
British Association of Chemists, 231
British Dyes Ltd., 186–7, 191, 194–5
British Dyestuffs Corporation, 195
British Institute of Preventive Medicine, 42
British Museum, 64, 73, 224
British Optical Society, 112
British Science Guild, 82, 92–6, 98, 112, 118, 124, 128, 131, 134, 136, 181, 183, 188, 207, 216, 231, 249
British South Africa Co., 48
Broadlands, 50
Brock, W.H., 70
Brocke, Bernhard vom, 5
Brunner, Sir John T., 38, 47, 52, 59, 124, 148, 185
Brunner Chair of Economic Science, 38
Brunner, Mond & Co., 38, 47
Buckmaster, C.A., 69–70, 131
Bunsen, Robert Wilhelm, 123

Cambridge, 24, 88–9, 143
Campbell-Bannerman, Sir Henry, 92, 124, 182
Canada, 65, 94, 153, 210
Cannon, Susan Faye, 86
Canterbury, Archbishop of, 25
Cape Colony, 46, 49
Capetown, 89
Cardiff, 139
Cardwell, D.S.L., 8, 30, 34, 62, 229
Carnegie, Andrew, 60
Carnegie United Kingdom Trust, 60
Cassel, Sir Ernest J., 46–7, 50–1, 166
Catholic University of Ireland, 33
Cavendish Laboratory, 24, 45, 143
Central Technical College, 167
Chadwick, James, 30
Chalmers, Robert, 180
Chamberlain, Austen, 35, 159–60, 167, 250–1, 254

Chamberlain, Joseph, 28, 39–40, 46–7, 53, 65, 92–3, 111, 130, 148, 159, 180–5
charities, 60
charity law, 60
Charles II, 14, 119
Charlottenburg, 140–1, 153, 156, 160–1, 163, 165, 171, 195
Charlottenburg Scheme, 155–6, 158–9, 168
chemical industry, 44, 52, 107, 111, 113–14, 132, 186, 194, 202, 213, 225, 242, 253
Chemical News, The, 78, 234
Chemical Products Supply Committee, 193–4
Chemical Research Laboratory, 211
Chemical Society, 15–17, 20, 30, 94, 121, 131, 193, 204, 217, 220, 225, 229–30, 234, 236, 239, 241–2
chemistry, 7, 21, 27, 61, 114, 133, 151–2, 221, 233–4, 239
chemists, 224, 233, 239, 241–3
China, 111
Christie, R.C., 56
Christie Library, 56
Churchill, Winston S., 191
City and Guilds Institute, 188
City of London, 50, 65, 92, 150, 166
civic universities, 28–9, 181
civil service, 27, 63, 234
Civil War, American, 133
Clare Market, 46
Clarendon Laboratory, 24
Clothworkers' Company, 39, 45
Coleman, D.C., 56, 115
College of Science and Technology, Glasgow, 152
Cologne, 50
Colonial Office, 6, 42, 64–6, 159, 173
Committee on Grants to University Colleges, 35, 208
Committee of the Privy Council for Scientific and Industrial Research, 97, 112, 206–10, 212
Congress of Vienna, 13
Congrès Scientifique de France, 86
Conjoint Board of Scientific Societies, 97–8, 189
cooperative research, 212
Copernicus, Nicolaus, 10
Crimean War, 99, 101, 103
Crookes, Sir William, 219, 223, 234
Crowther, J.G., 79
Crystal Palace, 100

284

Index

Currie, Donald, 41
Curtius, Ludwig, 135

Daily Mail, 104, 109
Daily Telegraph, The, 41, 114
Darmstadt, 49
Darwin, Charles, 10, 128, 225
Davy, Sir Humphry, 119
Davy–Faraday Laboratory, 42, 48
De Beers Consolidated Mines, 48
defence policy, 185
Denmark, 57
Department of Science and Art, 72
Department of Scientific and Industrial Research (DSIR), 5, 89, 97, 149, 176, 187, 200, 203–4, 208–12, 250, 254
Derby, Edward Stanley, Earl of, 53
Devonshire, William Cavendish, Duke of, 6, 45–6, 80, 90, 221
Devonshire Commission, 21, 61, 81, 84, 90, 99, 102, 117–18, 123, 126, 142, 150, 188, 207, 221, 235
Devonshire Report, 81
Dictionary of National Biography, 48
Diehl, Carl, 133
Disraeli, Benjamin, 102
DNA molecule, 10
Dohrn, Anton, 66, 90
dominions, 164, 210
Donation Fund, 20
dreadnoughts, 136
Dresden, 37
Dublin, 29

East Anglia, 238
Ecole Polytechnique, 217
economic backwardness, 99
economic decline, 105, 108, 113–16
economic growth, 3
Edinburgh, 29, 64, 88, 117
Edinburgh Journal of Science, The, 86
Education Act (1902), 158
Edward VII, 47, 50, 57, 65, 146, 159, 219, 248
Edwards, J. Passmore, 46
Egypt, 50
Egyptology, 38
Eidgenössische Technische Hochschule (Zurich), 149
Einstein, Albert, 10, 198
electrical engineering, 27
electrical industry, 44, 112, 124, 132
Elizabeth I, 33
Emancipation Act, Catholic (1829), 33

Empire Conference (1897), 109
Endowment of Research Movement, 83–5, 102, 120, 235
endowments 13–14, 20, 32, 34–6, 38, 40–2, 52, 55–8, 60, 65, 147, 154, 166, 174
Engineer, The, 78
Engineering, 78
engineering, 46, 151–2, 155, 159–60, 163–4
engineering industry, 44, 52, 113
English Mechanic, The, 78
Entomological Society of London, 58
entrepreneurs, 113, 115, 151, 196
Esperanto, 69
eugenics, 122
experimental physics, 45, 61

Fabian Society, 45, 152, 180
Fairbairn, Sir Andrew, 45
Faraday, Michael, 88, 112, 120, 144, 218
Ferber, Christian von, 230
First World War, 1, 5–6, 11, 20, 22, 26, 46, 52, 94–7, 103–4, 107, 118, 127, 133, 139, 147, 171, 174, 180, 186–7, 189, 191–202, 206, 209, 211, 213–14, 218–21, 223–4, 226–8, 231, 235, 237, 241, 243, 246–7, 250–1, 254
Firth College, 27
Fisher, H.A.L., 41–2, 199, 237
Fleming, A.P.M., 211
Fleming, J.A., 197, 221
Foreign Office, 114, 132
Foster, Sir Michael, 118, 188, 235–8, 254
France, 14, 78, 86, 100–1, 107–8, 111–12, 127, 141, 217
Franco-Prussian War, 107
Frankland, Edward, 84, 123
free trade, 111, 130
Furness, Sir Christopher, 124

Galton, Sir Douglas S., 140–2, 146, 179
Gassiot Fund, 147
Geikie, Sir Archibald, 135, 219, 246
General Board of Health, 173
Geological Society, 15
Geological Survey of Great Britain, 62–3, 135, 211, 224, 256
geology, 21, 90
George III, 79
George V, 145

Index

Germany, 1, 5, 7, 11, 13–14, 22, 26–7, 34, 47–8, 50–1, 57, 61–2, 73, 84, 86, 93, 97, 101, 103, 105–8, 111–12, 114–15, 120, 122, 124–7, 131–6, 140, 150, 153, 160–2, 165, 168, 171–2, 181, 186, 194–6, 198–200, 203, 213, 216–17, 219, 223, 225–7, 229, 233–4, 242, 244, 246–7, 251–2
Gesellschaft Deutscher Naturforscher und Ärzte (GDNÄ), 86–7
Giessen, 133, 234
Gilbert, Bentley B., 250
Gill, Sir David, 89, 126, 188
Gladstone, William E., 73, 81, 102, 182
Glasgow, 152
Glazebrook, Richard T., 134, 139, 141
Gmelin, Leopold, 234
Goldsmiths' Company, 46, 166
Gollancz, Sir Israel, 43, 59
Goschen, George J., 73
Göttingen, 133, 181
Government Chemist, 224, 240
Grant-in-Aid for Scientific Investigations, 21
Granville, George Leveson-Gower, Lord, 101
Great Depression, 36, 40, 106–7
Great Exhibition (1851), 64, 100–1, 121, 149, 159
Greenwich, 73, 120, 242
Gregory, Sir Richard A., 73, 95, 104
Grey, Sir Edward, 182
Griewank, Karl, 5

Haines, George, 133
Haldane, Richard B., 12, 46–7, 49, 59, 92, 94, 124–5, 131, 133–4, 148, 155–7, 160–2, 165–8, 171, 178–9, 181–5, 194, 199, 203, 205–7, 216, 219, 239, 241, 254–5
'Haldane Principle', 255
Hallé, Carl, 29
Hallé Orchestra, 29
Hamburg, 48–9
Harmsworth, Cecil, 187
Harnack, Adolf, 5, 132
Harrison, Heath, 52
Harrison, T. Fenwick, 52
Heath, Frank, 205
Heidelberg, 47, 234
Henderson, Arthur, 203, 206
Herschel, John, 218
Hertz, Henriette, 43
Hewins, William A.S., 210

Hicks Beach, Sir Michael E., 43, 73, 143
Hirst, Thomas Archer, 84
historiography of science, 7, 9–10, 115, 138, 200, 230
Hobsbawm, Eric J., 9, 109
Hofmann, August Wilhelm, 101, 110, 234
Hollenberg, Günter, 133
Home Office, 64
Hooker, Sir Joseph D., 84, 219
Hopkinson, Bertram, 208
House of Commons, 12, 67, 72, 134, 180, 184–91, 195, 197, 203, 205, 207, 220, 227, 243, 248
House of Lords, 92, 207
Howarth, O.J.R., 90
Huddersfield, 194
Huggins, Sir William, 219
Hughes, J.W., 52
humanities, 2, 53, 169, 215, 222
Humboldt, Alexander von, 86
Humboldt, Wilhelm von, 22, 140
Hutchinson, Henry Hunt, 45
Huxley, T.H., 66, 82–4, 128, 215, 225, 235

IG Farbenindustrie, 195
Imperial Chemical Industries (ICI), 47, 187, 195
Imperial College of Science and Technology, 12, 46, 49–50, 59, 91, 138, 154–61, 163–72, 176, 178–9, 184, 192, 216, 240, 249, 254
Imperial Institute (London), 64–5
Imperial Trust for the Encouragement of Scientific and Industrial Research, 212–13
imperialism, 109, 127, 129
India, 46, 65, 79, 153, 210
India Office, 173
indigo, 110–11
Industrial Revolution, 23–4, 106–7, 113–14, 173
industrialisation, 150, 253
Institute of Chemistry, 16, 95, 221, 224, 233
Institute of Physics, 231
Institution of Chemical Engineers, 231
Institution of Civil Engineers, 167
Institution of Professional Civil Servants, 231
international exhibitions, *see also* Great Exhibition, 39, 80, 99–100, 102, 110, 123

Index

Ipswich, 140
Ireland, 33, 68, 164, 185, 206, 247
Irish Home Rule, 83
Iron and Steel Institute, 101, 167
Isaac Wolfson Foundation, 60
Isaacs, Sir Rufus, 239
Italy, 107
Ivan Levinstein Ltd., 195
Iveagh, E.C. Guinness, Lord, 42

Jameson, Leander S., 245
Jameson Raid, 48
Jena, 112
Jenner Institute for Bacteriological Research, 42
Jenner Institute of Preventive Medicine, 42
Johannesburg, 89
John Rylands Library, 56
Jones, Alfred L., 55, 59
Jordan, W.K., 51
Journal of the British Science Guild, The, 95

Kaiser-Wilhelm Society, 132, 135–6
Kargon, Robert H., 7
Kassel, 47
Kelvin, William Thomson, Lord, 112, 117–19, 126, 139, 144, 219, 221
Kenyon, Sir Frederic G., 43
Keppel, Sir Henry, 219
Kew, 90, 142, 146
Kew Gardens, 64
Kew Observatory, 90, 142, 146–7
King Edward VII British–German Foundation, 51
King's College, London, 25–6, 151, 169
Kitchener of Khartoum, Lord, 219
Kuhn, Thomas S., 9–10

Laboratoire d'Essais, 141
laissez-faire, 1, 13, 61–2, 64, 74, 183, 247–8, 253
Landes, David S., 107
Lankester, Sir E. Ray, 97, 199
Larmor, Sir Joseph, 189
Lavoisier, Antoine, 10
Lawrence, Sir Joseph, 148
Lecky, W.E.H., 121, 219
Leeds, 27–8, 39, 44–5, 52, 101
Leicester, 89
Leipzig, 86, 133
Leverkusen, 194
Levinstein, Ivan, 195

liberal imperialists, 183
liberalism, 9
Liberal Party, 124
Liebig, Justus von, 25, 88, 101, 120, 133, 216, 222, 233–4
Linnean Society of London, 15, 58
Lister, Joseph Lister, Lord, 219
Lister Institute of Preventive Medicine, 42
Literary and Philosophical Society, 14
Liverpool, 14, 28, 44, 47, 54, 65–6, 142, 147
Liverpool Photographic Journal, The, 234
Livery Companies, 166–7
Lloyd George, David, 8, 11, 41, 128, 173–5, 177, 180, 199, 202–3, 205–6, 209, 254
Local Government Board, 64, 173, 175
Lockyer, Sir J. Norman, 34, 77, 81–2, 91–6, 104, 114, 123, 128–32, 136, 139, 196, 199, 207, 216, 225
Lodge, Sir Oliver J., 139, 144, 179
London, 15, 17, 21, 23, 29, 34–5, 39, 41, 45–6, 48–50, 52, 57, 64–5, 79, 82, 86–8, 90, 95, 100–2, 110, 119–21, 138, 149–56, 158–60, 162–7, 169–75, 179–80, 188, 221, 224, 230, 238, 240, 257
London Chamber of Commerce, 46
'London Charlottenburg', 157, 164, 166, 183
London County Council (LCC), 45, 152–5, 157–8, 160, 165–7, 180
London School of Economics and Political Science, 45–6, 50, 59, 157, 179, 186
London School of Tropical Medicine, 42, 65
Londonderry, Marquess of, 159, 166–7, 180
Lough, Thomas, 191
Low, Sidney, 221
Lowe, Robert, 67
Lunar Society, 14

McCormick, Sir William S., 208
MacDonagh, Oliver, 247
MacDonald, Ramsay, 182
Machinery of Government Committee, 254–5
McKenna, Reginald, 159, 171
MacLeod, Roy M., 8, 17, 68, 70, 204–5, 224
Macmillan (publishers), 82
Made in Germany, 104, 112

Index

Magnus, Sir Philip, 187–8
Mahan, Alfred Thayer, 129
Manchester, 14, 23, 27–8, 30, 36–7, 44, 56, 60, 70, 91, 151, 195
Marburg, 47, 123
Marine Biological Association of the UK, 66
marine biology, 66
Marine Laboratory (Plymouth), 66
Mason, Josiah, 28, 39, 51–3
Mason College, 27–8, 39, 44, 215
Massachusetts Institute of Technology, 150
Mathematical Society of London, 16
mathematics, 21, 90
Matthew, H.C.G., 182
Maxwell, James Clerk, 24, 112, 232
medical research, 175–7, 204
Medical Research Committee, 138, 172–8, 180, 204–5, 210–11, 240, 250, 254
Medical Research Council, 12, 174, 176, 209
Medical Research Fund, 173
medical science, 2
medicine, 7, 210, 222
Medick, Hans, 250
Meldola, Raphael, 30, 121, 131, 208, 216, 220, 236–7
Merchandise Marks Act, 112
Merton, Wilhelm, 36
metallurgy, 152
Meteorological Office, 70, 256
meteorology, 61
Michaelis, Maximilian, 166
Midlands, 23, 36, 45, 47, 53–4, 151, 165, 194, 224, 236
military research, 201–2, 243
Mill, John Stuart, 23, 25
Milner, Lord Alfred, 46, 182, 245
mining, 152
minister for science, 207
Ministry of Agriculture, 243
Ministry of Health, 158, 175–6, 187
Ministry of Munitions, 198, 202, 242
Ministry of Science, 73, 80
Ministry of Science and Industry, 209
Ministry of Technology, 210
molecular biology, 7
Mond, Sir Alfred, 187, 195
Mond, Ludwig, 38, 42–3, 47–9, 51, 57, 59, 187
Montreal, 89
Morant, Sir Robert L., 158–60, 164, 166, 180, 182, 254

Morning Post, 184, 198–9, 220
Morley, Sir John, 219
Morris dancing, 69
Moseley, H.G.J., 30
Moulton, John F., Lord, 175, 195
Mountbatten family, 50
Mowatt, Sir Francis, 41, 114, 158, 160, 162, 165, 180, 182
Munich, 48
Murray, Sir George, 180, 254
museums, 64
Musson, A.E., 23

Naples, 66, 90
National Bureau of Standards, 141, 146
National Chemical Laboratory, 211
'national efficiency', 27, 91, 93, 99, 118, 128, 177, 182, 184
National Gallery, 48–9
National Health Insurance Commission, 180
National Insurance, 177, 250
National Insurance Act, 173–5
National Physical Laboratory (NPL), 12, 49, 55, 134, 138–49, 158, 160, 172, 178, 185, 211, 227, 240, 243, 249, 254
National Union of Scientific Workers, 96, 231
National University of Ireland, 33, 241
National Vaccine Establishment, 63
Naturalists' Field Clubs, 17
Nature, 1, 5, 12, 34, 48, 66, 70–1, 73, 75, 78, 82–3, 94–5, 104–5, 107, 110, 114, 117–18, 120, 122, 124–6, 128, 134–6, 139, 150, 154, 161, 176–7, 181, 188–9, 192, 196, 199–201, 209–10, 215, 217, 219, 222, 226, 228–9, 231, 243, 246, 249, 252
Neglect of Science Committee, 97, 209, 220
Netherlands, The, 150
Newcastle-upon-Tyne, 91, 122
New College, Oxford, 40
Newton, Sir Isaac, 10, 198
New Zealand, 210
Nineteenth Century, The, 83, 221
Nineteenth Century and After, The, 153, 155
nitrogen synthesis, 194
Nobel Prize, 126, 143, 221, 225
Noble, Sir Andrew, 141
Norfolk and Norwich Naturalists' Society, 88

288

Index

Norwich, 80, 88, 90
nuclear physics, 7, 24
Nuffield Foundation, 60

optical glass, 95, 112, 201
optical industry, 107, 111–12
optical instruments, 112, 192
Order of Merit, 219–20, 234
organic chemistry, 30, 110
Ostwald, Wilhelm, 133
Owen, David, 41, 44, 53, 56–8
Owens, John, 37, 39, 44, 51, 53
Owens College, 27–8, 37, 56, 70, 151
Oxford, 24, 49, 88, 224

paradigm, 10
Paris, 39, 80, 100–2, 119, 123, 217, 234
Parliament, 12, 67, 92
Parris, Henry, 248
patent law, 95
patents, 110, 127, 194, 196, 212
pathology, 69, 236
patronage, 10, 13–14, 35–6, 38–41, 43–4, 47, 50–1, 56–62, 69–70, 73, 147, 154, 157, 178–9, 251, 253
Pattison, Mark, 102, 224
Payne-Townshend, Charlotte, 45
Pearson, Karl, 122
Pease, Sir Joseph A., 205–6, 212, 227
Peel, Sir Robert, 25, 218
Perkin, William Henry, 110, 124, 218
Pfetsch, Frank R., 4, 133, 230
pharmaceutical industry, 107
Pharmaceutical Society of Great Britain, 16
physical chemistry, 38
Physical Society of London, 15
physics, 21, 69, 90, 117, 139, 151, 221, 231
Physikalisch-Technische Reichsanstalt, 140–2, 144–6
Physiological Society of London, 16
physiology, 7, 236
Pilcher, Richard B., 233, 242
Pilgrim Trust, 60
Playfair, Lyon, 100–1, 121, 150, 216
Plymouth, 66
Polytechnische Hochschulen, 152
Poole, J.B., 7, 207
Portugal, 57
Post Office, 64
Pour le Mérite, 219
Pratt, Edwin A., 113
Preußische Akademie der Wissenschaften, 48
preventive medicine, 173
Price, Derek J. de Solla, 190
Primrose League, 127
Prince of Wales, *see* Edward VII
Privy Council, 64, 89, 176, 207
Procter, Richard, 81
professional associations, 145, 167, 170, 207, 231
professionalisation, 11, 228, 231–3
Prussia, 34, 78, 86, 100, 161, 219, 241
public lecture, 15
public schools, 215, 223

Quarterly Journal of Science, The, 78, 234
Queen's College Belfast, 33
Queen's University of Ireland, 33

Radium Institute, 50
Ramsay, Sir William, 197, 225, 237
Rayleigh, John William Strutt, Lord, 24, 97, 143, 146, 179, 181, 203, 208, 219, 221, 225
Read Holliday and Sons Ltd., 194
Reader, William J., 238
Reale Accademia dei Lincei, 48
Reay, Lord, 57, 214–16
Rendall, Gerald, 54
Research Associations, 212–13, 254
research laboratories, 196
Rhodes, Cecil, 47–9, 155, 166, 245
Rhodesia, 48–9
Ringer, Fritz K., 217, 241
Ripon, G.F.S. Robinson, Lord, 46
Roberts, Frederick Sleigh, Lord, 219
Robinson, Eric, 23
Rockefeller, John D., 59
Rockefeller Foundation, 174
Rome, 48
Rose, Frederick, 132
Rose, Hilary, 13, 61–2, 244
Rose, Steven, 13, 61–2, 244
Rosebery, Philip Archibald Primrose, Lord, 46–7, 59, 93, 153–8, 163, 165–6, 168–9, 171, 182
Rothschild, Lord Nathaniel, 45
Routh, Guy, 238
Royal Arsenal, 243
Royal Astronomical Society, 66, 78–9, 89
Royal College of Chemistry, 101, 110, 234
Royal College of Physicians, 176
Royal College of Science, 82, 91, 152,

Index

158, 160, 166–7
Royal College of Surgeons, 176
Royal Commission Appointed to Inquire into the Relations of Human and Animal Tuberculosis, 173
Royal Commission on Scientific Instruction and the Advancement of Science, *see* Devonshire Commission
Royal Commission on Technical Instruction, 103, 140
royal commissions, 12, 22, 80, 83, 132, 173
Royal Geographical Society, 15, 19, 66
Royal Institution of Great Britain, 42, 119–20, 143, 224
Royal Medal, 218
Royal Mint, 64
Royal Naval College, 242
Royal Observatory (Greenwich), 73, 120, 256
Royal School of Mines, 152, 158, 160, 166–7
Royal Scottish Meteorological Society, 19
Royal Society of Edinburgh, 14, 19
Royal Society of London, 12, 14–15, 18–21, 36, 42–3, 48, 56, 58, 66–7, 70–2, 77–9, 81–6, 89–91, 97, 101–2, 112, 117, 119, 123, 125–6, 135, 139, 140–6, 148–9, 167, 176, 179–80, 193, 203–4, 207–8, 211, 215, 218–20, 222–3, 225, 230, 232, 234, 241, 246, 248, 259
Royal University of Ireland, 33
Royal Zoological Society of Ireland, 19
Rücker, Arthur, 141, 157
Runciman, Walter, 34–5, 159, 193–4, 196–7, 203, 205, 241, 248
Russell, Bertrand, 45, 49
Rutherford, Sir Ernest, 24, 30
Rylands, John, 56

Salisbury, Robert Gascoyne-Cecil, Lord, 127, 142, 180, 188, 219, 221
Sanderson, Michael, 23, 32, 36, 39, 41, 44, 53–4, 125, 155
Schieder, Theodor, 76
Schofield, Robert E., 7, 23
School of Tropical Medicine (Liverpool), 55, 65
Schorlemmer, Carl, 30
Schroeder-Gudehus, Brigitte, 126
Schuster, Arthur, 70
science lobby, 76–7, 80, 95, 103–5,
116, 119–20, 128, 134, 136, 141, 148, 151, 176, 178, 180, 188–9, 201, 209, 220, 228, 231, 235, 238, 242–3, 246–7, 249, 252
science policy, 1–2, 4–5, 7–9, 11–12, 75, 79, 81, 103, 136–8, 144, 157, 166, 179–80, 182, 184–5, 187, 202–3, 207–8, 210, 249–55
Science Research Council (SRC), 210
scientific backwardness, 115, 196
scientific centres, 7
scientific community, 75, 117, 230
scientific expeditions, 62–3, 66–8
scientific experts, 63, 187
scientific instruments, 114, 213
scientific journals, 15, 82, 200–1, 229–30
scientific organisations, 10
scientific revolutions, 10
scientific societies, 5, 12, 14–21, 23–4, 34, 36, 43, 48, 58, 63, 65, 67–8, 77–9, 92, 96–8, 129, 144, 147, 179, 201, 204, 207, 211, 214, 229–30, 252, 256–7
Scotland, 32, 65, 88, 164, 206
Scott, Robert F., 66–7
Scottish Universities Trust, 60
Searle, G.R., 40, 182, 187
Second World War, 7, 60, 217, 243
secondary education, 68
Semmel, Bernard, 182
Seymour, Sir Edward, 219
Shackleton, Ernest H., 66
Shakespeare, William, 69
Shaw, George Bernard, 45
Sheffield, 27
Shils, Edward, 71
Siemens, Alexander, 141, 143
Siemens, Werner von, 140, 143
Sleeping Sickness Bureau, 65
smallpox, 63
Smith, Adam, 117
Snow, C.P., 177, 215, 217
Social Darwinism, 93, 127–9
social reforms, 182–3
social sciences, 2
Society of Arts, 73, 121, 217
Society of Chemical Industry, 48, 167
Solvay Process, 47
South Africa, 46–9, 89, 93, 99, 152–3, 158, 210
South Kensington, 64, 150, 158–9
South Pole, 66–7
Southport, 91, 129
Spectator, The, 117

290

Index

Spottiswoode, William, 84, 232
state intervention, 112, 183, 186, 195, 247–8, 251–3
statistics, 7
steel industry, 124
Strange, Sir Alexander, 77, 79–80, 100, 139, 235
'struggle for existence', 93, 127–30
Stuttgart, 114, 132
Sussex, Duke of, 18, 79, 85
Sweden, 50, 57
Switzerland, 57, 100, 111–12, 150, 194–5
synthetic dyes, 110–11, 186, 191, 195, 199

Taine, Hippolyte, 23
tariff reform, 130, 183–4
Tate, Sir Henry, 38
Taunton, Lord, 101
Taunton, Commission, 151
Taylor, Arthur J., 64, 247
technical colleges, 125, 163, 202, 241
technical education, 53, 100–1, 103, 105, 116, 123, 131, 150–1, 153–4, 158–62, 168, 172, 205
Technical Education Board, 152–3, 155
technical schools, 54
Technische Hochschulen, 26, 28, 34, 37, 134, 151–3, 161, 165, 168, 171–2, 198
Technische Universität Berlin, 171–2
Teddington, 138
textile industry, 192, 194
Thomson, Joseph J., 24, 219
Thompson, Silvanus P., 123–4, 218
Thorpe, Sir T. Edward, 91, 210
Threlfall, Sir Richard, 208
Tilden, Sir William A., 122, 224
Times, The, 57, 79, 81, 101, 103, 113, 123, 126, 128, 151, 153, 155–6, 189, 193, 197, 217–18
Toronto, 89
trade marks, 127
Transvaal, 48–9
Treasury, 6, 34–5, 40–1, 43, 63–4, 66, 71–3, 90, 105, 114, 135, 143, 145–8, 159, 166, 173, 175–6, 203, 209, 240, 242, 248, 254
Trinity College, Cambridge, 35
Trinity College, Dublin, 33
tropical diseases, 20, 55, 65, 173
Tropical Diseases Research Fund, 173
tuberculosis, 50, 173
Turner, Gerard, 61

Tyndall, John, 84, 225

United States, 4, 6–9, 11, 14, 26–7, 35–6, 42, 49, 60, 69, 71, 76, 82, 93, 100, 105–8, 111–13, 115, 124–5, 127, 131–2, 134, 141, 146, 150, 152–3, 162, 172, 190, 200, 203, 210–11, 228, 252–3
universal suffrage, 190
universities, 2, 15, 21–6, 28, 30, 33, 35, 164, 184, 186, 193, 202, 204, 207, 209, 211, 213, 215, 222, 224, 234, 236, 239, 242–3, 249, 251, 253–4
  American, 42
  British, 22, 26, 31, 51, 164–5, 181, 204, 225–6, 236, 240, 243
  French, 22
  German, 22, 25, 28, 32, 102, 133–4, 136, 153, 161, 168, 198
  Irish, 33
  Prussian, 22, 34, 241
  Scottish, 22–3, 30, 32–3, 65, 224
  South African, 49
University of
  Aberdeen, 31
  Berlin, 25, 241
  Birmingham, 28–9, 39–40, 52–4, 164
  Bonn, 25
  Bristol, 28–9, 40, 53
  Cambridge, 22–9, 33–5, 41, 45, 78, 84, 90, 99, 101, 118, 143, 222–4, 231, 234, 241, 244
  Chicago, 42
  Cornell, 42
  Dundee, 40
  Durham, 25, 27
  Edinburgh, 241
  Frankfurt, 36
  Glasgow, 151, 239
  Harvard, 42
  Heidelberg, 48
  Ireland, 33
  Leeds, 29, 42, 45, 164, 193, 241
  Liverpool, 29, 52, 164, 193, 241
  London, 23, 25–8, 32–3, 41–2, 46, 50, 54, 63, 65, 101, 118, 122, 153–4, 156–8, 160, 163–4, 167–71, 179, 187–8, 231, 234, 242
  Manchester, 29, 37–9, 53–4, 63, 164, 193, 239, 241
  Oxford, 22–3, 25, 27–9, 33–5, 41, 48, 84, 90, 99, 101–2, 164, 223–4, 231, 234, 241, 244
  St Andrews, 32

Index

Sheffield, 28, 175, 193
Southampton, 241
Strathclyde, 152
Wales, 31, 33, 248
Yale, 42
University College
  Aberystwyth, 32, 241
  Bangor, 32
  Bristol, 39, 45
  Cardiff, 32
  Cork, 33
  Dundee, 32
  Galway, 33, 241
  Liverpool, 28, 38–9, 53–5
  London, 25–6, 29, 41, 151, 169, 197, 224–5, 234, 241
  Nottingham, 40
  Reading, 40
university colleges, 32, 59, 63, 165, 223–4, 226, 231, 234, 236, 254
university constituencies, 180, 190
University Grants Committee, 35
University of London Act, 171
university students, 26–7, 29

Varcoe, Ian, 8
Verein Deutscher Ingenieure, 161
Versailles Treaty, 127
Victoria, Queen, 64, 75, 128, 218
Victoria University, 27–8, 32, 38–9
Vig, Norman J., 8
volcanism, 20

Wales, 31–2, 63, 164, 207
Walker, Sir Andrew B., 38, 53–4
Wallace, Alfred Russell, 219
War Office, 64, 66, 173, 193, 202
water pollution, 95
Watts, George Frederic, 219
Webb, Beatrice, 12, 45, 49, 55, 155–6, 158, 179, 182–4, 254
Webb, Sidney, 12, 45–6, 93, 151–3, 155–8, 160–1, 165–8, 180, 182–4
Weber, Max, 214
Weingart, Peter, 76
Weizmann, Chaim, 42, 241

Wellcome, Henry Solomon, 55, 59
Wellcome Chemical Research Laboratories, 55
Wellcome Physiological Research Laboratories, 55
Wellcome Trust, 55
Wellington, Arthur Wellesley, Duke of, 25
Wells, H.G., 220
Weltpolitik, 127
Wernher, Julius C., 46–9, 51, 55, 147, 154, 165–6
Wernher, Beit & Co., 46, 48, 154–5, 166
Wheatstone, Charles, 112
Whewell, William, 231–2
Whitworth, Sir Joseph, 56
Wiener, Martin J., 115
Wiese, Leopold von, 51
Williams, Ernest E., 104, 253
Williams, L. Pearce, 78
Williamson, Alexander William, 234
Wills family, 45
Wilson, Harold, 210
Winnipeg, 89
Wolseley, Lord, 219
Woolf, Virginia, 199
Woolwich, 243
Worcester College, Oxford, 243

X-Club, 83–4, 102, 232

Yarrow, Alfred F., 147
Year-Book of the Scientific and Learned Societies, 15, 17
York, 88
Yorkshire, 45
Yorkshire College of Science, 27–8, 39, 52, 101
Yorkshire Philosophical Society, 88

Zeiss Works, 112
zoology, 90
Zoological Station (Naples), 66, 90–1
Zurich, 149

THE LIBRARY
ST. MARY'S COLLEGE OF MARYLAND
ST. MARY'S CITY, MARYLAND 20686